# *Diffusion in Gases and Porous Media*

# Diffusion in Gases and Porous Media

## R. E. Cunningham

*Atanor S.A.M.*
*Buenos Aires, Argentina*
*and*
*University of La Plata*
*La Plata, Argentina*

and

## R. J. J. Williams

*University of Mar del Plata*
*Mar del Plata, Argentina*

SPRINGER SCIENCE+BUSINESS MEDIA, LLC

Library of Congress Cataloging in Publication Data

Cunningham, R   E
    Diffusion in gases and porous media.
    Includes bibliographies and index.
    1. Diffusion. 2. Porous materials. I. Williams, R. J. J., joint author. II. Title.
QC185.C86                          533'.63                          79-12120
ISBN 978-1-4757-4985-4      ISBN 978-1-4757-4983-0 (eBook)
DOI 10.1007/978-1-4757-4983-0

© 1980 Springer Science+Business Media New York
Originally published by Plenum Press, New York in 1980
Softcover reprint of the hardcover 1st edition 1980

We wish to express our deep appreciation to
Professor Mario Teruggi of the University of La Plata,
Argentina, for his advice on English style and
to Professor Rutherford Aris of the University of
Minnesota, for his encouragement.

# Preface

The world we live in exhibits, on different scales, many phenomena related to the diffusion of gases. Among them are the movement of gases in earth strata, the aeration of soils, the drying of certain materials, some catalytic reactions, purification by adsorption, isotope separation, column chromatography, cooling of nuclear reactors, and the permeability of various packing materials.

The evolution of the understanding of this subject has not always been straightforward and progressive—there has been much confusion and many doubts and misunderstandings, some of which remain to this day. The main reason for the difficulties in the development of this subject is, we now know, the lack of an understanding of the effects of walls on diffusing systems.

Textbooks usually treat diffusion on two levels: at the physicochemical or molecular level, making use of the kinetic theory of gases (which while a very rigorous and well-founded theory nevertheless is valid only for systems without walls), or at the level of a transport phenomenon, a level geared toward applications. The influence of walls is usually disregarded or is treated very briefly (for example, by taking account of the Knudsen regime or by introducing a transition regime of limited validity) in a way unconnected with previous studies. As a consequence, the extensive, generalized, and well-founded knowledge of systems without walls has often been applied without sound basis to real situations, i.e., to systems with walls.

Only recently has a unifying theory, the dusty gas model, been developed that correctly takes account of the influence of walls and which thus clarifies many of the problems and much of the confusion that has beset this subject. The dusty gas model shows why, for example, the fluxes of the species in a binary system are not always equal and opposite (and its predictions in this regard have been experimentally verified), it shows clearly that there is a difference between systems "without walls" and systems "without wall effects," and it clarifies the nature of the total diffusive flux of a system and the nature of the coupling between the diffusive and viscous fluxes.

In this book we present the subject of diffusion from the point of view of the dusty gas model, pointing out where the confusion and errors have arisen—and where they can and still unfortunately do arise if care is not taken. We will be more concerned with the foundations and preliminary questions than with the applications.

In transport theory, the solution of problems is usually obtained when a phenomenon is described by a differential equation, or equation of change, based on a conservation principle of a given extensive property; such a solution enables one to find the concentration or flux field of the given property. To construct the equation of change, we need to know the so-called constitutive equation, which provides the flux of the property in terms of coefficients and gradients. Our task is primarily to obtain this constitutive equation starting from fundamental principles, and four of the six chapters of this book are devoted to the achievement of this objective.

Our treatment of the subject proceeds by a sequence of increasingly realistic degrees of approximation. Thus, we begin in Chapter 1 with a phenomenological description which, in spite of its apparent simplicity, leads to a conceptual understanding of the phenomenon, both with and without walls, and leads to the correct anticipation of the properties and behavior of simple diffusing systems. With this description and the experimental evidence as a basis, constitutive equations of diffusion are immediately proposed.

Chapter 2 presents the kinetic theory of gases in equilibrium, and this makes possible a simplified kinetic theory for the elementary derivation of the transport coefficients.

The constitutive equations of diffusion are rigorously developed in Chapter 3 for multicomponent systems without walls by the use of the thermodynamics of irreversible processes and by the use of the more advanced kinetic theory of gases (using the Chapman–Enskog and Grad–Zhdanov developments).

In Chapter 4, the kinetic theory equations obtained in Chapter 3 are adapted to the dusty gas model, i.e., the effects of walls are incorporated in the equations, and with this we achieve our objective of obtaining the constitutive equations of diffusion for the most general case.

In Chapter 5, the dusty gas model constitutive equations are applied to the solution of real cases. This completes our theoretical–experimental treatment of the subject, though it does not, of course, exhaust it.

The last chapter (6) is devoted to a concise description and analysis of the evolution of the understanding of diffusion in gases from the time of Thomas Graham to the present. The objective here is to gather together and clarify the doubts, confusion, and misunderstandings which have arisen in this chapter of scientific history.

Throughout the book special emphasis is given to conceptual aspects rather than to tedious mathematical developments. Nevertheless, in every case the working hypotheses, the equations used, the mathematical framework, and the results obtained are given in some detail.

The general structure of the book makes it suitable for undergraduate and graduate courses. For the former, the first two chapters provide introductory material, while the more advanced methods and deeper insights of the remaining chapters will serve the latter.

Our main hope is to contribute to the understanding of the knowledge of diffusion in gases independently of its applications in the various branches of science and technology. We thus hope that the book will be of use to physicists, physical chemists, chemists, and to all types of engineers in their several specialities.

R. E. Cunningham
R. J. J. Williams
*La Plata, Argentina*

# Contents

## 2. Elementary Prediction of Transport Coefficients . . . . . . . 63

# Notation

The number of the equation in which a quantity is first used, together with the dimensions of the quantity (M = mass, L = length, t = time, and T = temperature), are given in square brackets at the ends of the definitions. Symbols which do not appear frequently are not listed.

| | |
|---|---|
| $a_i$ | Stoichiometric coefficient of species $i$ on a molar basis [mole/equiv] |
| $a^K$ | Coefficient in Knudsen's equation [(1.71), $L^3 t/M$] |
| $a_i^m$ | Stoichiometric coefficient of species $i$ on a mass basis [M/equiv] |
| A | Used to represent a species of gas |
| $A_G$ | Coefficient in Grad–Zhdanov's equations [(3.131)] |
| $b$ | Stoichiometric coefficient |
| $b^K$ | Coefficient in Knudsen's equation [(1.71), $L^2/t$] |
| $b$ | Impact parameter [(3.18)] |
| B | Used to represent a species of gas |
| $B$ | Generalized property (scalar, vector, or tensor) [(4.5)] |
| $B_k$ | Permeability [(1.68), $L^2$] |
| $B_k^*$ | $\equiv B_k p_0^2/\mu D_{AB} p$, dimensionless permeability [(5.8)] |
| $c_i$ | Molar concentration of species $i$ [mole/$L^3$] |
| $c_1^K, c_2^K$ | Coefficients in Knudsen's equation [(1.71), $Lt^2/M$] |
| $c_p$ | Specific heat at constant pressure [$L^2/Tt^2$] |
| $c_v$ | Specific heat at constant volume [$L^2/Tt^2$] |
| $C$ | Coefficient for calculation of $\Delta$ [(4.76)] |
| $C_i$ | ($i = 1, 2, 3, \ldots$) Proportionality coefficient |
| $C_r$ | Rugosity factor |
| $d$ | Distance [L] |
| $\mathbf{d}_i$ | Diffusion vector for species $i$ [(3.11), (3.74), $L^{-1}$] |
| $D_{ij}$ | Molecular diffusivity of species $i$ in $j$ or of $j$ in $i$ [$L^2/t$] |
| $D_{ij}^C$ | Configurational diffusivity [(5.69), $L^2/t$] |
| $D_{ij}^F$ | Diffusivity for a Fick's-type law [(5.96), $L^2/t$] |

| | |
|---|---|
| $[D_{ij}]_1, [D_{ij}]_2$ | First- and second-degree approximation diffusivities, respectively [(3.86), (3.89)] |
| $D_{ij,\,\text{eff}}$ | Effective diffusivity [(1.32a)] |
| $D_{i,\,\text{m}}$ | Binary type molecular diffusivity for a multicomponent mixture [(5.74), $L^2/t$] |
| $D_{AB}^K$ | Combined Knudsen diffusivity, $D_{AB}^K \equiv D_{AA}^K x_B + D_{BB}^K x_A$ $[L^2/t]$ |
| $D_{ii}^K$ | Knudsen diffusivity $[L^2/t]$ |
| $D^{K*}$ | $\equiv -D_{AA}^K/D_{BB}^K$, dimensionless Knudsen diffusivity |
| $D_{ij}^m$ | Multicomponent diffusivity $[L^2/t]$ |
| $D_{ij}^\circ$ | Pressure-independent diffusivity, $D_{ij}^\circ \equiv D_{ij}p$ $[ML/t^3]$ |
| $D_{sAB}$ | Slip diffusion coefficient [(4.65)] |
| $D_i^T$ | Thermal multicomponent diffusivity [(3.79, $M/Lt$] |
| $D^*$ | $\equiv D_{AA}^K p_0/D_{AB}p$, dimensionless diffusivity |
| $E_{AB}$ | Generalized coefficient of the viscous and molecular diffusive fluxes in binary mixtures [(4.52a), $L^2/t$] |
| $E_{AA}^K$ | Generalized coefficient of the viscous and Knudsen fluxes in binary mixtures [(4.52a), $L^2/t$] |
| Er | Relative error |
| $E_r$ | Relative kinetic energy [(3.83), $ML^2/t^2$] |
| $\mathbf{f}$ | Force $[ML/t^2]$ |
| $f^a$ | Distribution function after collision |
| $f^{(N_m)}$ | Distribution function for $N_m$ molecules |
| $f^{(0)}$ | Distribution function at equilibrium |
| $f(\mathbf{v})$ | Velocity distribution function |
| $F$ | Phase density distribution function |
| $\mathbf{F}_i$ | Molecular flux of species $i$ $[L^{-2}t^{-1}]$ |
| $F_{BA}$ | Molecular flux ratio, $F_{BA} \equiv F_B/F_A$ |
| $\mathbf{g}$ | Acceleration of gravity $[L/t^2]$ |
| $\mathbf{G}_i$ | Mass flux of species $i$ $[M/L^2t]$ |
| $G_f$ | Gibbs free energy $[ML^2/t^2]$ |
| $h$ | Planck's constant $[ML^2/t]$ |
| $h^*$ | Thiele modulus [(5.62)] |
| $H$ | Enthalpy $[ML^2/t^2]$ |
| $H^s$ | Sorption coefficient [(1.86), mole $t^2/ML^2$] |
| $I$ | Unitary tensor |
| $\mathbf{j}_i$ | Mass diffusive flux of species $i$ with respect to the velocity $\vartheta$ [(Table 1.3), $M/L^2t$] |
| $\mathbf{j}_{iM}$ | Mass diffusive flux of species $i$ with respect to the velocity $\vartheta_M$ [Table 1.3, $M/L^2t$] |
| $\mathbf{J}_i$ | Molar diffusive flux of species $i$ with respect to the velocity $\vartheta$ [Table 1.3, mole$/L^2t$] |
| $\mathbf{J}_i^m$ | Molecular diffusive flux of species $i$ with respect to the velocity $\vartheta$ [Table 1.3, $L^{-2}t^{-1}$] |

| | |
|---|---|
| $J_{iM}$ | Molar diffusive flux of species $i$ with respect to the velocity $\vartheta_M$ [Table 1.3, mole/$L^2t$] |
| $J_{iM}^m$ | Molecular diffusive flux of species $i$ with respect to the velocity $\vartheta_M$ [Table 1.3, $L^{-2}t^{-1}$] |
| $k$ | Boltzmann constant [$ML^2/t^2T$] |
| $k_T$ | Thermal diffusion ratio [(3.99)] |
| $k_V$ | Reaction rate coefficient [(5.55)] |
| Kn | $\equiv \lambda/\lambda_p$, Knudsen number |
| $L$ | Distance between boundaries [L] |
| $L_r$ | Average roughness [L] |
| $m_i$ | Molecular mass of species $i$ [M] |
| $m_{BA}$ | Molecular mass ratio, $m_{BA} \equiv m_B/m_A$ |
| $m_{ij}^+$ | Reduced molecular mass, $m_{ij}^+ \equiv m_i m_j/(m_i + m_j)$ [M] |
| $M_i$ | Molar mass of species $i$ [M/mole] |
| Ma | Mach number, ratio of the characteristic velocity and the velocity of sound, Ma $\equiv \vartheta_c/\vartheta_s$ |
| $M_{BA}$ | Molar mass ratio, $M_{BA} \equiv M_B/M_A$ |
| $n_i$ | Molecular concentration of species $i$ [$L^{-3}$] |
| $\mathbf{n}$ | Normal to a surface |
| $N_i$ | Molar flux of species $i$ [mole/$L^2t$] |
| $N_{Av}$ | Avogadro number [molecules/mole] |
| $N_{BA}$ | Molar flux ratio, $N_{BA} \equiv N_B/N_A$ |
| $N_M$ | Number of moles |
| $N^*$ | $\equiv (-N_{zA}/N_{zB})$, flux ratio in a binary system |
| $p$ | Pressure [$M/t^2L$] |
| $P$ | Variable-type pressure [(3.68), $M/Lt^2$] |
| $\mathbf{q}_i$ | Heat flux of species $i$ [$M/t^3$] |
| $Q$ | Obstruction factor |
| $Q_a$ | Molar heat of adsorption [$ML^2/t^2$] |
| $Q_m$ | Obstruction factor for molecular diffusion [(1.32a)] |
| $Q_M$ | Molar flow rate [mole/t] |
| $Q_p$ | Obstruction factor for Knudsen diffusion [(1.45a)] |
| $r$ | Radial coordinate [L] |
| $r_p$ | Radius of particle in the DGM [L] |
| $R$ | Radius [L] |
| $R$ | Radius of capillary [(1.67), L] |
| $\hat{R}$ | Molar reaction rate per unit mass [(5.2), mole/Mt] |
| Re | $\equiv 2\rho\langle\vartheta\rangle R/\mu$, Reynolds number |
| $R_e$ | Reaction rate per unit volume expressed in equivalents [equiv/$L^3t$] |
| $R_g$ | Gas constant [$ML^2/t^2T$ mole] |
| $R_v$ | Molar reaction rate per unit volume [mole/$L^3t$] |
| $s_{ij}$ | Transverse cross section for diffusion [(3.84)] |
| $S$ | Area of a cross section [$L^2$] |

| | |
|---|---|
| $S$ | Entropy $[ML^2/t^2T]$ |
| $S_{BET}$ | Surface area $[L^2]$ |
| $S_V$ | Surface area per unit volume $[L^{-1}]$ |
| $t$ | Time $[t]$ |
| $t_b$ | Average time between collisions $[t]$ |
| $t_m$ | Average time of a collision $[t]$ |
| $t_R$ | Retention time $[(1.77), t]$ |
| $T$ | Absolute temperature $[T]$ |
| $T$ | Shear tensor $[M/t^2L]$ |
| $U$ | Internal energy $[ML^2/t^2]$ |
| $\mathbf{v}$ | Velocity $[L/t]$ |
| $\bar{\mathbf{v}}_i$ | Average velocity of a large number of molecules of species $i$ $[L/t]$ |
| $\bar{v}_i$ | Average molecular speed of species $i$ $[L/t]$ |
| $\mathbf{v}_i^d$ | Diffusion velocity of species $i$ with respect to the velocity $\vartheta$ [Table 1.3, $L/t$] |
| $\mathbf{v}_{iM}^d$ | Diffusion velocity of species $i$ with respect to the velocity $\vartheta_M$ [Table 1.3, $L/t$] |
| $\mathbf{v}_i^D$ | Total diffusion velocity of species $i$ on a mass basis with respect to stationary coordinates [Table 1.4, $L/t$] |
| $\mathbf{v}_{iM}^D$ | Total diffusion velocity of species $i$ on a molar basis with respect to stationary coordinates [Table 1.4, $L/t$] |
| $\mathbf{v}^P$ | Peculiar velocity $[(3.37), L/t]$ |
| $V$ | Volume $[L^3]$ |
| $w$ | Pore coordinate $[L]$ |
| $w_i$ | Mass fraction of species $i$ |
| $W$ | Work $[ML^2/t^2]$ |
| $W_k$ | Tortuous porosity |
| $\mathbf{W}$ | Pore orientation |
| $\mathbf{x}$ | Rectangular coordinate $[L]$ |
| $x_i$ | Molar fraction of species $i$ |
| $X$ | Generalized driving force |
| $z$ | Rectangular coordinate $[L]$ |
| $Z$ | Frequency of molecular bombardment $[L^{-2}t^{-1}]$ |

## Greek letters

| | |
|---|---|
| $\alpha$ | Generalized, phenomenological coefficient of transport $[(3.1), L/t]$ |

| | |
|---|---|
| $\alpha'_{Ap}$ | Function of the molar fraction with limiting values $\alpha_{Lr}$ and $\alpha_{Qa}$ [(4.77)] |
| $\alpha_{ij}$ | Generalized thermal diffusivity [(3.98)] |
| $\beta$ | Slip friction coefficient [(1.69), $M/L^2t$] |
| $\gamma$ | Ratio of the specific heats, $\gamma \equiv c_p/c_v$ |
| $\Gamma_i$ | Generalized flux of property $i$ [(Property)$/L^2t$] |
| $\delta$ | Kronecker delta |
| $\boldsymbol{\delta}$ | Unit vector |
| $\Delta$ | Correction factor for the second-approximation diffusivity [(3.90), (3.128)] |
| $\varepsilon$ | Void fraction |
| $\varepsilon^s$ | Surface void fraction |
| $\zeta$ | Diffuse reflection coefficient [(2.40)] |
| $\eta$ | Effectiveness factor [(5.54)] |
| $\vartheta$ | Average velocity on a mass basis [(1.5), $L/t$] |
| $\vartheta_M$ | Average velocity on a molar basis [(1.4), $L/t$] |
| $\vartheta_s$ | Sound velocity [$L/t$] |
| $\Theta$ | Polar angular coordinate |
| $\Theta_s$ | Surface coverage |
| $\kappa$ | Tortuosity [(1.49a)] |
| $\lambda$ | Mean free path [L] |
| $\lambda_K$ | Thermal conductivity [$ML/t^2T$] |
| $\lambda_p$ | Distance between particles [L] |
| $\Lambda$ | Wavelength [L] |
| $\boldsymbol{\Lambda}$ | Driving force diffusion vector [(3.7), $L/t^2$] |
| $\mu$ | Viscosity [$M/Lt$] |
| $\xi$ | Collision function [(3.15), $ML^2$] |
| $\Xi$ | Energy [$ML^2/t^2$] |
| $\rho_i$ | Mass density of species $i$ [$M/L^3$] |
| $\sigma$ | Collision diameter [L] |
| $\sigma^s$ | Surface tension [$M/t^2$] |
| $\tau$ | Shear stress [$M/Lt^2$] |
| $\nu$ | Number of components |
| $\nu_{ij}$ | Collision frequency [$t^{-1}$] |
| $\nu_{ij}^+$ | Reactive collision frequency [$t^{-1}$] |
| $\varphi$ | Two-dimensional pressure [$M/t^2$] |
| $\psi$ | Molecular flux effectiveness [(1.45)] |
| $\psi_i$ | Generalized physical property |
| $\psi_p$ | Perturbation |
| $\omega$ | Momentum [$ML/t$] |
| $\Omega_{i,j}$ | Collision integral [(3.83)] |

## Subscripts

| | |
|---|---|
| A, B | refer to components in a binary mixture |
| A, B, C, | refer to components in a ternary mixture |
| a | macropore |
| c | characteristic |
| calc | calculated |
| eff | effective |
| g | gas |
| $i,j$ | refer to components in a multicomponent mixture |
| i | micropore |
| $L_r$ | Lorentzian |
| $L$ | boundary value ($z = L$) |
| m | molecule |
| $n$ | along the direction **n** |
| p | particle, pore, capillary |
| pot | potential |
| $Q_a$ | quasi-Lorentzian |
| $r$ | along the $r$ coordinate |
| $R$ | evaluated at the radius |
| S | surface |
| $w$ | along the $w$ coordinate |
| $x,y,z,$ | along the $x, y, z$ coordinate, respectively |
| 0 | boundary value ($z = 0$) |
| 1 | first approximation |
| 2 | second approximation |

## Superscripts

| | |
|---|---|
| D | total diffusive |
| d | diffusive |
| g | total nonsegregative |
| G | nonsegregative nondiffusive |
| K | Knudsen |
| N | nonequimolar |
| S | slip |
| s | surface |
| SV | slip plus viscous |
| tr | translational |

| | |
|---|---|
| V | viscous |
| V, ext | viscous extended |
| ′ | indicates that particles are also included (in the dusty gas model), e.g., $n' = n + n_p$ |
| * | indicates a dimensionless value |
| (0) | indicates a value at equilibrium |
| ° | indicates a reference value |

## Other symbols

Boldface symbols are vectors.

Boldface sans serif symbols are tensors.

[ ] indicate a matrix.

⟨ ⟩ indicate an average value over a cross section.

$\left.\begin{array}{c} ^- \\ ^\wedge \\ ^\sim \end{array}\right\}$ over a symbol means $\left\{\begin{array}{l} \text{average value} \\ \text{per unit mass} \\ \text{per mole} \end{array}\right.$

# Phenomenology of Diffusive and Viscous Fluxes

When we attempt to analyze a gaseous system at the molecular level, considering each molecule as an individual,[1] we are confronted by a chaotic motion of colliding molecules. If we analyze the system as a whole, i.e., macroscopically, and if there is a large enough number of molecules in the system, its behavior will no longer be chaotic. The properties of the system are then determined by the average of the contributions of the individual molecules.

A macroscopic analysis is useful since our senses and instruments usually perceive the behavior of the molecular population as a whole and not that of its individual components. This viewpoint implies that there are some properties (composition, pressure, temperature, density) inherent to the molecular population as a whole that will not be probabilistic, i.e., these properties will not be chaotic in spite of their origin in the molecular chaos. This is the basis of the *continuum hypothesis* used in the study of matter.

For example, if we imagine a plane in the system, molecules will cross it in both directions. The net flux depends on the probability of the molecules crossing the plane in either direction, and this probability depends on, among other factors, the concentration of the molecules on both sides of the plane. At the level of the individual molecules (i.e., microscopically), the flux is probabilistic, but at the molecular population level (macroscopically) it depends on concentration; when the concentration is not probabilistic (i.e., when the continuum hypothesis is valid) the phenomenon, while probabilistic in origin, can be described by causal laws.[†] When the number of individuals decreases to the extent that the

---

[†]An interesting analysis of the continuum hypothesis from the viewpoint of natural philosophy is given in References 64 and 67.

population as a whole also begins to behave chaotically, the continuum hypothesis is no longer applicable.

It seems plausible to expect to be able to predict the overall behavior of the population starting from a knowledge of the behavior of its individuals, the molecules. In what follows, we will attempt to do just this for the diffusion of gases. We will proceed by different and ever more rigorous degrees of approximation. In Part A (Description) of this chapter we will use the method of description, assuming the reader is familiar with the elementary kinetic theory of gases. This will provide some conclusions that we will attempt, in Part B (Analysis), to represent by equations.

In later chapters, we will deal with the molecular chaos using probabilistic language, and this will provide new equations to describe the same

Table 1.1. Some Systems Used to Study Diffusion

| | |
|---|---|
| 1. Systems used by Graham. Gas B in the tube diffuses through a porous plug into gas A (usually air) and A diffuses into the tube. In (b), the water level rises or drops depending on whether the molecular weight of B is less than or greater than that of A. (For more details see Section 6.1.1.1.) | 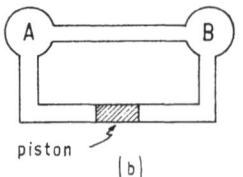 |
| 2. (a) A system used by Graham and Loschmidt. Gases A and B are initially kept separate in two bulbs and then (by, for example, turning a stopcock) are allowed to interdiffuse. A porous plug may be used in the joining tube.<br><br>(b) The gas chambers are joined by a second tube with a movable piston; then any change in pressure in the chambers resulting from the diffusion will be dissipated by movement of the piston. (For more details see Section 5.3.1.) | |

**Table 1.1**—*continued*

3. A system used by Stefan. The liquid B evaporates and the vapor diffuses through A (usually air) up the tube. (For more details see Section 5.3.2.)

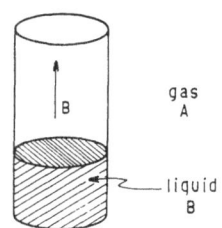

4. The diffusion cell. The gases A and B are forced to flow along two parallel tubes; they interdiffuse along the "working tube." The flow rates can be adjusted to produce any given pressure gradient along the working tube or to keep the pressure uniform. A porous plug may be used in the working tube. (For more details see Section 5.3.4.)

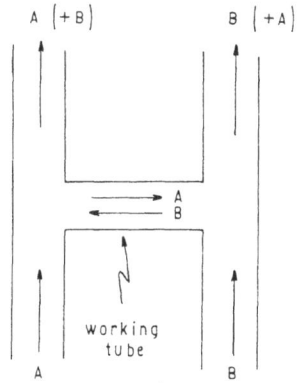

5. A core solid 1 gives off gas B which diffuses out through solid 2. The gas A surrounding the solids diffuses in toward solid 1. The solid 1 may, for example, react with A to produce B, or it may simply be wet and as it dries it gives off water vapor that diffuses out through the dry part. (For more details see Section 5.3.4.1.)

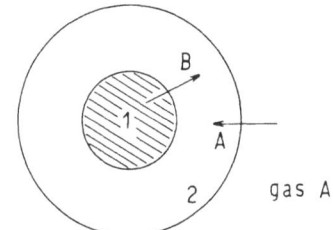

6. Schematic of a helium-cooled nuclear reactor. The core (of graphite and uranium) gives off the radioactive gas B, which diffuses out through the porous medium (graphite). Gas A (helium) diffuses into the chamber. (For more details see Section 5.3.4.)

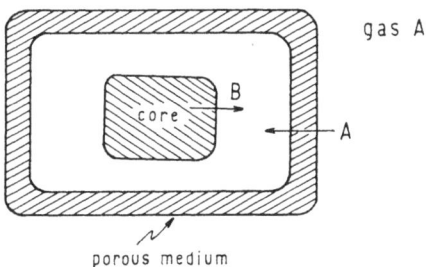

phenomenon. This approach can also be carried out to different degrees of approximation [what we will call the simplified kinetic theory (Chapter 2, Part B) and the rigorous kinetic theory (Chapter 3, Part B), respectively].

Table 1.1 shows some of the systems that have been used to study diffusion or in which diffusion plays an important part. System 2 in the table is an example of a *closed* system, i.e., the gases are completely enclosed. System 4, on the other hand, is an *open* system, i.e., there is an "infinite" source of gases A and B supplied to the ends of the working tube through which the gases interdiffuse. Systems 1, 3, 5, and 6 are *semiopen*, i.e., in these the source of one of the gases, B, is confined, and B itself diffuses out into an "infinite" unconfined gas A. In any of these systems there may be a chemical reaction of the gases, e.g., with each other or one gas may react with the source of the other (System 5); in some cases a porous medium may act as a catalyst for a reaction (see Section 5.3.3.1). All of the systems shown in Table 1.1, as well as others, will be described in more detail in later chapters.

## PART A.   DESCRIPTION

The molecules of a gas collide frequently and so are constantly exchanging kinetic energy. The reader will recall that for a gas in equilibrium, or for a gas not far from equilibrium, all the molecules have on the average the same kinetic energy (this average kinetic energy depends on the temperature of the gas). Since the kinetic energy is proportional to the mass of a particle, it follows that on the average the lighter molecules have greater speeds than the heavier molecules.

The pressure a gas exerts on a surface (real or imaginary) is the average effect of the collisions of the molecules with the surface (if the surface is imaginary then we consider one portion of the gas as exerting pressure on the adjacent portion). The more frequently collisions occur, the greater the pressure, so that the pressure increases as the number of molecules per unit volume increases (but recall that the pressure does not depend on the mass of the molecules).

Unless the molecular concentration is very high, the most probable type of collision is that of two molecules, usually called a *binary* or *simple collision*.

## 1.1.   Simple Molecular Collision

In a collision of free molecules, momentum is conserved. If the collision is elastic—and we shall restrict ourselves to elastic collisions—

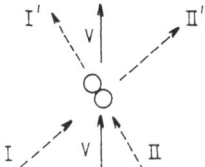

Figure 1.1. Simple collision between two like molecules. The dashed arrows represent the velocities of the molecules before (I, II) and after (I′, II′) the collision; v is the resultant of I and II (as well as of I′ and II′).

then the kinetic energy of the molecules is also conserved. These two conservation principles relate the velocities of the molecules before and after the collision. For example, if the colliding molecules belong to the same species, both masses are the same, and the conservation of momentum and kinetic energy implies conservation of the resultant velocity[†] (indicated by v in Figure 1.1) and conservation of the magnitude of the relative velocity. In a collision between two unlike molecules with different molecular masses, the relationship between the initial and final velocities is more complicated but is still determined by the conservation principles.

## 1.2. Systems without Walls

Consider a system of molecules that occupies a volume large enough for it to be considered devoid of walls (i.e., wall effects such as viscous drag can be ignored and the gas can be considered as unconfined). This volume can move as a whole (e.g., there may be a flow of gas) at a given overall velocity, and any such uniform motion is superimposed on the chaotic, intrinsic motion of the molecules.

If we wish, we can make the system independent of such an overall velocity by choosing a coordinate frame moving at that velocity, and we can define molecular fluxes with respect to this frame. We shall call this coordinate frame a *reference coordinate frame*[‡] and the velocity of this frame the *reference velocity*.

Consider a system composed of molecules of a single species and with a uniform molecular concentration (uniform pressure). In the reference coordinate frame the system will be motionless as a whole since the net

[†]Obtained by the parallelogram construction before and after the collision.

[‡]We are here using a reference coordinate frame that is moving at the average velocity of the molecules of the gas. Later (in Part B of this chapter) we will also make use of another coordinate frame, one which moves at the same velocity as the center of mass of the gas. These two frames may be different (see Section 1.3.2.7), and it is purely a matter of convenience in any particular problem which frame is chosen. For systems with walls the reference coordinate frame is usually fixed relative to the wall. It is important, of course, to specify which frame is being used in each case. We shall use the first-mentioned frame, the one moving at the average velocity of the molecules of the gas, throughout this descriptive section (Part A of this chapter) for systems devoid of walls.

molecular flux through any plane moving at the reference velocity will be zero. In the absence of external forces, any molecular concentration gradient that happens to arise will immediately dissipate, since such a gradient gives rise to a molecular flux which restores the uniform pressure. The dissipation of the concentration gradient occurs without loss of momentum since there are no walls to which the molecules can transfer momentum (e.g., via a viscous drag).

We thus have the following: In a one-component system without walls and with a uniform pressure, the only way there can be a permanent net molecular flux across a plane is to have the system move as a whole. Thus in the reference coordinate frame there will never be a permanent net molecular flux.

Consider now a system consisting of two or more species. If a concentration gradient exists for one of the species, there will be a net molecular flux of the molecules of that species from the more concentrated region to the less concentrated one. In general, there will be different fluxes for the different species. These fluxes contribute to any overall flux of the system.

When the molecules of a given species in a gas mixture possess an overall velocity[†] different from the reference velocity, the molecules of that species are said to be diffusing and we refer to the flux of the species as a *diffusive flux*. The overall velocity of a species relative to the reference coordinate frame is called the *diffusion velocity* of the species.[‡] We can define a *diffusive flux vector* for a given species: Consider a plane perpendicular to the diffusion velocity and moving at the reference velocity; then the magnitude of the diffusive flux vector is the net number of molecules of the given species that cross unit area of that plane in unit time and the direction of the diffusive flux vector is that of the diffusion velocity.

From the definition of the diffusive flux vector, it follows that the vector sum of the diffusive flux vectors of all the species is zero, since otherwise there would be an overall flux relative to the reference coordinate frame. For a binary system, the diffusive flux vectors are equal and opposite in the reference coordinate frame.

Diffusion corresponds, in a sense, to "segregation of species," for a diffusing species has a velocity different from that of the gas as a whole,

---

[†] When we refer here to the overall velocity of a given species, we mean the average velocity of a very large number of molecules; this average velocity should not be confused with the average thermal speed of an individual molecule.

[‡] The diffusion velocity as defined here is the difference between the average velocity of the given species and the average velocity of the gas. It can also be defined relative to the velocity of the center of mass of the gas. Again, it is important in any given situation to specify which definition is used.

and, under the right conditions, that difference in velocity will tend to separate the species (e.g., in a separation tube). Even though, then, we often think of diffusion as producing a more complete mixture of components, i.e., as integration of species, we find it convenient to emphasize the "segregative" aspect.

Thus, we will refer to the flux corresponding to a diffusion velocity as a *segregative* (or *separative*) flux, and we will refer to a flux which carries the different species with the same velocity as a *nonsegregative* flux.

If the masses of the diffusing species are different, which is the most common case, there will be an initial increase in concentration in the region toward which the lighter, faster molecules move. In a system maintained at constant pressure, this gradient of the total molecular concentration (pressure gradient) will instantaneously dissipate by generating a nonsegregative flux opposite to the flux of the lighter molecules. Since the pressure is constant, the total number of molecules of all species per unit volume must remain constant, and the net molecular flux of the whole system across any plane is therefore zero (when measured in the reference coordinate frame). Thus, we see that the nonsegregative flux spontaneously generated in the system is such that the overall flux is zero.

Further, from the point of view of an observer (measuring, e.g., the concentration), only one phenomenon is detectable, namely a net molecular flux for each species, and the sum of the net fluxes for all species is zero for the whole system. The two subphenomena—the initial increase in concentration in the region toward which the lighter, faster molecules move, and the dissipation of this pressure gradient by a nonsegregative flux opposite to the direction of the flux of the lighter molecules—which give rise to the one observed phenomena are naturally coupled and are inseparable. We shall thus say that at the level of observable phenomena there are only diffusive fluxes in this system.[†]

The diffusion of a given component of a gas is resisted by collisions of the molecules of that species with the other molecules. In consequence, the diffusive flux of a given component in a system without walls depends on two factors: (1) the number of molecules of the given component per unit volume (the molecular flux of a given component increases as its concentration increases); (2) the number of molecules per unit volume of the other species (as the concentration of the other species increases, the collision frequency increases, the molecular flights between collisions become shorter, and so there is more resistance to diffusion and a lower diffusive flux of the given species). However, if the increase in the total molecular concentration (i.e., in the pressure) is such that there is the same

---

[†] The diffusion of two species in opposite directions is referred to as *countercurrent diffusion* or *interdiffusion*.

relative increase of the molecular concentration of every species, the diffusive flux of the given species tends to increase because of factor (1) above and to decrease because of factor (2) in the same relative proportion. We conclude, therefore, that *at the level of rigor with which we are now analyzing the problem, the diffusive flux is independent of pressure*.

Further, since there is conservation of momentum in molecule–molecule collisions, and there are no walls to which momentum can be transferred, the diffusion takes place with conservation of the total amount of momentum.

Because this kind of diffusion proceeds through molecule–molecule collisions (in contrast to molecule–wall collisions, which we will consider below), it is called *molecular diffusion*.

## 1.3.  Systems with Walls

### 1.3.1.  Molecule–Wall Collision

The presence of a wall gives rise to a new type of collision, the molecule–wall collision. The interaction between a molecule of the gas and a solid wall is more complex than that between two molecules in a simple collision: The molecule may be retained by adsorption on the wall, and/or the wall surface may be smooth or rough.

Consider first what happens when the molecule collides with the wall without being adsorbed. We then have a typical elastic collision between a moving element and a body at rest; this results in a *specular collision* with a surface element of the wall, the angle of incidence being the same as the angle of reflection, as shown in Figure 1.2. There is conservation of momentum in the direction tangent to the surface element, and the normal component of the momentum is exactly reversed. If the surface is smooth, this description is valid for the whole surface. The situation is in a way identical to that of a system devoid of walls, since for a system with smooth walls there is no transfer of momentum in the direction of flow of a gas along a wall (the only difference between a system without walls and one with smooth walls is that in the latter the molecular velocity component normal to the wall is reversed).

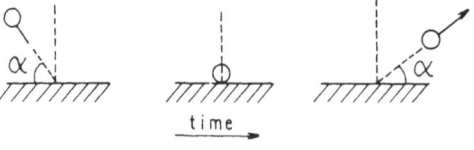

time

Figure 1.2. Molecule–wall elastic collision on a smooth surface. The angle of incidence ($\alpha$) is equal to the angle of reflection.

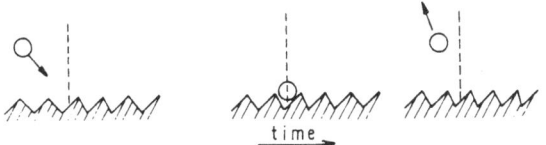

Figure 1.3. Molecule–wall elastic collision on a rough surface.

If the surface is rough, the orientation of a surface element does not necessarily coincide with the macroscopic orientation of the surface as a whole. Macroscopically, then, the collision is no longer specular: there is no conservation of momentum along the direction tangent to the overall surface and the angle of incidence is no longer related to the angle of reflection. A small but macroscopic element of the rough surface will rebound the molecules chaotically as shown in Figure 1.3, and the reflection, while specular on the microscopic scale, is *diffuse* on the macroscopic scale. Thus in contrast to the situation with smooth walls, the gas will lose momentum as it flows along a rough wall.

Now consider the case in which the molecule is adsorbed by the wall surface. On adsorption, the molecule transfers momentum to the wall. If the binding energy (the amount of energy released on adsorption) is not too great, the molecule can migrate over the surface, colliding with wall molecules and giving rise to the phenomenon of *surface diffusion*. When the adsorbed molecule acquires a momentum normal to the wall surface large enough to exceed the attractive force of the wall, it will be emitted. The direction of emission is independent of the original direction of incidence and the "reflection" is therefore both microscopically and macroscopically diffuse, as shown in Figure 1.4. On a macroscopic scale, it is therefore impossible to distinguish between a reflection that is specular on a rough surface and a reflection that is really an adsorptive collision.

A surface will never be smooth on the molecular level, so that in practice we will always be dealing with diffuse reflections.

Diffuse reflection has an interesting property if the wall is an isotropic medium, for then the flux from the surface must be the same in every direction, i.e., the flux through the equal elemental areas $dS$ in Figure 1.5

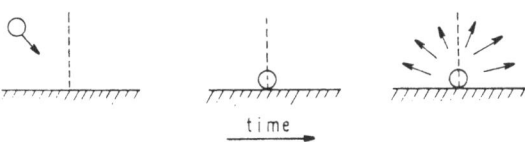

Figure 1.4. Molecule–wall collision with adsorption.

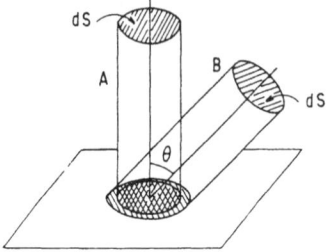

Figure 1.5. For diffuse reflection, the flux from the surface through the equal elements of area $dS$ shown is the same for all values of $\theta$. The cross-sectional area cut out on the surface by a cylinder inclined at angle $\theta$ is $dS/\cos\theta$.

must be the same. The molecules emitted at the angle $\theta$ and passing through the cylinder B come from a larger surface area than those emitted normal to the surface and passing through the vertical cylinder. This means that the number of molecules emitted per unit time per unit area of the surface is greatest in the direction normal to the surface (compare Lambert's law in optics).

### 1.3.2.  Mechanisms of Flux

In order to study the flux mechanisms in a system with walls it is convenient to represent the walls as suspended particles (like giant molecules) distributed throughout the rest of the gas molecules. For convenience we assume the "wall" particles are spherical and that the reference coordinate frame is fixed upon them, i.e., they are at rest in this frame.

This system of molecules and particles displays three characteristic lengths: (1) the mean distance between two molecular collisions, called the mean free path $\lambda$; (2) the average distance $\lambda_p$ between two neighboring particles; (3) the particle radius $r_p$. A fourth characteristic length, the molecular size, also arises; however, we assume for now that the molecular size is negligible in comparison to the other lengths.

In this system the molecules can collide with particles or with each other. We shall assume that the molecule–particle collision is macroscopically diffuse.

#### 1.3.2.1.  Viscous Flux

To analyze viscous flux we need to consider only the interaction between the moving fluid and one of the particles. The picture will be the same for all the other particles.

If $\lambda \ll \lambda_p$ and $r_p \gg \lambda$, there will be an enormously large number of molecules in the neighborhood of a particle. Some of these molecules will be colliding with other molecules and some will be colliding with the particle. But, under the given conditions, we can assert that *all* molecules rebounding from the particle will collide with molecules in the immediate vicinity of the particle. Our use of the word *all* here is important to our discussion, and because *all* of the rebounded molecules interact (via collisions) with the molecules surrounding the particle, we shall say that there is *total interaction* between the rebounded and surrounding molecules.

The rebounded molecules carry the properties of macroscopically diffuse collisions, the result of the molecule–particle collisions, and those molecules that have not collided with the particle "learn" about its existence from the rebounded molecules.

Consider now the evolution of this system of gas and particles from the moment a pressure gradient is applied. At first the layer of gas surrounding the particle moves at some velocity with respect to the particle. Because of the gradient in the total molecular concentration, the particle is bombarded more intensively on the high-concentration (high-pressure) side, and if the particle is to remain motionless (remember that we assumed that the particles are at rest in the reference coordinate frame) a force must be applied to it to balance out the force the gas exerts on it. Momentum must therefore be transferred from the gas to the particle as a whole.

The rebounded molecules are diffusely reflected and so have lost, on the average, the component of their momentum along the direction of the flow of gas produced by the pressure gradient. These molecules, through collisions, tend to decrease the average velocity of the molecules surrounding the particle. There is thus a damping effect on the boundary layer surrounding the particle, and the damping continues until the boundary layer reaches the same velocity as the particle.

As a consequence, *it is necessary to apply a constant pressure gradient in order to keep the flow of the gas constant*. This is not the case for a system devoid of walls or for a system in which the molecule–wall collisions are specular on a smooth surface.

In summary, we have seen that by virtue of two factors—the macroscopically diffuse molecule–wall collision and the total interaction between rebounded and adjacent molecules—all of the molecules (independently of the species to which they belong) reach the same velocity in the region immediately surrounding a particle.

This phenomenon differs from molecular diffusion in two respects: the gas molecules lose momentum as a whole and there is no segregation of species. We can distinguish between molecular diffusion and the viscous flux just described above by the following: In molecular diffusion there is segregation of species, there is no loss of momentum, and it is resisted by molecule–molecule collisions with other species. In the case of a viscous flux, there is no segregation of species, there is a loss of momentum, and it is resisted by molecule–wall collisions.

When *all* molecules rebounded from a wall collide with the surrounding molecules, we shall say that the molecule–wall collision is a *viscous collision*. When the momentum exchange is progressively and smoothly transferred to the bulk of the fluid, we have the so-called *viscous regime* of flow. If this progressive transmission is interrupted by vortexes, we have the so-called *turbulent regime* of flow.

The possibility of building up either the viscous or turbulent regime depends on the feasibility of dissipation of momentum, and this, in turn, depends on the intrinsic resistance of the fluid to this dissipation, the degree of influence of the wall (wall surface per unit volume of system), and the flow rate; the higher the first and second of these and the lower the third, the greater the possibility of achieving the viscous regime, which is the most common case in porous media. Consequently, we shall restrict our considerations of this type of flow to the viscous regime.

### 1.3.2.2.  Knudsen Flux

Let us now turn our attention to the opposite condition, namely $\lambda \gg \lambda_p$. In terms of the so-called Knudsen number, $\mathrm{Kn} \equiv \lambda / \lambda_p$, this condition is $\mathrm{Kn} \gg 1$. In this case the probability of a molecule–molecule collision is negligible compared to that of a molecule–wall collision. Those molecules rebounded or re-emitted by the wall do not collide with other molecules and the particle can thus only influence those molecules that collide with it—the molecules that have not collided with the wall cannot "learn" about its existence from the rebounded molecules. This kind of molecule–wall interaction will be called a *Knudsen collision* and the corresponding phenomenon will be called the *Knudsen regime*.

If in a multicomponent system with Knudsen flux there is a concentration gradient for a given component, there will be a net flux of the molecules of that component. As in a system devoid of walls, the species segregate, but collide with only one other species—the particles. A given species cannot "learn" about the existence of the other species since there are no molecule–molecule collisions. Then, because pressure affects only the concentration of the molecular species, *the Knudsen flux of a given component will be directly proportional to the pressure of the system.*

Further, we see that the Knudsen flux is determined by the thermal chaotic velocity of the molecules; this is in contrast to the viscous flux, where the molecules, while colliding chaotically among themselves, are being transported in a given direction at a given overall velocity.

For a one-component system under a pressure gradient, a nonseparative flux occurs and, as in a multicomponent system, the Knudsen collisions determine the flux.

We were able to distinguish molecular diffusion from viscous flux using dynamical considerations (p. 11), i.e., one occurred with, the other without, loss of momentum, and the mechanisms resisting the fluxes were different. We cannot do the same for the separative Knudsen flux of a multicomponent system and the nonseparative Knudsen flux of a pure system. For both there is loss of momentum and the mechanism resisting the flow is the same. The only difference between the two types of Knudsen flow is that for a pure system a pressure gradient is needed in order to develop a flux, while in a multicomponent system there can be isobaric Knudsen fluxes.

In summary, for a system in the Knudsen regime, there are as many individual fluxes present as there are species (as in molecular diffusion), and these fluxes are independent of each other (in contrast to molecular diffusion).

### 1.3.2.3.  Nonequimolar Flux

A system with walls exhibits another important difference from one without walls, as shown in the following.

Imagine a closed system, i.e., a system in which there is no entrance or exit of mass. We shall assume that there are neither sources nor sinks of matter within the system, that it consists of two components that are initially separated, and that the masses of the molecules of the two components are different. System 2a of Table 1.1 is an example of the type of system we have in mind.

As in the system devoid of walls, the lighter molecules fly faster than the heavier ones, creating a higher concentration of molecules in the region toward which they flow. A pressure gradient thus develops and, since there are walls, it will dissipate with loss of momentum. The pressure gradient supports a nonsegregative viscous flux that tends to restore the uniform pressure, while the diffusion tends to maintain the pressure gradient—the diffusive and the nonsegregative fluxes are thus clearly coupled.

At some point a quasi-steady-state will be reached and the pressure gradient will remain constant. The net molecular flux (diffusive plus the overall nonsegregative flux) must then be zero, for if it were not the pressure gradient would continue to vary, indicating that the assumed quasi-steady-state had not yet been reached.

Thus, in this quasi-steady-state the net molecular flux, as in a system with walls, is zero, but in contrast to a system without walls there is now a pressure gradient. The pressure gradient supports a flux that is balanced out by the diffusive fluxes.

The flux due to the pressure gradients can obviously be eliminated if the system is forced to remain at uniform pressure. This can only be achieved by external means and the net molecular flux will then no longer be zero[†]; further, once the flux due to the pressure gradient has died out, only the diffusive component of the flux will remain. However, we now have an apparent paradox: We expect the diffusive fluxes to be equal and opposite in a binary system (in the reference coordinate frame), but we have here a binary system with diffusive fluxes exclusively and yet the net flux is not zero. In order to make these two properties compatible, we have to propose that a nonsegregative diffusive[‡] flux has developed; we shall call this flux a *nonequimolar flux* because it arises as a result of a difference in concentration (nonequimolarity) of species.[¶] The more commonly used name for this flux is *diffusion slip flux*.

We reach the same conclusion regarding the existence of the non-equimolar flux by considering our original closed system in a quasi-steady-state, with its pressure gradient and zero net flux. The pressure gradient supports a nondiffusive nonsegregative flux, so the remaining fluxes, which are diffusive and opposite have to be different in order to balance out the pressure-gradient-induced flux.

The nonequimolar flux arises because (1) the species have different molecular masses—the different masses of the molecules results in a different average velocity of the molecules and this difference in velocity is what tends to produce a pressure gradient; and (2) the system has walls. If the molecular velocities did not depend on the molecular masses, the nonequimolar flux would not exist.

We thus see that in a system with walls the total diffusive flux of a given species consists of two components: the segregative one and the nonequimolar, nonsegregative one. The sum of these two components gives the *total diffusive flux* of the species: diffusive flux + nonequimolar flux = total diffusive flux.[§]

In Chapter 2 we will find characteristic parameters (i.e., the molecular diffusion coefficients) for the segregative diffusive flux for a system *without walls*, a system in which (as we saw on p. 7) two inseparable phenomena take place. Now in a system *with walls*, the total diffusive flux of a given species is, as we have just seen above, solely due to the different masses of the molecules (resulting in the first of the two subphenomena previously described); for such systems the segregative diffusive flux is given by the

---

[†]We could use, for example, System 2b of Table 1.1. If the mass of a molecule of B is less than that of a molecule of A, there will be a net flux of gas counterclockwise through the system—a net flux that is *not* the result of a pressure gradient.

[‡]We refer to this flux as diffusive even though it is nonsegregative because it is the result of a difference in concentration of species (its "driving force" is the same as that for the segregative diffusive flux), and it accompanies the segregative diffusive flux.

[¶]For example, in System 2b of Table 1.1, if the molecular mass of B is less than that of A, the nonequimolar flux is from B to A in the top cross tube.

[§]In the limit of a pure Knudsen regime, the Knudsen flux is the only flux to be considered.

same formula used for systems without walls, so the unobservable non-equimolar flux has to be added to the segregative diffusive flux term in order to obtain the correct value for the total diffusive flux. Further, the nonequimolar flux is nonsegregative because it balances out the nonsegregative flux (the second of the two subphenomena) that arises naturally in a system without walls.

In the most general case that we will consider there can be a diffusive flux (segregative), nonequimolar flux (diffusive, nonsegregative), and viscous flux (nondiffusive, nonsegregative). Any theory of diffusion has to make the calculation of these fluxes possible.

There is considerable confusion in the literature regarding nonsegregative fluxes. The following terms have been used with no attempt to distinguish between what we have called the nonsegregative diffusive and the nonsegregative nondiffusive fluxes: hydrodynamic,[59] massic,[57] nondiffusional.[29] The nonequimolar flux, as we noted above, is also called "slip diffusion."[55]

We shall from now on use the term *diffusive flux* to refer only to the segregative diffusive flux, and the term *total diffusive flux* will be used only when we are referring to the sum of the segregative diffusive flux and the nonequimolar flux. We will sometimes refer to the nonequimolar flux as a *nonsegregative diffusive flux*.

In summary, the phenomenon of diffusion in a system with walls is one in which, provided the molecular masses of the diffusing species are different, (a) there is a net molecular flux when pressure gradients are absent, and (b) there must be a pressure gradient when there is no net molecular flux.

Further, we can say that diffusion is a phenomenon in which the molecules of a given species transfer momentum to the molecules of other species (if the transfer is only from molecule to molecule the diffusion is called molecular diffusion) or to the walls (if the transfer is only to the walls the diffusion is called Knudsen diffusion) and there is in general a different flux for each species; a nonsegregative flux is one in which the molecules interchange momentum and transfer some to the walls with the same fluxes for all species.

A more detailed analysis of the problem leads us to the following: Every kind of flux (molecular, Knudsen, viscous) will be characterized by properties of the system through, for example, different *constitutive equations*, i.e., equations such as: flux of a property = coefficient × gradient of the property concentration, with a different coefficient for each type of flux.

In accordance with what we have previously said, the nonequimolar flux will depend, not on properties inherent to the viscous flux, but on properties corresponding to the molecular and/or Knudsen diffusive fluxes.

But since the nonequimolar flux exists because there are walls, we expect that the Knudsen diffusion coefficient will appear in the equations for both diffusion regimes. There must therefore be some property of the non-equimolar flux that depends only on the Knudsen coefficient. We shall see what this property is in the next section (Section 1.3.2.4).

On the other hand, when walls are absent, the constitutive equation for diffusion will involve two subphenomena (segregative and nonsegregative fluxes), and there is no nonequimolar flux to be described by a corresponding constitutive equation.[†]

### 1.3.2.4.    Relationships between Fluxes

When the pressure is uniform, the molecular bombardment of the walls must be such that the net component of momentum transferred to the walls in the direction of the flow, summed over all species, is zero. This represents a condition that the total diffusive fluxes of all the species must satisfy. This relationship must depend on the molecular masses of the species, as we can see from a particular, simple example: In the Knudsen regime, the diffusive flux of a given species is proportional to the molecular velocity of that species and is thus inversely proportional to its mass. Since, further, this proportionality also holds for the free molecular flux through an opening in an extremely thin plate (see Section 1.3.3.2), the condition on the total diffusive fluxes of all species must depend on the molecular masses even when the frequency of molecule–wall collisions is negligible.

### 1.3.2.5.    Surface Flux

Until now, we have dealt with the movement of molecules in the gaseous phase. However, as we have seen, there is the possibility that a molecule will be adsorbed by the walls of the system and held for a given retention time $t_R > 0$. Only when the adsorbed molecule acquires an energy exceeding the energy barrier (equal to the energy $Q_a$ developed during the adsorption) will it be able to return to the gaseous phase. When this happens, the molecule escapes from the surface in a random direction.

---

[†]An illustrative and simple picture of these phenomena is given by the following: Imagine a train with two locomotives, A and B, one at each end, pulling on the train in the same or opposite directions. If A (at least initially) exerts more force than B, the train will move in the direction in which A is pulling and B's force favors or resists this motion. The passenger flux inside the train involves the velocity of the passengers with respect to the train and corresponds to the diffusive flux. The movement of the train corresponds to the nonsegregative fluxes: its velocity can be considered as the resultant of two components $v_A$ and $v_B$ (one for each locomotive), which represent the nonequimolar and viscous fluxes, respectively.

Figure 1.6. Surface of a solid with adsorbed molecules: the atoms or molecules of the solid are indicated by the large circles and the adsorbed molecules by small circles. A molecule can hop, as indicated by the arrows, from one adsorption site to another.

Let us see what happens during the time the adsorbed molecule is retained by the surface. Imagine a surface packing of the molecules of the wall such as the one depicted in Figure 1.6a, b. The large circles represent molecules of the solid while the small ones indicate adsorbed molecules. The location of the latter in the sites of higher coordination indicates that adsorption is due to van der Waals forces.

The distribution of energy on the solid surface is represented by the lower curve in Figure 1.7. Valleys correspond to sites of higher coordination energy while peaks represent lower coordination energy sites (we have assumed two coordination sites; the generalization to more is simple). The energy barrier between like coordination sites is $\Delta Q_a = \frac{1}{2} Q_a$ for the solid of Figure 1.6a and $\Delta Q_a = \frac{1}{3} Q_a$ for the solid of Figure 1.6b. Both values correspond to physical adsorption. For localized bonds (chemisorption) $\Delta Q_a \simeq Q_a$, while for delocalized bonds (e.g., on metal surfaces) $\Delta Q_a$ will be a small fraction of $Q_a$.

The picture of the surface flux is then as follows. A gas molecule is adsorbed on a low-energy site. After vibrating for a certain time, the molecule acquires an amount of energy $\Delta Q_a$ sufficient to allow the molecule to hop to another adsorption site, where it will begin to vibrate again. The molecule makes a certain number of hops until it acquires an amount of energy (greater than $Q_a$) sufficient to escape from the surface, thus returning to the gaseous phase. The existence of a net flux on the surface is thus a probabilistic phenomenon that depends on the existence of a surface concentration gradient of the hopping molecules. On the other hand, since the velocity of the surface migration is very much lower than the exchange rate between the gaseous and adsorbed phases, it can be assumed that equilibrium is reached locally at every site.

Figure 1.7. Variation of the surface energy of a solid assuming there are two types of coordination sites. The lower curve shows the energy variation along the solid surface; to escape from the surface, a molecule must acquire an energy equal to or greater than $Q_a$.

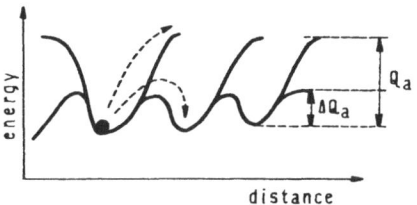

According to this picture we might expect the surface diffusion to increase rapidly with temperature. However, we must not forget that as the temperature increases, the concentration of adsorbed molecules decreases. Consequently, when the energy of the thermal agitation is greater than the adsorption energy $Q_a$, the surface concentration of adsorbed molecules as well as the surface flux will be practically zero. On the other hand, at low temperatures the probability is small that an adsorbed molecule will gain enough energy to overcome the energy barrier $\Delta Q_a$. Since the surface flux is very small at both high and low temperatures, the surface flux as a function of temperature must have a maximum. Obviously, the possibility of actually observing the maximum will depend on the relative values of $Q_a$ and $\Delta Q_a$.

In general, we can say that the flux within the body of the system enclosed by the wall (i.e., the volumetric flux) and the surface fluxes are additive. This is true to the extent that the adsorbed layer does not significantly alter the transverse section for the volumetric flux. If, however, the phenomenon takes place in the interior of micropores, the coupling between the two kinds of fluxes can be much more complex.

### 1.3.2.6.  Transition Regime

As we have seen, the diffusion velocity of a given species is limited by the two resistances to diffusion, namely molecule–molecule collisions[†] and molecule–wall collisions. In the molecular diffusion regime, the resistance due to molecule–wall collisions is extremely low, while in the Knudsen regime, it is the resistance due to molecule–molecule collisions that is extremely low. In the transition between these two extremes, both resistance mechanisms play an important role.

If the system flows as a whole and there are no walls, i.e., if $r_p = 0$, there will be no resistance to the flow, i.e., no viscous flux. This flux is also absent in an isobaric system with walls or in the Knudsen regime.

When $\lambda \simeq \lambda_p$, then for the nonsegregative nondiffusive flux the interaction between the rebounded molecules and the adjacent ones is not complete; the boundary layer surrounding the particle is not completely stopped and a steady state is reached in which the layer moves with respect to the particle or, as is usually said, the layer slips on the wall. This transition flux regime is usually called *slip flux*. This is not a new mechanism but a transition between the two extremes of complete interaction

---

[†]An idea of the magnitude of the resistance due to molecule–molecule collisions is obtained from the fact that, in general, the diffusion velocity is 1/10,000 that of the thermal, chaotic velocity of the molecules.

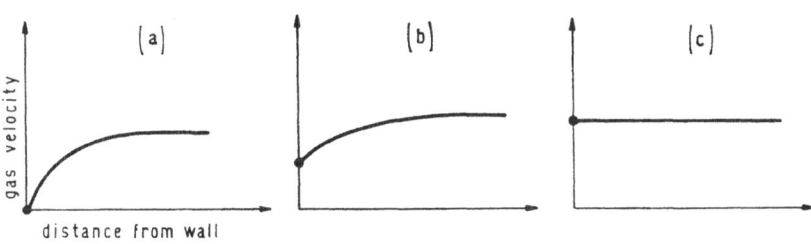

Figure 1.8. Velocity profiles near a wall: (a) viscous flow; (b) slip flow; (c) Knudsen flow.

between rebounded and adjacent molecules (viscous flux) and no interaction (Knudsen flux). In a viscous flux, the velocity of the gas decreases progressively toward the wall and is zero on it (Figure 1.8a); in the slip flux the velocity is not zero at the wall (Figure 1.8b); and in the Knudsen flux the velocity of the gas is independent of the distance from the wall (Figure 1.8c). We cannot foresee, at this stage of our knowledge, whether the nonequimolar flux will have such a velocity profile. This will be considered in Chapter 4.

We should point out that there can be another kind of flux whenever $r_p \rightarrow 0$. In such a case, there are not enough rebounded molecules to collide with all of the surrounding ones. As in the slip flux, the interaction is not complete, but for different reasons; in this regime, which is a transition between systems with walls and those devoid of them, the flux tends to increase as the particle size decreases for a given pressure gradient. This flux regime should be called *pseudo-Knudsen's*.

Finally, we must remember that we originally assumed that the molecular size was negligibly small compared to the other characteristic

Table 1.2. The Various Segregative and Nonsegregative Fluxes

| Condition | Regime | Nonsegregative | Segregative |
|---|---|---|---|
| $\lambda \ll \lambda_p$; $\lambda \ll r_p$ | Viscous | × | |
| $\lambda \gg \lambda_p$; $\lambda \ll r_p$ | Knudsen | × | × |
| $\lambda \simeq \lambda_p$; $\lambda \ll r_p$ | Slip | × | × |
| $\lambda_p \ll \lambda$; $r_p \ll \lambda_p$ | Pseudo-Knudsen | × | × |
| $\lambda \ll \lambda_p$; $r_p > 0$ | Molecular | | × |
| $\lambda \ll \lambda_p$; $r_p \rightarrow 0$ | Inviscid | × | |
| Molecular size $\simeq \lambda_p$ | Configurational | | × |
| Different molecular masses; $r_p > 0$ | Nonequimolar | × | |

lengths of the system. Whenever the molecular size is of the order of $\lambda_p$, another mechanism of diffusion appears, the so-called *configurational diffusion* (see Sections 1.3.3.3, 4.3.1.1, and 5.4.1).

The various segregative and nonsegregative flux regimes are summarized in Table 1.2.

### 1.3.2.7.  Molar and Mass Fluxes

In a nonsegregative flux, the direction in which the molecules, or moles, move is the same as the direction in which the mass (i.e., the center of mass of the gas as a whole) moves. However, if there is diffusion, conditions may be such that the molecular flux vector of the whole system is in the direction in which the lighter molecules diffuse, while the center of mass of the whole system moves in the direction in which the heavier molecules diffuse. It is thus possible for the molecular flux and the mass flux to be opposite in direction, even though they describe the same phenomenon.

It is important, then, that in speaking of the diffusive flux of a given component, we specify whether the flux is with respect to the molar velocity (i.e., the average velocity of the molecules of the gas) or with respect to the mass velocity (i.e., the velocity with which the center of mass of the gas moves).

### 1.3.3.  Other Wall Effects

### 1.3.3.1.  Permeability and Obstruction Effects

We shall now analyze the effect of wall geometry on the various flux mechanisms.

In molecular diffusion, it is obvious that the resistance due to molecule–molecule collisions is not modified by the wall geometry. While we thus do not expect the wall geometry to appear in the equations describing this type of flux, it does in general appear, as we will immediately see, but only because of the need to work with equations involving simple expressions.

If we consider diffusing molecules passing through a given cross section, the particles generating wall effects are merely obstacles to the diffusing molecules, decreasing the free area of the cross section and increasing the diffusion path. If, when calculating the diffusive flux, we take the true cross section and length, the calculation generally proves to be rather cumbersome. On the other hand, if we "ignore" the existence of the particles when computing cross sections and lengths, a correction factor in general appears that gives the effect of the particles on these quantities. When the container is a straight tube of uniform cross section, the wall does not affect either the diffusion cross section or the diffusion path, i.e., the correction factor for wall obstruction is then simply equal to one.

For Knudsen and viscous fluxes, the wall affects the intrinsic mechanism of the flux. For example, in a straight tube of uniform cross section, the flow resistance per unit length depends on the size and shape of the cross section; as the cross-sectional area increases and the perimeter increases, the resistance to flow decreases (i.e., the *permeability*[†] of the tube increases). But, since the resistance mechanisms for the Knudsen and viscous fluxes are different, the wall geometry will affect them differently. A correction for obstruction also has to be made, and this correction, too, is different for each flux mechanism.

### 1.3.3.2.  Influence of Pressure on Flow

Let us first consider a pure gas enclosed in a container that has as one wall a very thin plate with a small opening. We assume that the diameter of the opening is very small compared to the mean free path of the molecules, a condition that may be attained by using very low pressures, i.e., $p \to 0$. The net flux is determined by the probability that a molecule inside will cross the orifice and the probability that a molecule outside will cross it. Because the wall with the opening is assumed to be very thin, every molecule with a trajectory that falls inside the opening will pass through the orifice. The *molecular flux effectiveness*, which is the ratio of the number of molecules that completely pass through the opening to the number of molecules that enter it, will be 100%.

Let us now increase the thickness of the plate so that the opening is converted into a short tube. Now, some of the molecules entering the tube will collide with the tube wall and a fraction of these will be rebounded back through the inlet if the molecule–wall collision is macroscopically diffuse. Other molecules will pass through the tube without touching its walls, and a significant fraction of the exiting molecules will have collided with the walls. The molecular flux effectiveness will now be less than 100%.

These effects are enhanced if the tube length is increased; the enhancement is drastic if the tube is made tortuous as in the case of a real pore (see Section 1.3.3.3) in a solid medium. It is then impossible for a molecule to leave the pore without having collided with its walls, and the molecular flux effectiveness is extremely low.

---

[†] The resistance to flow is related to the permeability in roughly the same way that the resistance to an electric current is related to the electrical conductivity. See Section 1.7.2.1 for a more precise definition of permeability.

From this we see that the number of molecules with a probability of a long flight is directly proportional to $\lambda/L$, where $L$ is the tube length. But we must remember that all this is valid only for the Knudsen regime, e.g., for $\lambda \gg \lambda_p$ ($\lambda_p$ is the tube diameter).

Let us now see what happens when the pressure is increased, with the driving force per unit length, $\Delta p/L$, kept constant. The pressure increase has two consequences: (1) the ratio $\lambda/L$ decreases, i.e., the pressure increase is equivalent to an increase in $L$; (2) molecule–molecule collisions appear and begin to compete with molecule–wall collisions. This competition is inversely proportional to the Knudsen number, $Kn \equiv \lambda/d$, where $d$ is the tube diameter, and we see that as the Knudsen number decreases, a new flux mechanism, the viscous flux, appears.

A pressure increase thus has two opposite effects on the flux: Because of its effect on the length of molecular flights, it tends to decrease the flux, but because of its effect on molecule–molecule collisions (i.e., on the molecular concentration) it tends to increase the flux. If the increase in pressure begins from $p \simeq 0$, then in a capillary where $L > d$, the effect which predominates at first is the decrease of $\lambda/L$, and the flux therefore will initially tend to decrease. At higher pressures, the second effect prevails and the flux tends to increase. This means that there must be a minimum in the flux per unit driving force per unit length, usually called the *specific flux*, as a function of pressure (see Figure 1.14 below).

In porous media with tortuous pores, long molecular flights are not expected; thus, the decrease of flux with increase in pressure will be absent and the minimum will not be observed. This will be discussed in more detail in Chapter 4.

The minimum in the specific flux was observed for the first time in 1909 by Martin Knudsen[50] and its absence in porous media was also verified,[23, 41, 52, 66] although certain porous media show abnormal behavior.

### 1.3.3.3.  Influence of Pore Diameter on Diffusion

We saw in Table 1.2 that the mean distance between particles, or its equivalent, the pore diameter, influences the diffusion regime. Imagine a porous medium under uniform pressure containing a multicomponent mixture, and let us vary the mean distance between particles, or the pore diameter, keeping the total void (or pore) section and the total void (or pore) length constant, i.e., keeping the obstruction factor constant. When the pore diameter is large enough for molecular diffusion to prevail, the diffusive flux will be independent of the pore diameter. However, as the pore diameter or the pressure decreases, the resistance due to molecule–wall collisions appears, and the flux tends to decrease. Since in the Knudsen

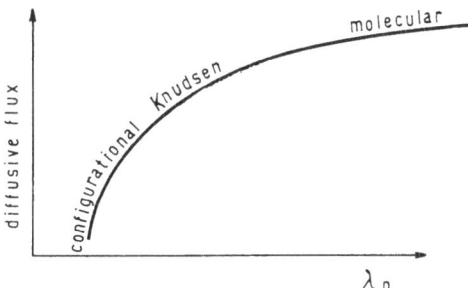

Figure 1.9. Type of diffusion regime
as a function of the pore diameter.

regime the flux is proportional to the length of the molecular flights, it seems reasonable to expect that the Knudsen flux will be directly proportional to the pore diameter.

If the pore diameter is decreased to the point at which the molecular size is no longer negligible compared with the pore diameter, another resistance appears, and Knudsen diffusion gives way to configurational diffusion. Consequently, we can represent the behavior of the system as depicted in Figure 1.9.

Finally, the reader should observe that the description given in Section 1.3.3.2 and in this section can be expressed, for both pure and multicomponent systems, in terms of two main parameters: the Knudsen number, and the ratio of the molecular size to the pore diameter.

### 1.3.3.4. Additivities of Momenta and Fluxes

It is obvious that the total amount of momentum that is lost by a given species is equal to the sum of the portions of its momenta transferred to the other molecular species plus the portion of its momentum transferred to the walls. In other words, additivity of momenta is valid.

*Since in molecule–molecule collisions the molecules simply exchange momenta and the total momentum is conserved, the total momentum lost by all species through molecule–molecule collisions is zero. On the other hand, the total momentum lost by the system through molecule–wall collisions is zero only when the pressure is uniform.*

This has an important consequence. When the diffusion regime is molecular it is customary and correct to state that *wall effects are negligible for diffusion*. But this is not to be confused with the *absence of walls*. In fact, in the molecular diffusion regime, the amount of momentum that molecules of a given species give to the walls is negligible in comparison with the amount of momentum given to the other species, but even though the amount lost to the walls may be very small, it is nevertheless a finite number and not zero. From this point of view, the walls do exist.

On the other hand, and by virtue of its definition, the total flux of a given species is the sum of its diffusive flux, nonequimolar flux, and viscous flux. The diffusive flux is zero for uniform composition, the nonequimolar flux is zero when the species have the same molecular mass, and the viscous flux is zero if the pressure is uniform. Thus in the approximation to reality with which we are now concerned, additivity of the fluxes of a given species is valid.

This has a further consequence. If the system is flowing as a whole with simultaneous diffusion, momentum is transferred to the walls, with the diffusive flux and the viscous flux each contributing to the total amount of momentum transferred to the walls. This means that the total diffusive flux of the system (i.e., the sum of the total diffusive fluxes for all the species) or, what is the same, the total nonequimolar flux of the system (sum of the nonequimolar fluxes for all the species) is related to the viscous flux through the total amount of momentum transferred to the walls.

### 1.3.3.5.  Distribution of Velocities

As a consequence of the chaotic behavior of the molecular population, the molecules do not all have the same velocity. There is a distribution of velocities, and the particular distribution occurring in a gas depends on the conditions of the system.

In a system devoid of walls, this distribution is determined by the molecule–molecule collisions exclusively. The presence of walls perturbs this velocity distribution since then there are molecule–wall collisions. It is to be expected, then, that the velocity distribution in the bulk of the fluid will in general be different from that in the neighborhood of the walls. This is not true, however, in the Knudsen regime, where there is only one kind of collision (molecule–wall), and the velocity distribution is the same for the whole system.

# PART B.   ANALYSIS

## B.1.   DEFINITIONS

# 1.4.   Diffusive and Nonsegregative Fluxes

## 1.4.1.   Preliminaries

In the gaseous systems that we will consider, the properties of the gas, in particular the composition, will vary throughout the gas. To analyze such a system, we will consider it to be made up of small elements of volume, with each volume element macroscopic in size in the sense that it contains a large number of molecules, but at the same time small enough so that the properties of the gas are approximately constant[†] in each element. The requirement that the volume elements be macroscopic ensures that the definitions of properties such as density, pressure, and composition will be meaningful for these elements, while the limitation on the size of the volume elements permits us to use the techniques of the calculus.[‡]

If we are dealing with a gas in which the properties vary along only one direction, e.g., the $z$ axis, the system can be considered one-dimensional, and the element of volume is a thin slice of the gas, i.e., a section of the gas formed by two parallel planes perpendicular to the $z$ axis and separated by a small distance $\Delta z$. For the more general case, the volume element may be a small cube.

The volume elements are generally taken as at rest relative to the walls of the gas container. This is not necessary, however, and we are free to choose any convenient coordinate frame. For example, for a gas flowing through a tube, it may prove convenient to use a frame at rest relative to the gas rather than a frame at rest relative to the tube.

In Part A, we considered quantities such as the average velocity of the molecule of a given species, diffusion velocities, diffusive flux vectors, etc. Each such quantity in general varies throughout a gas, and while many of them can be defined for the gas as a whole, we will define them for a representative volume element in order to be able to take into account their variation throughout the gas.

We will at first specifically refer to the volume elements in giving definitions, but we will later drop such references when it is obvious that a definition is given for a volume element and not for the gas as a whole.

The properties of the system will in general also vary with time. The quantities defined are therefore functions of time and are defined for a given time $t$. Of course, if the system reaches a steady state, its properties will no longer depend on time.

### 1.4.2. Definitions

We begin by defining the average velocity of the molecules of species $i$ in a given volume element: It is equal to the sum of the velocities of all of

---

[†]Actually, we assume that the variation of these quantities is "infinitesimal," i.e., if $f(\mathbf{r})$ is the value of a given property at $\mathbf{r}$, then for an infinitesimal displacement $\Delta \mathbf{r}$ we have $f(\mathbf{r} + \Delta \mathbf{r}) = f(\mathbf{r}) + (\nabla f) \cdot \Delta \mathbf{r}$.

[‡]This is in fact the continuum hypothesis, which is still valid for the Knudsen regime and breaks down only at extremely low pressures, i.e., those of open, cosmic space.

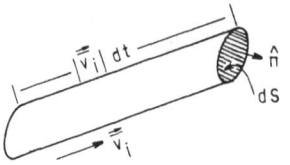

Figure 1.10. The number of molecules of species $i$ passing through the element of area $dS$ in time $dt$ is equal to the number of molecules of $i$ contained in the cylinder; $\bar{\mathbf{v}}_i$ is the average velocity of species $i$ and $\hat{\mathbf{n}}$ is the normal to $dS$.

the molecules of species $i$ in the volume element, divided by the total number of molecules of $i$ in that element.

Consider now a (plane) element of area $dS$ within a given volume element, and assume $dS$ has an orientation specified by a normal $\hat{\mathbf{n}}$ as in Figure 1.10. Let $\bar{\mathbf{v}}_i$ be the average velocity of species $i$ in the volume element relative to the coordinate system in which $dS$ is at rest. In time $dt$ the number of molecules of $i$ crossing $dS$ is the number contained in a cylinder with base $dS$ and length $\bar{\mathbf{v}}_i dt$.[†] The volume of the cylinder is $\bar{\mathbf{v}}_i dt \cdot \hat{\mathbf{n}} dS$, and if $n_i$ is the molecular concentration (number of molecules per unit volume) of species $i$ in the volume element, then the number of molecules of $i$ crossing the section $dS$ per unit area per unit time is $n_i \bar{\mathbf{v}}_i \cdot \hat{\mathbf{n}}$. Since we have used the average velocity here, this quantity is the *net molecular flux* of species $i$ across the section. The quantity

$$\mathbf{F}_i = n_i \bar{\mathbf{v}}_i \tag{1.1}$$

is called the *molecular flux vector for species i*. In general, $\mathbf{F}_i$, $n_i$, and $\bar{\mathbf{v}}_i$ vary throughout the gas, i.e., they are functions of the position vector $\mathbf{r}$, and to emphasize this we could write equation (1.1) as $\mathbf{F}_i(\mathbf{r}) = n_i(\mathbf{r})\bar{\mathbf{v}}_i(\mathbf{r})$.

The total molecular flux vector, i.e., that for all species of the gas, is obtained by summing expression (1.1) over all $i$.

We can in a similar fashion define the *molar flux vector* for species $i$, using the molar concentration $c_i$ of species $i$,

$$\mathbf{N}_i = c_i \bar{\mathbf{v}}_i \tag{1.2}$$

and we can define the *mass flux vector for species i*, using the mass concentration (density) $\rho_i = m_i n_i$, where $m_i$ is the mass of a molecule of species $i$:

$$\mathbf{G}_i = \rho_i \bar{\mathbf{v}}_i \tag{1.3}$$

---

[†]The increment $dt$ is assumed short enough so that the entire cylinder is contained in the volume element.

These three flux vectors are all related, since $n_i/c_i = N_{Av}$, where $N_{Av}$ is Avogadro's number, and $\rho_i = m_i n_i = c_i M_i$, where $M_i$ is the molar mass (usually called the molecular weight) of species $i$. The quantities $c_i$, $\rho_i$, $\mathbf{N}_i$, and $\mathbf{G}_i$ may be, of course, functions of the position vector $\mathbf{r}$.

The overall velocity of the molecules in the volume element can also be defined on a molecular, molar, or mass (i.e., center of mass) basis. The overall molecular and molar velocities are identical:

$$\vartheta_M = \frac{\sum\limits_{i=1}^{\nu} n_i \bar{\mathbf{v}}_i}{\sum\limits_{i=1}^{\nu} n_i} = \frac{\sum\limits_{i=1}^{\nu} c_i \bar{\mathbf{v}}_i}{\sum\limits_{i=1}^{\nu} c_i} \tag{1.4}$$

The overall velocity on the mass basis (i.e., the velocity of the center of mass of the molecules of the volume element) is given by

$$\vartheta = \frac{\sum\limits_{i=1}^{\nu} \rho_i \bar{\mathbf{v}}_i}{\sum\limits_{i=1}^{\nu} \rho_i} = \frac{\sum\limits_{i=1}^{\nu} n_i m_i \bar{\mathbf{v}}_i}{\sum\limits_{i=1}^{\nu} \rho_i} \tag{1.5}$$

where $\nu$ is the number of components in the system. Equations (1.4) and (1.5) give the velocities at which the moles and the mass of the system are moving, respectively. In a single-component system, these two velocities will be the same, but in a multicomponent system they can differ (see our earlier discussion in Section 1.3.2.7 and the example given in Section 1.4.3).

These velocities are defined, of course, relative to some particular coordinate frame, usually the frame in which the walls of the gas container are at rest.

The definitions of these overall velocities can be extended to include the whole gas. If, for example, the gas is in a closed container and the velocities are measured relative to the container, then $\vartheta_{Mt}$ (the subscript t indicates this is the velocity of the total gas) is zero, but $\vartheta_t$ may not be zero, e.g., as two gases with different molecular masses interdiffuse, the center of mass of the gas will shift.

If the average velocity of a given species is different from either of the velocities (1.4) or (1.5), it is said to be diffusing. We can define two diffusion velocities, one with respect to $\vartheta_M$, the other with respect to $\vartheta$:

$$\mathbf{v}_{iM}^d = \bar{\mathbf{v}}_i - \vartheta_M; \qquad \mathbf{v}_i^d = \bar{\mathbf{v}}_i - \vartheta \tag{1.6}$$

Table 1.3. Definitions of Diffusive Fluxes

| Flux | Defined with respect to $\vartheta$ | Defined with respect to $\vartheta_M$ |
|------|-------------------------------------|---------------------------------------|
| Molecular | $J_i^m = n_i(\bar{v}_i - \vartheta) = n_i v_i^d$ | $J_{iM}^m = n_i(\bar{v}_i - \vartheta_M) = n_i v_{iM}^d$ |
| Molar | $J_i = c_i(\bar{v}_i - \vartheta) = c_i v_i^d$ | $J_{iM} = c_i(\bar{v}_i - \vartheta_M) = c_i v_{iM}^d$ |
| Mass | $j_i = \rho_i(\bar{v}_i - \vartheta) = \rho_i v_i^d$ | $j_{iM} = \rho_i(\bar{v}_i - \vartheta_M) = \rho_i v_{iM}^d$ |

Corresponding to these two ways of defining the diffusion velocity and to the three types of flux (molecular, molar, and mass), there are six ways to define a diffusive flux vector; these definitions are listed in Table 1.3. Physicists seem to prefer to work with $j_i$ and $j_{iM}$, while chemists and chemical engineers use $J_{iM}$.[20] We shall use $j_i$ and $J_{iM}$.

It follows from their definitions that

$$\sum_{i=1}^{\nu} J_{iM} = \sum_{i=1}^{\nu} J_{iM}^m = \sum_{i=1}^{\nu} j_i = 0 \tag{1.7}$$

If there are only two components A and B of the gas, we have $J_{AM} = -J_{BM}$, $J_{AM}^m = -J_{BM}^m$, and $j_A = -j_B$, i.e., the molar, the molecular, and the mass diffusive fluxes of one component are equal and opposite to those of the other component.

### 1.4.3.  Example

Before we continue our discussion of fluxes, let us apply some of the above ideas to a relatively simple example: the one-dimensional interdiffusion[†] of two gases A and B. This might be approximated experimentally by starting with the gases in two separate containers divided by a diaphragm at $z = 0$ (Figure 1.11a), with the containers long in the $z$ direction; when the diaphragm is removed (gently enough so as not to disturb the gas), diffusion will begin. The number densities $n_A$ and $n_B$ for such an experimental setup are shown as a function of $z$ in Figure 1.11b, c. Figure 1.11b shows the number densities before the diaphragm is removed, and Figure 1.11c shows them at some time $t$ after diffusion has proceeded. The curves in Figure 1.11c are not intended to be exact, but merely to show what we expect of the number densities during the diffusion process: (i) at $z = \pm L$ (the location of the end walls of the container; we imagine $L \to \infty$), the

---

[†]Also often referred to as *counterdiffusion*.

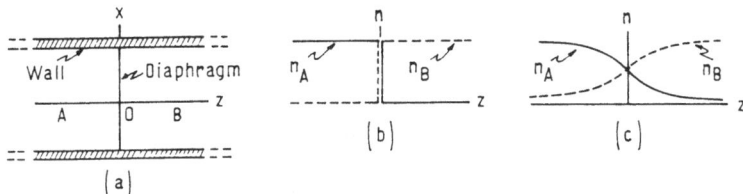

Figure 1.11. The interdiffusion of two gases A and B originally kept separate by a diaphragm at $z = 0$ as shown in (a). The curves in (b) give the number densities of A (solid curve) and B (dashed curve) before the diaphragm is removed. The curves in (c) give the number densities after diffusion has proceeded for some time.

number densities have their initial values; (ii) the number density of each gas drops off as we leave the end wall of the container in which it was originally confined and approach $z = 0$; (iii) the number density of A in B's original container is no longer zero, and vice versa.

We will assume that the experimental conditions are such that for at least the portion of the gas not too near any wall we can use as a volume element a slice of gas as shown in Figure 1.12. This volume element, centered on an arbitrary point $z_0$ and of width $\Delta z$, must satisfy the requirements discussed in Section 1.4.1, i.e., it must contain many molecules, but be small enough so that the properties of the gas do not vary much throughout it. For example, for the net molecular flux vectors at $z_0 - \frac{1}{2}\Delta z$, $z_0$, and $z_0 + \frac{1}{2}\Delta z$, we have

$$F\left(z_0 - \tfrac{1}{2}\Delta z\right) = F(z_0) - \left(\frac{\partial F}{\partial z}\right)_{z_0}\left(\tfrac{1}{2}\Delta z\right)$$

$$F\left(z_0 + \tfrac{1}{2}\Delta z\right) = F(z_0) + \left(\frac{\partial F}{\partial z}\right)_{z_0}\left(\tfrac{1}{2}\Delta z\right)$$

i.e., the variation of the properties is at most "infinitesimal" throughout the volume element. In fact, by virtue of the continuum hypothesis, the molecular structure of the gas is no longer taken into account and

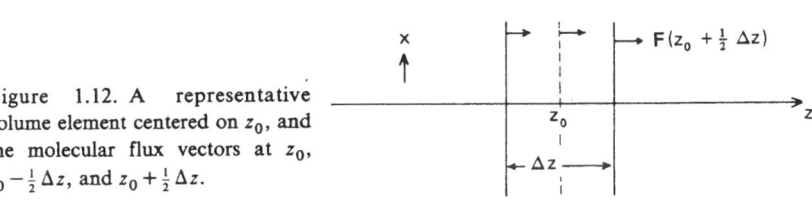

Figure 1.12. A representative volume element centered on $z_0$, and the molecular flux vectors at $z_0$, $z_0 - \frac{1}{2}\Delta z$, and $z_0 + \frac{1}{2}\Delta z$.

properties are defined for an arbitrary given point. This means that $\Delta z \to 0$ and fluxes are defined for the arbitrary plane at $z = z_0$ (the fact that a one-dimensional example is being considered enables us to define properties for a plane rather than for a point).

Let us assume that a quasi-steady-state has been reached by the system (cf. our discussion in Section 1.3.2.3). In this case there will be no net molecular flow across the plane at $z = z_0$, i.e., the average molecular velocity $\vartheta_M = 0$. It follows from equations (1.1), (1.2), and (1.4) that $\mathbf{F} = 0$ and $\mathbf{N} = 0$. From their definitions, it follows that $\mathbf{F}_i = \mathbf{J}_{iM}^m$, $\mathbf{N}_i = \mathbf{J}_{iM}$, where $i = A, B$, and we can verify that, in agreement with equation (1.7), we have $\mathbf{J}_{AM}^m = -\mathbf{J}_{BM}^m$, i.e., the molecular diffusive flux vectors are equal and opposite. Similarly, $\mathbf{J}_{AM} = -\mathbf{J}_{BM}$.

If the masses of the two components are the same, then the velocity $\vartheta$ of the center of mass of the volume element is also zero, and $\mathbf{G} = 0$. If the masses differ, then $\vartheta$ is not necessarily zero. For example, suppose $m_A = \frac{1}{2} m_B$. From the definition of $\vartheta_M$ we have $0 = n_A \bar{\mathbf{v}}_A + n_B \bar{\mathbf{v}}_B$, so $\vartheta = (m_A n_A \bar{\mathbf{v}}_A + m_B n_B \bar{\mathbf{v}}_B)/(m_A n_A + m_B n_B) = n_B \bar{\mathbf{v}}_B/(n_A + 2n_B)$, and $\vartheta$ is in the direction of $\bar{\mathbf{v}}_B$, the average velocity of the heavier molecules. For the mass flux vectors, we have $\mathbf{G}_A = m_A n_A \bar{\mathbf{v}}_A = -\frac{1}{2} m_B n_B \bar{\mathbf{v}}_B = -\frac{1}{2} \mathbf{G}_B$, and $\mathbf{j}_A = -\mathbf{j}_B$.

Let us for this same system now switch to a coordinate frame moving at the same velocity as the center of mass of the volume element. The superscript $+$ will be used to denote quantities measured in this new frame. We have, since $\vartheta_M = 0$,

$$
\begin{aligned}
\vartheta_M^+ &= \vartheta_M - \vartheta = -\vartheta, & \vartheta^+ &= \vartheta - \vartheta = 0, \\
\mathbf{F}_i^+ &= n_i \bar{\mathbf{v}}_i^+ = n_i(\bar{\mathbf{v}}_i - \vartheta) = \mathbf{F}_i - n_i \vartheta, & \mathbf{F}^+ &= -n\vartheta, \\
\mathbf{N}_i^+ &= \mathbf{N}_i - c_i \vartheta, & \mathbf{N}^+ &= -c\vartheta, \\
\mathbf{G}_i^+ &= \mathbf{G}_i - m_i n_i \vartheta, & \mathbf{G}^+ &= 0, \\
\mathbf{F}_i^+ &= \mathbf{J}_i^m, \quad \mathbf{N}_i^+ = \mathbf{J}_i, & \mathbf{G}_i^+ &= \mathbf{j}_i
\end{aligned}
\tag{1.8}
$$

The fluxes $\mathbf{F}^+$ and $\mathbf{N}^+$ correspond, as to be expected, to an overall (in the volume element) flow of gas at the velocity $\vartheta$. For the diffusive fluxes we still have, of course, $\mathbf{J}_{AM}^m = -\mathbf{J}_{BM}^m$, $\mathbf{J}_{AM} = -\mathbf{J}_{BM}$, $\mathbf{j}_A = -\mathbf{j}_B$.

### 1.4.4.  Relationships between Fluxes

It follows from their definitions that

$$
\mathbf{N}_i = \mathbf{J}_{iM} + x_i \sum_{j=1}^{\nu} \mathbf{N}_j
\tag{1.9}
$$

$$
\mathbf{F}_i = \mathbf{J}_{iM}^m + x_i \sum_{j=1}^{\nu} \mathbf{F}_j
\tag{1.10}
$$

$$
\mathbf{G}_i = \mathbf{j}_i + w_i \sum_{j=1}^{\nu} \mathbf{G}_j
\tag{1.11}
$$

where $x_i = c_i/\Sigma_i c_i$ is the molar fraction and $w_i = \rho_i/\Sigma_i \rho_i$ is the mass fraction of species $i$. Using $\Sigma_i x_i = \Sigma_i w_i = 1$, we can again obtain equation (1.7) from equations (1.9)–(1.11). We have

$$\sum_{i=1}^{\nu} \mathbf{j}_i = \sum_{i=1}^{\nu} m_i n_i \mathbf{v}_i^d = \sum_{i=1}^{\nu} M_i c_i \mathbf{v}_i^d = \sum_{i=1}^{\nu} m_i \mathbf{J}_i^m = \sum_{i=1}^{\nu} M_i \mathbf{J}_i = 0 \quad (1.12)$$

Equations (1.9)–(1.11) give the relationship between the total flux of a given species and its diffusive flux. Let

$$\mathbf{N}_i^g \equiv x_i \sum_{j=1}^{\nu} \mathbf{N}_j \quad (1.13)$$

This is the second term on the right-hand side of equation (1.9), and it represents the contribution of species $i$ to the total nonsegregative flux [nonsegregative diffusive (nonequimolar) plus nondiffusive]. We can consider $\mathbf{N}_i^g$ as made up of two parts,

$$\mathbf{N}_i^g = \mathbf{N}_i^N + \mathbf{N}_i^G \quad (1.14)$$

where $\mathbf{N}_i^N$ is the nonequimolar flux of species $i$, and $\mathbf{N}_i^G$ is the total nonsegregative nondiffusive flux of species $i$. As we noted earlier (Section 1.3.2.1), the only nondiffusive nonsegregative contribution to the flux that we will consider is the viscous one; so we take $\mathbf{N}_i^G = \mathbf{N}_i^V$. The total diffusive flux $\mathbf{N}_i^D$ of species $i$ is then

$$\mathbf{N}_i^D = \mathbf{J}_{iM} + \mathbf{N}_i^N \quad (1.15)$$

and we can write for the total flux $\mathbf{N}_i$

$$\mathbf{N}_i = \mathbf{J}_{iM} + \mathbf{N}_i^N + \mathbf{N}_i^V = \mathbf{N}_i^D + \mathbf{N}_i^V \quad (1.16)$$

We have separated the total flux of species $i$ into two parts (the total diffusive and viscous) for, as we shall see later (in the introductory paragraphs to Chapter 4), we can then write separate constitutive equations for these two contributions to the total flux. Relationships similar to equation (1.16) can be written for the fluxes $\mathbf{F}_i$ and $\mathbf{G}_i$.

If we sum over all species in equation (1.15) and use equation (1.7), we obtain

$$\mathbf{N}^D = \mathbf{N}^N \quad (1.17)$$

The nonequimolar flux, like the viscous flux, is nonseparative and so we

Table 1.4. Velocities Associated with Different Fluxes

| Molar flux | Mass flux |
|---|---|
| $J_{iM} = c_i v_{iM}^d$ | $j_i = \rho_i v_i^d$ |
| $N_i = c_i \bar{v}_i$ | $G_i = \rho_i \bar{v}_i$ |
| $N_i^D = c_i v_{iM}^D$ | $G_i^D = \rho_i v_i^D$ |
| $N_i^N = c_i \vartheta_M^N$ | $G_i^N = \rho_i \vartheta^N$ |
| $N_i^V = c_i \vartheta^V$ | $G_i^V = \rho_i \vartheta^V$ |

have $N_i^N = N^N x_i$, and it then follows from equation (1.17) that $N^N x_i = N^D x_i$. We can therefore write equation (1.15) as

$$N_i^D = J_{iM} + x_i \sum_{j=1}^{\nu} N_j^D \tag{1.18}$$

The reader should observe that the diffusive fluxes of the different species are related by the conditions given in equations (1.7) and (1.12), and the relationship between the nonseparative fluxes for species $i$ and $j$ is $N_i^N / N_j^N = N_i^V / N_j^V = x_i / x_j$.

On the other hand, similar equations may be written in terms of velocities rather than fluxes. Table 1.4 gives the definition of the velocities associated with different fluxes. It should be noted that since the viscous and nonequimolar fluxes are nonsegregative, the velocities $\vartheta^V$, $\vartheta_M^N$, and $\vartheta^N$ are the same for all species. On the other hand, the velocity $\vartheta^V$ corresponding to the viscous flux is the same on both the molecular and mass bases, while $\vartheta_M^N$ and $\vartheta^N$ may in general be different from each other.

From equations (1.16)–(1.18) and Table 1.4 the following relationships arise:

$$\bar{v}_i = v_{iM}^D + \vartheta^V \tag{1.19}$$

and

$$v_{iM}^D = v_{iM}^d + \vartheta_M^N \tag{1.20}$$

where

$$\vartheta_M^N = \left( \sum_{i=1}^{\nu} c_i v_{iM}^D \right) / c$$

Table 1.5. Different Types of Fluxes

| Flux | Viscous | Nonequimolar | Total diffusive | Diffusive |
|---|---|---|---|---|
| Total of the system | $\mathbf{N}^V$ | $\mathbf{N}^N$ | $\mathbf{N}^D$ | $0$ |
| Of component $i$ | $\mathbf{N}_i^V = \mathbf{N}^V x_i$ | $\mathbf{N}_i^N = \mathbf{N}^N x_i$ | $\mathbf{N}_i^D$ | $\mathbf{J}_{iM}$ |

The reader can verify that

$$\boldsymbol{\vartheta}_M = \boldsymbol{\vartheta}_M^N + \boldsymbol{\vartheta}^V$$

$$\boldsymbol{\vartheta} = \boldsymbol{\vartheta}^N + \boldsymbol{\vartheta}^V$$

$$\mathbf{v}_i^D = \mathbf{v}_i^d + \boldsymbol{\vartheta}^N = \mathbf{v}_{iM}^D$$

$$\boldsymbol{\vartheta}^N = \left( \sum_{i=1}^{\nu} \rho_i \mathbf{v}_i^D \right) / \rho \tag{1.21}$$

For the relative velocity of two species, we can write any of the following:

$$\bar{\mathbf{v}}_i - \bar{\mathbf{v}}_j = \mathbf{v}_{iM}^d - \mathbf{v}_{jM}^d = \mathbf{v}_i^d - \mathbf{v}_j^d = \mathbf{v}_{iM}^D - \mathbf{v}_{jM}^D \tag{1.22}$$

Table 1.5 gives a summary of the different fluxes discussed.

## 1.4.5. Relationships between Diffusive Fluxes

According to their definitions (Table 1.3), there are two sets of directly related diffusive fluxes:

$$j_i / J_i = j_{iM} / J_{iM} = M_i; \qquad j_i / J_i^m = j_{iM} / J_{iM}^m = m_i \tag{1.23}$$

If we write for a binary mixture

$$M = M_A x_A + M_B x_B; \qquad m = m_A x_A + m_B x_B; \qquad x_A \equiv n_A / n \tag{1.24}$$

and use the definitions in Table 1.3, then after some algebra we obtain

$$\mathbf{j}_{AM} M_B = \mathbf{j}_A M; \qquad \mathbf{j}_{AM} m_B = \mathbf{j}_A m \tag{1.25}$$

and from equations (1.23) and (1.25) we obtain

$$\mathbf{j}_A = (M_A M_B / M) \mathbf{J}_{AM}; \qquad \mathbf{j}_A = (m_A m_B / m) \mathbf{J}_{AM}^m \tag{1.26}$$

## B.2. EQUATIONS OF CHANGE

## 1.5. Conservation Principles

A simple phenomenological description has shown us that the study of the phenomenon concerning us here is related to at least two extensive properties: the amount of momentum exchange and the amount of mass exchange. We shall see later that the amount of thermal energy exchange is also related to the process.

The use of conservation principles for these three extensive properties leads to *equations of balance* for them in the form of well-known *equations of change*. Since these equations of change are extensively analyzed in related literature[19, 21, 62, 71, 82] and their study is not our main objective, we shall discuss them here only briefly, merely writing them down and explaining the meaning of the various terms (we will derive some of them in Chapter 3).

The equation of balance for a particular extensive property is obtained by taking account of the different ways in which that property can be changed. There are different mechanisms of change for the three extensive properties but there are two transport mechanisms which are the same for all of them: the convective contribution due to the macroscopic flow of the system as a whole, and the molecular contribution due to molecular interaction.

The equation of balance is written for a differential volume element.

### 1.5.1. Momentum

To the molecular and convective contributions to the momentum balance we have to add the forces acting on the system; for a fluid these are, usually, the pressure $p$ of the fluid and gravity. The equation of change usually obtained is known as the *Navier–Stokes equation*, which is written as follows:

$$\frac{\partial}{\partial t}\rho\vartheta = -\nabla\cdot\rho\vartheta\vartheta - \nabla p - \nabla\cdot\tau + \sum_{i=1}^{\nu}\rho_i g_i \qquad (1.27)$$

where $t$ is time, $g_i$ is an external force acting on species $i$ (if gravity is the only external force to be considered, then $g_i = g$), and $\tau$ is the shear stress tensor corresponding to viscous transport.† The left-hand side of equation (1.27) is the rate of increase of momentum per unit volume; on the right-hand side, the first term is the rate of increase of momentum per unit volume due to convection, the second term is the force per unit volume

---

† $\tau$ is given by Newton's law of viscosity [Section (1.7.2)].

due to the fluid pressure on the volume element, the third term is the increase of momentum per unit volume due to molecular (or viscous) transport, and the fourth term is the external force on the volume element per unit volume.

The scalar product of equation (1.27) and the velocity $\vartheta$ gives the equation of change for the mechanical energy.

## 1.5.2. Energy

### 1.5.2.1. Total Energy

If we do not take into account energy forms such as nuclear and electromagnetic, which do not usually vary in the phenomena that we are concerned with, the total energy of the system is the sum of the internal energy and the kinetic energy of any overall motion. In writing down the energy balance we must include the work done by or on the system. The equation of change finally obtained is

$$\frac{\partial}{\partial t}\rho\left(\hat{U}+\tfrac{1}{2}\vartheta^2\right) = -\nabla\cdot\rho\vartheta\left(\hat{U}+\tfrac{1}{2}\vartheta^2\right) - \nabla\cdot\mathbf{q} + \rho(\vartheta\cdot\mathbf{g}) - \nabla\cdot p\vartheta - \nabla\cdot(\tau\cdot\vartheta)$$

(1.28)

where $\hat{U}$ is the internal energy per unit mass, and $\mathbf{q}$ is the flux of thermal energy due to molecular transport. The left-hand side of equation (1.28) is the rate of increase of energy (internal plus kinetic) per unit volume; on the right-hand side, the first term is the rate of increase of energy per unit volume due to convection, the second term is the rate of increase of energy per unit volume due to molecular transport, the third term represents the rate of work done by gravity on the fluid per unit volume, the fourth term is the rate of work done on the fluid per unit volume by pressure forces, and the last term is the rate of work done on the fluid per unit volume by viscous forces.

### 1.5.2.2. Thermal Energy

If from equation (1.28) we subtract the mechanical energy contribution, the equation of change for the thermal energy is obtained:

$$\frac{\partial}{\partial t}\rho\hat{U} = -\nabla\cdot\rho\hat{U}\vartheta - \nabla\cdot\mathbf{q} - p\nabla\cdot\vartheta - \tau:\nabla\vartheta$$

(1.29)

The last two terms arise from the last two terms of equation (1.28) on subtracting the mechanical energy. Both terms represent the conversion of mechanical energy into internal energy. The first of these terms corresponds to a reversible transition, while the second corresponds to an irreversible one (mechanical energy degradation into heat).

Equation (1.29) can also be written with the enthalpy or temperature as the dependent variables.[21]

### 1.5.3.  Mass

#### 1.5.3.1.  Total Mass

The only net mass contribution to the total mass is the convective one; therefore, the equation of change of the total mass is[†]

$$\frac{\partial \rho}{\partial t} = -\nabla \cdot \rho \vartheta \tag{1.30}$$

#### 1.5.3.2.  Mass of a Component

The rate of change of the mass of species A of a multicomponent system is equal to the net flux $-\nabla \cdot G_A$ of A into the volume element plus the rate of reaction, if any, of the component A:

$$\frac{\partial \rho_A}{\partial t} = -\nabla \cdot G_A + \sum_{j=1}^{q} a_j^m R_{ej} \tag{1.31}$$

where $R_e$ is the reaction rate per unit volume expressed in equivalents, $a^m$ is the stoichiometric coefficient on mass basis (positive for products, negative for reactants), the subscript $j$ denotes the $j$th reaction, and $q$ is the number of linearly independent reactions in which A participates. The sum of equation (1.31) over all species yields, of course, equation (1.30).

### B.3.    CONSTITUTIVE EQUATIONS

To express the problem of the transport properties in a complete analytical form, it is necessary to write down the constitutive equations for the fluxes $\tau$, $q$, and $G_A$. In doing this, we split, as in equation (1.16), the

---

[†]Care must be taken regarding the validity of equation (1.30). If the gas is diffusing through a porous solid, equation (1.30) represents the equation of change of total mass in the gas, provided the solid is not reacting or decomposing; otherwise, the rate of the reaction of the gas with the reactive solid must be added to the right-hand side of equation (1.30).

mass flux $G_A$ into the nonsegregative nondiffusive contribution (which, for us, will be $G_A^V$) and the total diffusive contribution ($G_A^D$). Our main purpose is to study the constitutive equation for $G_A^D$ (or its molar analog $N_A^D$, which we will use more frequently). We shall see that a particular solution of equation (1.27) will provide the constitutive equation for $G^V$ (or $N^V$).

We now want to obtain the constitutive equation for the mass (or molar) diffusive and viscous fluxes. To do this, we will assume (i) that the only gradients in the system are concentration gradients of either a given species or of the whole system, and (ii) that there are no external forces acting on the system that can produce segregation of the different species.[†]

## 1.6. Diffusive Fluxes

### 1.6.1. Binary Mixture. Diffusion in the Molecular Regime at Constant Pressure. Fick's Law

Our phenomenological description showed that the diffusion of a given species depends directly on the concentration gradient of that species: the larger the gradient, the greater the diffusive flux. Consequently, we can postulate that, in an *isobaric binary mixture*,

$$J_{AM}^m = -D_{AB} \nabla n_A \tag{1.32}$$

where $D_{AB}$ is a proportionality constant that we will call the *binary molecular diffusivity of A in B*. We obviously know nothing yet about the dependence of $D_{AB}$ on the state variables of the system or on the molecular masses of A and B.[‡]

Equation (1.32) is known as *Fick's law* or as *Fick's first law*. When this law was first formulated and for a long time afterward, there was no precise statement as to which flux (diffusive or total diffusive) was to be set equal to $D_{AB} \nabla n_A$. A systematic analysis of this problem was first made by Bird.[20]

---

[†] For example gravitation, which tends to segregate the heavier and lighter components.

[‡] If we add (wall) particles to the system, a correction factor due to the obstruction by the walls appears (cf. our earlier discussion in Section 1.3.3.1). The diffusivity $D_{AB}$ is then replaced by an effective coefficient of the form

$$D_{AB, \text{eff}} \equiv D_{AB} Q_m \tag{1.32a}$$

where $Q_m$ is the *obstruction factor*.

In principle, it would also have been correct to postulate that $\mathbf{J}_A^m = -D_{AB}\nabla n_A$, which provides a different definition of $D_{AB}$. However, it is more consistent to use equation (1.32) since in that equation the molecular flux used is referred to $\vartheta_M$ and not to $\vartheta$, and so both sides of the equation (the right-hand side through $n_A$) are on a molecular basis.

We can write equation (1.32) so that it is "on a mass basis" as follows. Since we have assumed the system is isobaric, we have $\nabla n = 0$, and so equation (1.32) can be written as

$$\mathbf{J}_{AM}^m = -D_{AB}\, n\, \nabla x_A \tag{1.33}$$

Using equation (1.24) and $w_A = \rho_A/\rho$ to relate $\nabla x_A$ and $\nabla w_A$, we obtain

$$\nabla x_A = \left( M^2/M_A M_B \right)\nabla w_A \tag{1.34}$$

If we substitute equation (1.34) into equation (1.33) and use equation (1.26) we finally obtain

$$\mathbf{j}_A = -\rho D_{AB}\nabla w_A \tag{1.35}$$

Equation (1.35) is the mass flux of A with respect to $\vartheta$. The diffusivity $D_{AB}$ in equation (1.33) is the same as that in equation (1.35).

As we found in the phenomenological description, at the level of approximation with which we are now working, $\mathbf{J}_{AM}^m$ is independent of the pressure and hence of $n$ (see the end of Section 1.2). Thus from equation (1.33), for a given $\nabla x_A$, we have

$$D_{AB}n = \text{const} \tag{1.36}$$

i.e., $D_{AB}$ depends on $n$ (and thus on the pressure); it may also depend on the relative composition and the temperature.

Using the ideal gas law ($n = p/kT$, where $k$ is the Boltzmann constant and $T$ is the absolute temperature) together with the principle of partial pressures ($p = p_A + p_B, n_i = p_i/kT, i = $ A, B), and assuming a constant temperature, we have $\nabla n_A = \nabla p_A/kT$. Then from equation (1.18) written on a molecular basis and from equation (1.32) we obtain

$$-\frac{\nabla p_A}{kT} = \frac{F_A^D x_B - F_B^D x_A}{D_{AB}} \tag{1.37}$$

Now using Table 1.4 together with equation (1.22), we can write

$$F_A^D x_B - F_B^D x_A = \frac{n_A n_B}{n}\left(v_{AM}^D - v_{BM}^D\right) = \frac{n_A n_B}{n}\left(v_{AM}^d - v_{BM}^d\right)$$

$$= \frac{n_A n_B}{n}\left(\bar{v}_A - \bar{v}_B\right) \tag{1.38}$$

By multiplying both sides of equation (1.37) by $kT$ we obtain a momentum balance for species A. The left-hand side may be regarded as the rate at which the momentum of A decreases per unit volume, and since A loses momentum only through collisions with B, the right-hand side is the rate, per unit volume, at which momentum of A is being transferred to species B by collisions. It thus follows from equation (1.38) that the momentum lost by A is proportional to the difference of the diffusion (or total) velocities of A and B and to the molecular concentrations of A and B. This reasoning, as well as equation (1.37), was developed by Maxwell[56] and Stefan[73, 74] and equations such as (1.37) are usually known as *Stefan–Maxwell equations.*[21, 47]

Later [equation (4.12)] it will be shown that equation (1.37) is also valid for a *nonisobaric* isothermal binary system in the molecular diffusion regime.

*For a system devoid of walls*, we have seen (Section 1.2) that in the reference coordinate frame $\mathbf{F}_A^D = -\mathbf{F}_B^D$ and equation (1.37) can thus be written as

$$\mathbf{F}_A^D = \mathbf{J}_{AM}^m = -D_{AB}\nabla n_A \qquad (1.39)$$

Analogous equations can be written on a molar basis.

It follows from their definitions that

$$\mathbf{J}_{AM}^m = -\mathbf{J}_{BM}^m \qquad (1.40)$$

so that $D_{AB} = D_{BA}$. Further, since equation (1.12) is valid whether or not there are walls, the following holds:

$$\mathbf{J}_A^m / \mathbf{J}_B^m = -m_B / m_A \qquad (1.41)$$

It is interesting to note that considerable confusion exists in the literature regarding the relationship between the diffusive fluxes of species in a system devoid of walls. Some authors give equation (1.40) as the relationship between the molar diffusive fluxes in a binary mixture, while others quote equation (1.41). We see here that both assertions are correct, depending on whether we take the velocity $\vartheta_M$ or $\vartheta$ as the reference velocity.

## 1.6.2. Multicomponent Mixtures

### 1.6.2.1. Knudsen Regime

We shall deal with the Knudsen regime through the two pictures of it developed in Sections 1.3.2.2 and 1.3.3.2.

1.6.2.1.a.   *Collisions between Gas and Giant Molecules.* Equation (1.37) can be adapted to this case using equation (1.38) and can be written as follows:

$$-\frac{\nabla p_A}{kT} = \frac{n_A n_p}{n' D'_{Ap}}\left(v_{AM}^D - v_{pM}^D\right) = \frac{n_A n_p}{n' D'_{Ap}} v_{AM}^D \qquad (1.42)$$

The last form holds since velocities are taken relative to the particles, i.e., $\bar{v}_p = 0$, implying $v_{pM}^D = 0$; $D'_{Ap}$ is the diffusion coefficient for the molecule–particle collision; the prime indicates that the properties correspond to a system including particles, so $n' = n + n_p$, and the subscript p refers to particles.

Since as we noted above, equation (1.37) is also valid for nonisobaric conditions, equation (1.42) must be valid for both a constant or variable total pressure.

It is to be noted that there is no viscous flux (i.e., $\vartheta^V = 0$) in the Knudsen regime, in spite of the existence of a pressure gradient. For this reason, either $v_{AM}^D$ or $\bar{v}_A$ can be used on the right-hand side of equation (1.42), a result that follows directly from equation (1.22).

Rearranging equation (1.42) and again using Table 1.4, we obtain

$$F_A^D = -D_{AA,\,eff}^K \nabla n_A \qquad (1.43)$$

where $D_{AA,\,eff}^K \equiv n' D'_{Ap}/n_p$ is the diffusion coefficient for this type of flux and is called the *effective Knudsen diffusivity.* Just as for $D_{AB}$, we know nothing yet about the variables on which $D'_{Ap}$ depends.

Since we are dealing with a Knudsen flux, equations (1.42) and (1.43) are independent of the presence of other species. The total flux is therefore the sum of the individual, independent, noncoupled fluxes:

$$F^D = \sum_{i=1}^{\nu} F_i^D = \sum_{i=1}^{\nu} \left(-D_{ii,\,eff}^K \nabla n_i\right) \qquad (1.44)$$

This property of the independence of species in the Knudsen regime has some important consequences: (i) $F_A^D$ is independent of the number of species, so equation (1.43) holds also for a pure gas (in which case $\nabla n_A = \nabla n$); (ii) equation (1.43) applies for both isobaric *and* nonisobaric systems, as already noted.

1.6.2.1.b.   *Knudsen Flux through an Orifice.* If on one side of the orifice there is a gas at a certain pressure and on the other side there is a vacuum, the flux will be proportional to the speed $\bar{v}_A$ of the molecules,

$$F_{zA}^D \propto \psi n_A \bar{v}_A \qquad (1.45)$$

where $\psi$ is the molecular flux effectiveness[†]; $\psi = 1$ when the thickness of the orifice tends to zero and $\psi \to 0$ as the thickness increases. For $\psi = 1$ the flux is in fact a free molecular one (no collisions with the orifice walls take place).

If a pressure drop exists between the two sides of the orifice, equation (1.43) holds.[‡] By comparison with equation (1.45) we infer that

$$D_{AA,eff}^{K} \propto \bar{v}_A \tag{1.46}$$

### 1.6.2.2. Transition Regime

In a multicomponent nonisobaric mixture in the transition regime, the molecules of a component A collide with all the other species, including the walls. In every one of these collisions, the species interchange momentum, so that the total amount of momentum lost by one species is equal to the total amount gained by all the other species. Consequently, the multicomponent generalization of equation (1.37) will have as many addends as species which collide with A, including the walls,

$$-\frac{\nabla p_A}{kT} = \frac{n_A n_B}{n' D_{AB}' Q_m}\left(\bar{v}_{AM}^D - \bar{v}_{BM}^D\right) + \frac{n_A n_C}{n' D_{AC}' Q_m}\left(\bar{v}_{AM}^D - \bar{v}_{CM}^D\right)$$
$$+ \cdots + \frac{n_A}{D_{AA}^K Q_p}\bar{v}_{AM}^D \tag{1.47}$$

where $\bar{v}_p = 0$, $\vartheta_p^V = 0$ (though $\vartheta_i^V \neq 0$, $i \neq p$) and, from equation (1.19), $v_{pM}^D = 0$.

Again, it is worth pointing out that the first term of equation (1.47) could have been written using total velocities, $\bar{v}_i$, instead of the total diffusive flux velocities, $v_{iM}^D$, through equation (1.22). For the general nonisobaric case we find ourselves compelled, however, to select one of the two possibilities because they are not equivalent for the last term, a fact that has caused several errors and misunderstandings in the subject of diffusion. The choice must be $v_{iM}^D$ for reasons to be discussed in Chapter 4.

---

[†]See Section 1.3.3.2 for a definition of this quantity.

[‡]If the orifice has a uniform cross section, it is nonsense to speak about an obstruction factor, but in the most general case (nonuniform cross section, tortuous path), we can propose

$$D_{AA,eff}^{K} \equiv D_{AA}^{K} Q_p \tag{1.45a}$$

where $D_{AA}^{K}$ is the Knudsen diffusivity for a straight tube with a uniform cross section, and $Q_p$ is the corresponding obstruction factor [cf. equation (1.32a) in the footnote on p. 37].

Of course, for isobaric conditions $\vartheta^V = 0$ and $\bar{v}_i = v_{iM}^D$, so no choice need be made in that special case.

Equation (1.47) can be written more concisely as

$$-\frac{\nabla p_i}{kT} = \sum_{\substack{j=1 \\ j \neq i}}^{\nu} \frac{n_i n_j}{n D_{ij,\,\text{eff}}} \left( v_{iM}^D - v_{jM}^D \right); \qquad v_{pM}^D = 0 \qquad (1.48)$$

where the particles are regarded as one of the $\nu$ components and we have taken into account equation (1.36), generalized to the other species[†]:

$$D_{ij} n = D'_{ij} n' = \text{const} \qquad (1.49)$$

Taking equation (1.38) into account, we can write equation (1.48) as

$$-\frac{\nabla p_i}{kT} = \sum_{\substack{j=1 \\ j \neq i \\ j \neq p}}^{\nu} \frac{F_i^D x_j - F_j^D x_i}{D_{ij} Q_m} + \frac{F_i^D}{D_{ii}^K Q_p} \qquad (1.50)$$

This is the constitutive equation for diffusion in the transition regime for a multicomponent mixture. The method we used to obtain it was suggested by Mason,[54, 55] who also proposed the following analogy with an electric current to visualize the phenomenon: The total diffusive flux is considered as a current that must flow through a set of resistances in series, each type of collision (molecule–molecule between one species and each of the other species, and molecule–wall) corresponding to a resistor.

Despite this and similar attempts to derive the constitutive equations from a momentum balance,[63, 68, 72] the transformation of equation (1.37) into equation (1.50) is a generalization without a rigorous foundation, even though it seems plausible. We cannot therefore assert that the diffusivities appearing in equation (1.50) are the same as those in equation (1.37), although we can assume that they will be the same if the molecule–molecule collisions are simple, i.e., ternary collisions do not take place.

---

[†]We observe that, when working with effective coefficients for porous media, we are in fact applying the continuum hypothesis to the porous medium. In other words, the operator $\nabla$ in equations (1.47) and (1.48) involves coordinates that are not necessarily the same as those of the real diffusion path and the fluxes are for a total cross section (solid plus voids) and not just the void section. This can be taken into account by an obstruction factor of the form

$$Q = \varepsilon^s \kappa \qquad (1.49a)$$

where $\varepsilon^s$ is the surface porosity of the porous medium and $\kappa$ is a tortuosity factor.
The degree of approximation with which we are now working makes it impossible for us to give here a more rigorous demonstration of equation (1.49a). This is left for Chapter 4.

Although we have obtained equation (1.50) in a straightforward way, its solution is very complex since the fluxes of the various species are coupled (see Chapter 5 for a more detailed analysis of this), so that the concentration gradient of one species influences the flux of the other species. This explains why we can observe: (1) a diffusive flux of a given species with no concentration gradient of that species, a phenomenon called *osmotic diffusion*[44]; (2) no diffusive flux of a given species even when there is a concentration gradient of that species—the result of the existence of a *diffusive barrier*[76]; (3) a diffusive flux opposite the concentration gradient of the species, a phenomenon called *inverted diffusion*.[33]

We remind the reader that the term $\nabla p_i / kT$ in the above equations originates from $\nabla n_i$ for isothermal conditions [see the discussion leading to equation (1.37)]. If it is assumed that $\nabla T \neq 0$, we cannot obtain the constitutive equation for diffusion starting from the equation valid only for the isothermal case.

### 1.6.2.3. Relationship between Total Diffusive Fluxes

The sum over $i$ of the first term on the right-hand side of equation (1.50) is zero since the double sum then changes sign on interchanging $i$ and $j$ and all the terms cancel out:

$$\sum_{i=1}^{\nu} \sum_{\substack{j=1 \\ j \neq i \\ j \neq p}}^{\nu} \frac{F_i^D x_j - F_j^D x_i}{D_{ij} Q_m} = 0 \tag{1.51}$$

This result reflects the fact that momentum is conserved in molecule–molecule collisions. The sum of equation (1.50) over $i$ yields, using equation (1.51),

$$-\nabla n = \sum_{i=1}^{\nu} \frac{F_i^D}{D_{ii,\text{eff}}^K} \tag{1.52}$$

Equations (1.51) and (1.52) are valid for both isobaric and nonisobaric systems; further, equation (1.52) is valid for any diffusion regime because equation (1.51) always holds.

When the pressure is uniform, equation (1.52) yields

$$\sum_{i=1}^{\nu} \frac{F_i^D}{D_{ii,\text{eff}}^K} = 0 \tag{1.53}$$

Equations (1.52) and (1.53) show what we foresaw (Section 1.3.3.4) in the phenomenological description: (1) Equation (1.52) shows that the relationship between the total diffusive fluxes depends on molecule–wall collisions; (2) since $D^{K}_{ii,\,\text{eff}} \propto \bar{v}_i$ [equation (1.46)] and $\bar{v}_i \propto m_i^{-1/2}$, it follows from equation (1.53) that, even when the pressure is uniform, there will be a net molecular flux (the nonequimolar flux) whenever the molecular masses of the diffusing species are different; this is due both to the presence of walls and to the nonequimolarity of the diffusing species, i.e., the origin of the driving force of such a flux is diffusive and not fluid-dynamic.

## 1.7.  Viscous Flux

### 1.7.1.  Probability of Specular vs. Diffusive Molecule–Wall Collisions

We have seen that, because of the macroscopically diffuse molecule–wall collisions, the wall exerts a damping effect on a flowing fluid and this makes it necessary to maintain a pressure gradient to keep the fluid flowing.

The mere existence of a rough surface is sufficient to produce macroscopically diffuse collisions, but the effect of the surface roughness on the flow depends on the ratio of the number of macroscopically diffuse collisions to the number of specular collisions. We can get a quantitative estimate of this as follows.

The necessary condition for a specular reflection of a light beam (more generally of any plane transverse wave) by a surface is

$$2 L_r \cos \alpha < \Lambda \tag{1.54}$$

where $L_r$ is the average roughness of the surface (i.e., the root mean square deviation of the surface from a plane surface), $\alpha$ is the angle of incidence with respect to the normal, and $\Lambda$ is the wavelength of the incident beam. For a de Broglie wave of a molecule of species $i$, the wavelength $\Lambda_i$ is

$$\Lambda_i = h / m_i \bar{v}_i \tag{1.55}$$

where $h$ is Planck's constant. It follows from equations (1.54) and (1.55) that the necessary condition for specular reflection of an incident molecule by a surface with an average roughness $L_r$ is given by

$$\cos \alpha < h / 2 L_r m_i \bar{v}_i \tag{1.56}$$

For an extremely well-polished surface with an average roughness $L_r \simeq 10^{-5}$ cm and for the lightest molecule (hydrogen), at 20°C specular reflection will occur for

$$\alpha > 89°58' \tag{1.57}$$

Under these conditions, the probability of specular reflection of a molecule is extremely small. The only way to have an important fraction of the collisions specular is to use a perfect monocrystal or a liquid surface. In both cases, the average roughness is of the order of the atomic diameters: $L_r \simeq 10^{-8}$ cm. For such a case, equation (1.56) gives, for hydrogen at 20°C,

$$\alpha > 55°42' \tag{1.58}$$

which indicates that in this case the fraction of specular reflections can be important.

It is to be noted that the cause of a diffuse reflection cannot be experimentally observed, i.e., experiment cannot distinguish between a rough surface and a retention time different from zero. The only observable fact is that the rate of emitted molecules is the same in any direction or that the Lambert–Knudsen law holds (Lambert[21] deduced it for thermal radiation and Knudsen[50] applied it to molecules incident on a surface). This law, also known as the cosine law, states that for any isotropic surface flux phenomenon (i.e., for a flux such that the amount of property emitted by the surface per unit time per unit area perpendicular to the direction of emission is independent of the direction of emission), the amount of that property emitted per unit area of the surface must be proportional to the cosine of the angle between the direction of emission and the normal to the surface:

$$\mathbf{F} \propto \cos \alpha \tag{1.59}$$

We foresaw this law earlier (Section 1.31, Figure 1.5).

### 1.7.2.  Newton's Law of Viscosity

We now obtain an expression for the momentum transfer between the layers of a fluid flowing over a surface.

Imagine two large parallel and horizontal plates, separated by a very small distance $\Delta z$, with a gas at rest between them (Figure 1.13). If the lower plate is moved at velocity $\vartheta_{x0}$, it will give momentum to the gas, accelerating it until a steady state is reached. In the steady state, the velocity of the gas at the wall will be the same as the velocity of the wall, will decrease in the positive $z$ direction, and will be zero at the upper plate.

The motion of the bottom plate causes the gas to flow and transfer momentum to the upper plate. To keep the gas flowing at a steady rate it is

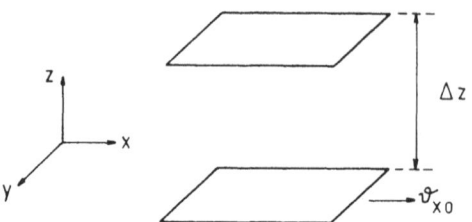

Figure 1.13. A gas contained between two large flat parallel plates a distance $\Delta z$ apart, with the bottom plate moving in the positive $x$ direction at the constant velocity $\vartheta_{x0}$. Momentum is transferred from the bottom plate to the gas and from the gas to the top plate.

necessary to keep the lower plate moving at a constant velocity $\vartheta_{x0}$, and this requires application of a constant force $f_x$. The higher the velocity and plate area and the smaller the distance between plates, the larger the force that must be applied,

$$\frac{f_x}{A} = \frac{\mu \vartheta_{x0}}{\Delta z} \tag{1.60}$$

where the proportionality coefficient $\mu$ is the gas viscosity.

The gas can be considered as a series of thin layers parallel to the plates, each layer pulling in the positive $x$ direction on the layer of gas above it. The force per unit area exerted in the $x$ direction by one layer on the layer above it is the shear stress tensor component $\tau_{zx}$ (the first subscript $z$ indicates the force is perpendicular to the $z$ axis, and the second subscript indicates the force is in the $x$ direction); $\tau_{zx}$ is thus the rate per unit area at which momentum is transferred from one layer of gas to the next, and so it is the momentum flux along $z$. Writing equation (1.60) in terms of $\tau_{zx}$ at any point of the gas, we obtain Newton's law of viscosity:

$$\tau_{zx} = -\mu \frac{d\vartheta_x}{dz} \tag{1.61}$$

Equation (1.61) shows that the direction of the momentum flux is opposite to that of the velocity gradient just as the direction of the diffusive flux is opposite to that of the concentration gradient.[†]

---

[†] We may point out that a similar development leads to Fourier's law of heat transfer,

$$q_z = -\lambda_K \frac{dT}{dz} \tag{1.61a}$$

where $\lambda_K$ is the thermal conductivity.

### 1.7.3.  Viscous Flux

#### 1.7.3.1.  Laminar Flow in a Capillary

The flow of an ideal gas in a straight capillary is described by the Navier–Stokes equation [equation (1.27)] written in cylindrical coordinates. We introduce Newton's law of viscosity into that equation and assume a steady state (so that, e.g., $\partial \vartheta_z / \partial t = 0$). We further assume that the pressure $p$ (and hence $\rho$) varies only along the $z$ axis (which is taken as the capillary axis) and that only the $\tau_{rz} = \tau_{zr}$ components of $\tau$ are nonzero. From equation (1.27) together with the equation of continuity of the total mass [equation (1.30)] we obtain

$$\rho \vartheta_z \frac{\partial \vartheta_z}{\partial z} = -\frac{\partial p}{\partial z} + \frac{\mu}{r}\left[\frac{\partial}{\partial r}\left(\frac{\partial \vartheta_z}{\partial r}r\right)\right] \tag{1.62}$$

We will also need to use the equation of state for the gas,

$$\rho = \frac{pM}{R_g T} \tag{1.63}$$

Observe that, if the pressure drops as $z$ increases, equation (1.63) predicts a corresponding decrease in the mass density and, consequently, equation (1.30) predicts a velocity increase along $z$. We thus see that the force due to the pressure is partially transferred to the walls and partially accelerates the fluid [equation (1.62)].

We are interested in the solution of equation (1.62) in a laminar regime where the inertial forces [the left-hand side of equation (1.62)] are negligible compared to the other ones (i.e., for a regime with Reynolds number $\mathrm{Re} \equiv 2\langle \vartheta_z \rangle \rho R / \mu < 2100$; the inertial-force term is only important at high Mach numbers). In this case, integration of equation (1.62) with the boundary conditions

$$d\vartheta_z / dr = 0 \text{ at } r = 0 \qquad \vartheta_z = 0 \text{ at } r = R \tag{1.64}$$

leads to

$$\vartheta_z \equiv \vartheta_z^V = -\frac{R^2}{4\mu}\left[1 - \left(\frac{r}{R}\right)^2\right]\frac{dp}{dz} \tag{1.65}$$

The average velocity in a capillary cross section obtained from equation (1.65) is

$$\langle \vartheta_z^V \rangle = -\frac{R^2}{8\mu}\frac{dp}{dz} \tag{1.66}$$

The molar viscous flux for the laminar flow of gases in capillaries is thus given by

$$N_z^V \equiv c\langle \vartheta_z^V \rangle = -\frac{R^2}{8\mu} \frac{p}{R_g T} \frac{dp}{dz} \tag{1.67}$$

Equation (1.67) is the *Hagen–Poiseuille law*,[60, 61] and has been widely verified experimentally for $Kn \ll 1$.

The form of equation (1.67) shows that it is the constitutive equation for the molar flux in the viscous regime for laminar capillary flow. We observe, however, that unlike Fick's and Fourier's laws, the statement of a conservation principle is required to obtain the Hagen–Poiseuille law.

### 1.7.3.2. Darcy's Law

When we are dealing with a porous medium rather than a simple capillary, we must take account of the obstructing effect of the porous medium in order to calculate the flux; the geometric factor $R^2/8$ in equation (1.67) is then replaced by an effective coefficient $B_k$, called the *permeability*, and we obtain[†]

$$N_z^V = -\frac{B_k}{\mu} \frac{p}{R_g T} \frac{dp}{dz} \tag{1.68}$$

Equation (1.68) is known as *Darcy's law*.[31]

In practice, the systems studied are usually in the steady state. For this case the number density will at each point have a constant value (although it will in general vary with $z$); the fluxes into and out of any cross-sectional slice of the capillary must therefore be equal, i.e., $N_z^V$ is the same across each cross section. It then follows from equation (1.67) that $p(dp/dz) =$ constant and integration of this equation from $z = 0$ to $z = L$ yields $(p_L^2 - p_0^2)/2 = L \cdot \text{constant}$ ($p_L$ and $p_0$ are the pressures at $z = L$ and $z = 0$, respectively), or $\bar{p}(\Delta p/\Delta z) = \text{constant}$, where $\bar{p} = (p_L + p_0)/2$, $\Delta p = p_L - p_0$, and $\Delta z = L$. Equations (1.65)–(1.68) can all be written in terms of $\bar{p}$ and $\Delta p/\Delta z$ and are usually written in that form in applications.

---

[†]Almost a century later, Brinkman[22] proposed a more complete equation than Darcy's:

$$\vartheta^V = B_k[\nabla^2 \vartheta^V - (1/\mu)\nabla p] \tag{1.68a}$$

The Darcy and Brinkman constitutive equations are analyzed in Chapter 4.

### 1.7.4.  Transition Regime

We have seen that when some of the molecules that have rebounded from the wall do not collide with adjacent molecules of the gas, the fluid layer on the wall has a certain velocity with respect to the wall, a phenomenon called *slip flux*.

Since the slip flux is a transition flux, we can treat it either as an extension of a pure viscous flux or as an extension of a Knudsen flux. The first of these possibilities is the simplest one since it merely implies the modification of one of the boundary conditions of equation (1.62) and we shall therefore consider it now, leaving the other possibility for Chapter 4.

The boundary condition at the wall is now given by

$$\tau_{rz} = \beta \vartheta_{z0} = -\mu \frac{d\vartheta_z}{dr} \qquad \text{at } r = R \qquad (1.69)$$

where the coefficient $\beta$, usually known as the *slip friction coefficient*, determines the value of the fluid velocity at the wall. When $\beta \to \infty$, the velocity at the wall goes to zero ($\vartheta_{z0} \to 0$), and we have the viscous flux described by the Hagen–Poiseuille law, which is thus a particular case of the more general law arising from equation (1.69). Using equation (1.69) in integrating equation (1.62) gives

$$N_z^{V,\text{ext}} = -\left( \frac{R^2}{8\mu} + \frac{R}{2\beta} \right) \frac{p}{R_g T} \frac{dp}{dz} \qquad (1.70)$$

The second term on the right-hand side of equation (1.70) represents the slip flux and $N_z^{V,\text{ext}}$ is the "extended" viscous flux. According to equation (1.70) the flux $N_z^{V,\text{ext}}$ is greater than the flux predicted by the Hagen–Poiseuille law, a result verified by the experiments of Kunt and Warburg[53,79] and Christiansen.[30]

Equation (1.70), while an extension from the pure viscous regime into the transition regime, does not correctly predict the behavior of the system when $Kn \gg 1$, i.e., it does not contain the Knudsen flux as a particular case.

We cannot, with the degree of approximation we are using here, obtain a theoretical expression valid for the viscous and Knudsen regimes and for the entire transition region between these two extremes. An attempt will be made to do this in Chapter 2 and again, with different arguments, in Chapter 4. Here we will merely cite and briefly discuss two expressions suggested as valid for all regimes.

The first of these was proposed by Knudsen[50] in an attempt to explain his experimental data. Figure 1.14 gives some of his data, in the

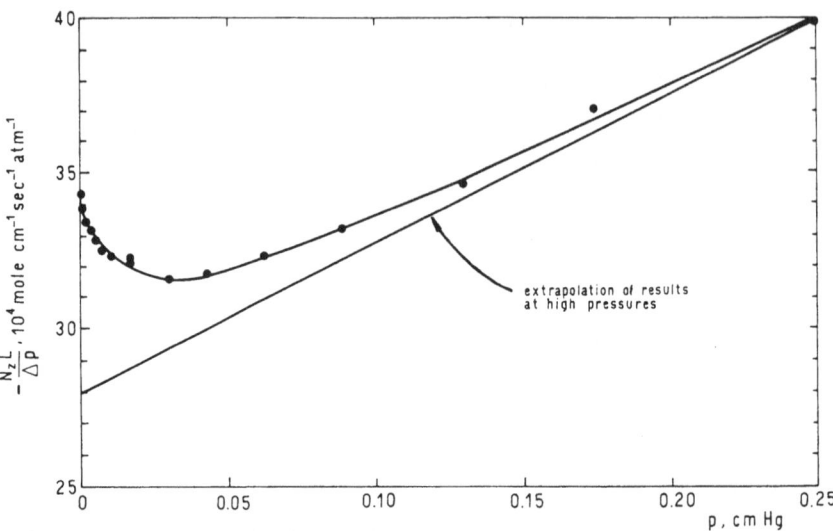

Figure 1.14. Some experimental results of Knudsen[50] for the laminar flow of a gas through capillaries.

form of a plot of $N_z L/\Delta p$ as a function of the pressure $\bar{p}$, for the flow of carbon dioxide at 25°C through a bundle of 24 capillaries each 2 cm long and with a radius of $3.33 \times 10^{-3}$ cm. The curve shows a minimum corresponding to $Kn = 1.02$. Such a minimum is generally found for flow in capillaries, with the location of the minimum depending on the nature of the gas and of the capillaries and on other conditions such as the temperature. The semiempirical expression proposed by Knudsen to describe such curves is

$$N_z = -\left( a^K \bar{p} + b^K \frac{1 + c_1^K \bar{p}}{1 + c_2^K \bar{p}} \right) \frac{1}{R_g T} \frac{\Delta p}{\Delta z} \tag{1.71}$$

where $a^K$, $b^K$, and the $c_i^K$ are coefficients that depend on the nature of the gas and the capillaries. At high pressures (viscous regime) the plot of $N_z L R_g T/(p_0 - p_L)$ as a function of the mean pressure $\bar{p}$ gives a straight line with slope $a^K$ and intercept $b^K c_1^K/c_2^K$. From a comparison with equation (1.70), $a^K = R^2/8\mu$ is the viscous flux coefficient and $b^K c_1^K/c_2^K = R\bar{p}/2\beta$ is the slip flux coefficient. The coefficient $b^K$ is the Knudsen diffusivity, as can be seen by considering the case $\bar{p} \to 0$. Since $b^K$ and the $c_i^K$ are constants, it follows that the slip friction coefficient $\beta$ must be proportional to $\bar{p}$.

Table 1.6. Values of the ratio $c_1^K/c_2^K$ [1, 2]

| Gas | Capillary material | $c_1^K/c_2^K$ |
|---|---|---|
| Hydrogen | Silver | 0.89 |
| Hydrogen | Aluminum | 0.81 |
| Hydrogen | Copper | 0.79 |
| Hydrogen | Iron | 0.73 |
| Hydrogen | Glass | 0.91 |
| Acetylene | Glass | 0.88 |
| Propane | Glass | 0.90 |

Knudsen predicted theoretically the coefficients $a^K$ and $b^K$, but not the coefficient $\beta$. He used the values of $c_1^K$ and $c_2^K$ obtained by applying the method of least squares to his experimental results:

$$c_1^K = 2.00(8/\pi)^{1/2} R/\mu \bar{v} \qquad (1.72)$$

$$c_2^K = 2.47(8/\pi)^{1/2} R/\mu \bar{v} \qquad (1.73)$$

From these values we find $c_1^K/c_2^K = 0.81$.

Adzumi[1, 2] calculated $c_1^K/c_2^K$ using experimental values for several gases flowing in capillaries of various materials. His results are shown in Table 1.6 and are similar to Knudsen's.

Another attempt to predict theoretically the flux in the transition regime was made by Weber,[81] who obtained the following expression:

$$\mathbf{N} = \mathbf{N}^V + \tfrac{4}{3}\mathbf{N}^S(1 - \Phi_w) + \mathbf{N}^K \Phi_w \qquad (1.74)$$

where $\mathbf{N}^S$ is the molar slip flux, $\Phi_w$ is a flux modulus defined by

$$\Phi_w \equiv Kn/(1 + Kn) \qquad (1.75)$$

and $\mathbf{N}^K$ is the molar Knudsen flux. Equation (1.74) is correct in the limiting cases and predicts the minimum observed by Knudsen (it is to be observed that $Kn \propto p^{-1}$).

## 1.8. Relationship between Total Diffusive and Viscous Fluxes

From the phenomenological description we know (Section 1.3.3.4) that to a first approximation the viscous and diffusive fluxes occur simultaneously and independently under the same pressure gradient. To use an electrical analogy it is as if both fluxes were in parallel. Therefore, for flow in a capillary the total flux $N_i$ of a given component is the sum of that obtained from the solution of equation (1.50) for $N_i^D$ and from equation (1.67) for $N_i^V$.

From equations (1.52) and (1.68) we obtain the following relationship between the viscous and total diffusive fluxes for flow in a porous medium:

$$-\nabla n = \frac{\mu}{B_k p} \mathbf{F}^V = \sum_{i=1}^{\nu} \frac{\mathbf{F}_i^D}{D_{ii,\,\text{eff}}^K} \tag{1.76}$$

## 1.9. Surface Flux

### 1.9.1. Retention Time

In the phenomenological description given in Section 1.3.2.5 we saw that the molecules adsorbed by a surface can migrate over the surface during the time $t_R$ that they are retained. Let us see what variables influence the retention time.

In 1924, Frenkel[32] derived the following theoretical expression for the retention time,

$$t_R = t_{R0} \exp(Q_a/R_g T) \tag{1.77}$$

where $t_{R0}$ is the period of oscillation of the molecules in the adsorbed state (vibrations normal to the surface) and $Q_a$ is the molar heat of adsorption. Equation (1.77), which has been experimentally verified for many pairs of adsorbent–adsorbate, predicts an increase of the retention time on going from physical adsorption to chemisorption, for this transition corresponds to an increase in $Q_a$.

The period of oscillation $t_{R0}$ is usually between $10^{-12}$ and $10^{-14}$ sec. It can be calculated using statistical mechanics, since the period of oscillation is related to the entropy lost by the adsorbed molecule. We shall not go into the details of the calculation here, but merely give the results. When the sorbed molecule can migrate on the surface, only one degree of freedom (corresponding to linear translation perpendicular to the surface) is lost and $t_{R0}$ is given by

$$t_{R0} = h/kT = 1.6 \times 10^{-13} \text{ sec} \tag{1.78}$$

at room temperature. The period of oscillation is greater than $1.6 \times 10^{-13}$ sec when a fraction of the lost entropy is transformed into vibrational entropy (thus giving rise to an excited vibration in the direction normal to

Table 1.7. Retention Times for Different Heats of Adsorption[a]

| $Q_a$, kcal/mole | $t_R$, sec |
|:---:|:---:|
| 0.1 | $1.2 \times 10^{-13}$ |
| 1.5 | $1.3 \times 10^{-12}$ |
| 3.5 | $4.0 \times 10^{-11}$ |
| 4 | $10^{-10}$ |
| 10 | $3.2 \times 10^{-6}$ |
| 15 | $1.8 \times 10^{-2}$ |
| 20 | $10^2$ |
| 25 | 1 week |
| 30 | 1 century |
| 147 | $10^{85}$ centuries |

[a]Reprinted from de Boer, [32] by permission of the Oxford University Press, Oxford.

the surface). If there is no surface migration, in which case the three translational degrees of freedom are lost, $t_{R0}$ can be as small as $10^{-16}$ sec.

De Boer[32] has calculated retention times as a function of the molar heat of adsorption at room temperature. The values obtained are given in Table 1.7. The last value in the table corresponds to the adsorption of oxygen on tungsten, and becomes 1 sec at 2,200°C.

## 1.9.2. Evidence for the Mobility of Adsorbed Molecules

The movement of molecules over surfaces can be demonstrated experimentally in several different ways. For example, the dispersion of drops of two immiscible liquids, the wetting of a solid by a liquid, and the ascent of crystals up the walls of a beaker involve surface migration. In the last case, the solution first migrates and then evaporation occurs. More recently, the surface migration of different chemisorbed gases on tungsten has been observed with the field emission microscope.

The mobility of physically adsorbed molecules was first demonstrated by Volmer and Adhikari[77] for molecules of benzophenone. Later, the same authors[78] showed that the benzophenone migration on glass was the result of a surface concentration gradient. Bangham and Fakhoury[8] pointed out the existence of a two-dimensional pressure (see Section 1.9.3) in the two-dimensional "gas" of adsorbed molecules.

Further evidence for a surface flux arose from the observation of gas fluxes higher than those predicted for the viscous, transition or Knudsen regimes. For example, Wentworth[83] observed that the measured fluxes were proportional to $(\nabla p)^n$ with $n > 1$. Later, Wicke[84,85] studied the permeation of hydrogen, methane, nitrogen, and argon through sintered glass and activated carbon, finding that the results agreed with those predicted by the theories for either the viscous or Knudsen fluxes, depending on the working pressure; for the flux of $n$-butane through sintered glass, however, the following deviations from the theoretical predictions were observed:

$$T, °C \quad 0 \quad -35 \quad -80$$
$$\text{Deviation, \%} \quad 5.3 \quad 9.8 \quad 12$$

Other studies of this phenomenon were made at about the same time.[37,42] More systematic and better-founded works began to appear after 1950.[†]

It should be pointed out that a nonadsorbable gas such as helium cannot diffuse superficially, as shown by Ash, Barrer, and Lowson[5] for the temperature range $-195$ to $300°C$. This behavior is important in connection with the variation of the Knudsen flux with temperature, as we shall see in Chapter 4.


### 1.9.3.  Two-Dimensional Pressure

Just as a gas exerts pressure (expressed in $dynes/cm^2$) on the walls of its container, a two-dimensional gas exerts a two-dimensional pressure (expressed in $dynes/cm$) on its border.

For a fluid, the two-dimensional pressure $\varphi$ is related to the surface tension $\sigma^s$. Langmuir, in 1917, designed a balance for measuring the two-dimensional pressure on a liquid surface: a surface barrier is used as a two-dimensional analog of a piston.

Imagine a liquid surface, a fraction of which is contaminated by a sorbate which forms a surface film. The stippled area in Figure 1.15 represents this contaminated surface, the rest being free of sorbate. If we displace the rod of length $L$ through a distance $dz$, the work done by the film on the rod is

$$W = \varphi L \, dz \tag{1.79}$$

which is the work involved in the generation of the new contaminated surface (or the removal of the uncontaminated one).

On the other hand, work done against the surface tension of a surface is the work per unit area necessary to generate that surface. This work is done against the forces of molecular cohesion since, for a molecule on the

---

[†]See References 3–18, 24–29, 32, 35, 36, 38–40, 43, 48, 49, 58, 65, 80.

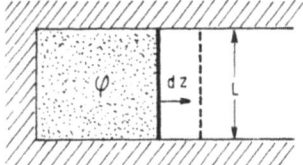

Figure 1.15. The measurement of the work done in moving the rod (heavy solid line) a distance $dz$, thereby stretching the surface film (the stippled area), permits the determination of the two-dimensional pressure in the film.

surface, the resultant of the cohesive forces is not zero and is directed into the bulk of the fluid.[†]

Let the surface tension of the pure liquid be $\sigma_0^s$ and the surface tension of the contaminated liquid be $\sigma_1^s$; then the work done by the system as the rod is moved a distance $dz$ is

$$W = (\sigma_0^s - \sigma_1^s) L \, dz \tag{1.80}$$

where $\sigma_0^s > \sigma_1^s$ since the adsorption of foreign molecules partially compensates the resultant of the cohesive forces on a surface molecule. A comparison of equations (1.79) and (1.80) yields

$$\varphi = \sigma_0^s - \sigma_1^s \tag{1.81}$$

For a solid, the two-dimensional pressure represents the difference between the superficial free energy of the solid in vacuum and the same energy at the pressure of the adsorbate.

In 1878, Gibbs obtained, by thermodynamic considerations, the following equation[(32)] relating the two-dimensional pressure $\varphi$ to the three-dimensional pressure $p$ for ideal gases:

$$N_M^s = \frac{p}{R_g T} \frac{d\varphi}{dp} \tag{1.82}$$

where $N_M^s$ is the number of moles of sorbate per unit surface area of sorbent,

$$N_M^s = N_M / S_{BET} \tag{1.83}$$

Here $N_M$ is the number of moles of sorbate and $S_{BET}$ is the surface area of the sorbent.[‡]

---

[†] We refer the reader to any elementary physics text for a discussion of surface tension and molecular cohesion forces.

[‡] $S_{BET}$ is measured by the Brunauer–Emmett–Teller method for determining the surface area of a porous solid.

The superficial molar concentration of sorbate (i.e., the number of moles of sorbate per unit volume of sorbent) is

$$c^s = S_V N_M^s \tag{1.84}$$

where $S_V$ is the surface area of sorbent per unit volume. By substituting equation (1.82) into (1.84) we obtain

$$c^s = \frac{pS_V}{R_g T} \frac{d\varphi}{dp} \tag{1.85}$$

If the adsorption isotherm $c^s(p)$ is linear, i.e.,

$$c^s = H^s p \tag{1.86}$$

where $H^s$ is a constant[†] (called the sorption coefficient), equation (1.85) leads to

$$\frac{d\varphi}{dp} = \frac{H^s}{S_V} R_g T \tag{1.87}$$

By integrating equation (1.87) with the boundary condition $\varphi = 0$ for $p = 0$ and then replacing $H^s$, $S_V$, and $N_M^s$ by using equations (1.86), (1.84), and (1.83), we finally obtain

$$\varphi S_{BET} = N_M R_g T \tag{1.88}$$

Equation (1.88) is the ideal gas law for two-dimensional gases. A linear adsorption isotherm (at low coverages, practically all isotherms are linear) thus leads to ideal gas behavior.

### 1.9.4.  Fluid Dynamic Model of the Surface Flux

By analogy with the volumetric flux, we can write a constitutive equation for the surface flux in terms of the two-dimensional pressure gradient. This idea was originally applied by Babbit[7] to the diffusion of gases in nonporous solids, was extended by Gilliland,[39,40] and applied

---

[†]In the general case (a nonlinear isotherm), $H^s$ may be defined as

$$H^s = dc_s/dp \tag{1.86a}$$

independently by Flood.[35-38] We write[†]

$$\mathbf{N}^s = -\frac{c^s}{R^s}\nabla\varphi \tag{1.89}$$

where $R^s$ is a resistance coefficient that is independent of the surface concentration and is a function of the temperature:

$$R^s = R_0^s T^{1/2}\exp(\Delta Q_a / R_g T) \tag{1.90}$$

As we saw in Section 1.3.2.5, $\Delta Q_a$ is the activation energy for the surface flux.

Using $\nabla\varphi = (d\varphi/dp)\nabla p$ and equation (1.85) solved for $d\varphi/dp$, equation (1.89) becomes

$$\mathbf{N}^s = -\frac{R_g T}{R^s S_V}\frac{[c^s(p)]^2}{p}\nabla p \tag{1.91}$$

Observe that the application of Gibbs' equation [through equation (1.85)] in the development leading to equation (1.91) implies the assumption that equilibrium has been reached between the gaseous and adsorbed phases. This is plausible since the exchange rate between the gaseous and sorbed molecules is much higher than the rate of surface migration.

### 1.9.5. Fick's-Type Law Model

The surface flux can also be defined in terms of a Fick's-type law,

$$\mathbf{N}^s = -D^s\nabla c^s \tag{1.92}$$

where $D^s$ is the surface diffusivity. Using equation (1.86a) we can write equation (1.92) in terms of the coefficient $H^s(p)$,

$$\mathbf{N}^s = -D^s H^s(p)\nabla p \tag{1.93}$$

---

[†]Equation (1.89) can also be arrived at by assuming there is a balance of the viscous forces $\mathbf{f}^V$ and the pressure forces $-\nabla\varphi$ in the adsorbed layer, i.e.,

$$\mathbf{f}^V = -\nabla\varphi \tag{1.89a}$$

Assuming that the viscous forces are directly proportional to the surface velocity $\boldsymbol{\vartheta}^s$ of the adsorbed molecules, we have

$$R^s\boldsymbol{\vartheta}^s = -\nabla\varphi \tag{1.89b}$$

where $R^s$ is the proportionality constant. Use of $\mathbf{N}^s = c^s\boldsymbol{\vartheta}^s$ in equation (1.89b) leads to (1.89).

Expressions similar to equation (1.93) were used by Barrer,[3,9-17] Carman,[25,27,29] and later authors. However, the first quantitative development was made by Taylor and Langmuir,[75] who measured the rate of dispersion of a deposit of cesium on a tungsten surface. The measured diffusivity was $D^s \simeq 10^{-6}$ cm$^2$/sec at 700°K.

It is important to emphasize that both equation (1.89), which is based on a fluid-dynamic picture of a two-dimensional fluid phase, and equation (1.92), which is a conventional diffusion law, are simply assumed to be valid, and each contains a coefficient ($R^s$ and $D^s$, respectively) to be determined experimentally. A comparison of equations (1.91) and (1.93) and use of equation (1.90) gives the relationship between the two phenomenological coefficients:

$$D^s = D_0^s \exp(-\Delta Q_a / R_g T) \tag{1.94}$$

where

$$D_0^s = R_g T^{1/2} [S_V R_0^s H^s(p)p]^{-1} [c^s(p)]^2.$$

The use of the surface diffusivity $D^s$ is inconvenient because of its variation with pressure (or, what is the same, with the degree of coverage). This variation was shown by Gilliland[40] by observing that the surface diffusivity of hydrocarbons on Vycor glass, when plotted as a function of coverage, first had a maximum and then a minimum. Despite this difficulty, the empirical relationships used to calculate surface fluxes (see Section 1.9.6) are expressed in terms of surface diffusivities.

De Boer[32] as well as Weaver[80] have developed mechanistic models based on the theory of the hopping molecule in order to predict surface diffusivities. The application of these models requires, however, the knowledge of parameters of the adsorbed-molecule–surface interaction, the determination of which is almost impossible.

The validity of equations (1.89) and (1.92) has been investigated experimentally by Horiguchi.[48] His results indicate that the equations provide correct results for high coverages in the monolayer and in the multilayer region. For lower coverages, the surface diffusivity must be multiplied by a coverage function $f(\Theta_s)$. For energetically uniform surfaces in which the heat of adsorption does not vary with coverage (e.g., for Graphon), the coverage function is given by

$$f(\Theta_s) = (1 - \Theta_s)^{-1} \qquad \text{for } \Theta_s < 0.6 \tag{1.95}$$

An expression similar to equation (1.95) was obtained theoretically by Higashi[45,46] by assuming that as the coverage increases there is an increase in the distance traveled in one hop of a migrating molecule.

## 1.9.6.  Empirical Relationships for Calculating Surface Diffusivities

Because of the lack of suitable and useful theoretical methods for calculating surface diffusivities, attempts have been made to obtain a "universal" empirical relationship for estimating surface diffusivities.

Using data given in the literature and his own results, Sladek[69,70] proposed the following relation:

$$D^s = 1.6 \times 10^{-2} \exp(-0.45 Q_a / m R_g T) \text{ cm}^2/\text{sec} \qquad (1.96)$$

where $m$ is a coefficient that depends on the type of sorbent–sorbate bond, as shown in Table 1.8. The table shows that values of $m > 1$ correspond to surface migration on conductors (except for the nonpolar adsorbates) where electronic delocalization decreases the activation energy, as we saw in Section 1.3.2.5 (Figure 1.6).

The relation (1.96) is based on values of $D^s$ that range over 11 orders of magnitude, values of the activation energy that range from 0.3 up to 200 kcal/mole, and temperatures that range from $-230$ to $600°C$. However, the estimation has an error margin of $\pm 75\%$.

Horiguchi[48] has proposed the following empirical relationship, which is valid for physical adsorption on some solids:

$$D^s = D_0^s \exp(-0.57 Q_a / R_g T) f(\Theta_s) \text{ cm}^2/\text{sec} \qquad (1.97)$$

where $D_0^s = 0.8$ for Graphon, 0.08 for Vycor glass, and 0.008 for Carbolac and silica-alumina. Equation (1.97) with these constants gives estimates of surface fluxes with an error margin of $\pm 50\%$.

Table 1.8.  Different Types of Gas–Solid Bonds[69,70]

| Bond | Nature of solid | $m$ | Example |
|---|---|---|---|
| van der Waals | | | |
|   polar adsorbate | Conductor | 2 | Sulfur dioxide–carbon |
| | Insulator | 1 | Sulfur dioxide–glass |
| | | | Ammonia–glass |
|   nonpolar adsorbate | Conductor | 1 | Argon–tungsten |
| | | | Nitrogen–carbon |
| | Insulator | 1 | Krypton–glass |
| | | | Ethylene–glass |
| Ionic | Conductor | 2 | Cesium–tungsten |
| | Insulator | 1 | Barium–tungsten |
| Covalent | Conductor | 3 | Hydrogen–metals |
| | | | Oxygen–tungsten |
| | Insulator | 1 | — |

It should be noted that $D_0^s$ is a function of the surface area per unit volume and of the tortuosity.

For the nonisothermal surface flux, Engel[34] and Gilliland[39] proposed the following relationship:

$$\mathbf{N}^s = - \frac{c^s}{R_0^s} \nabla \left[ \varphi T^{-1/2} \exp(-\Delta Q_a / R_g T) \right] \qquad (1.98)$$

which has been verified experimentally for several systems.

Krückels[51] has observed that for surface diffusion of water on nonporous silica-gel, the surface diffusivity depends on the surface concentration gradient; this is similar to the variation of viscosity with the velocity gradient and can be explained by means of Eyring's[47] rate process theory.

Before concluding this discussion of surface flux, let us point out that all the models of surface flux assume local equilibrium between the gaseous and adsorbed phases and additivity of the volumetric and surface fluxes. The condition of additivity of fluxes can be applied whenever the volume of the adsorbed phase is negligible compared to the pore volume.


# PART C.   STUDY SYSTEMATICS

The phenomenological analysis has enabled us to establish certain relationships, usually known as constitutive equations between fluxes and concentration gradients; these equations involve proportionality coefficients—the so-called transport coefficients. However, the phenomenological method of analysis has told us nothing about the transport coefficients, i.e., what quantities they depend upon. To do this it is necessary to formulate new, more sophisticated models. The more elaborate the model, the higher the degree of approximation between it and the real phenomenon.

We will first (in Chapter 2) use a simplified kinetic theory method and then (in Chapter 3) use more rigorous methods (such as the thermodynamics of irreversible processes and the more advanced kinetic theory) to obtain the constitutive equations of diffusion for multicomponent systems without walls. The influence of walls on the equations so developed will then be considered (in Chapter 4) and this step will lead to the most rigorous constitutive equations presently available for the description of the diffusion of gases in porous media. These constitutive equations, together with the conservation equations (dealt with in Chapter 5), enable us to predict the behavior of the systems that are our main concern in this book.

We shall be proceeding along a course of increasing complexity. Our presentation obviously does not correspond to the chronological evolution of the subject, which followed a much more tortuous and confused course, as the reader will see from the historical analysis given in Chapter 6.

# 2

# Elementary Prediction of Transport Coefficients

The constitutive equations for the molecular transport of momentum, heat, and mass can be derived from the kinetic theory of gases. The primary assumption of this theory is that any observable macroscopic property of a gas may be obtained from knowledge of the internal structure of its molecules and of the forces acting between them.

If it were possible to specify the instantaneous values of the position vector $\mathbf{x}$ and the momentum vector $\omega$ of each of the molecules of the system (in fact this violates the uncertainty principle) for all time $t$, any physical property $\Psi$ could be expressed as a function $\Psi(\mathbf{x}, \omega)$ in $(\mathbf{x}, \omega)$ space (the phase space) and would be completely determined from the information available. Classically such a description is in principle possible, but in practice it is impossible because of the huge number of molecules and the corresponding huge number of coupled equations of motion. A similar difficulty prevents a complete quantum mechanical description.

To get around this difficulty, it is customary to use a distribution function $F^{(s)}(\mathbf{x}_1, \mathbf{x}_2, \ldots, \mathbf{x}_s, \omega_1, \omega_2, \ldots, \omega_s, t)$ which is the probability that, in a system of $s$ molecules at time $t$, molecule 1 is at the point $(\mathbf{x}_1, \omega_1)$ of phase space, molecule 2 at $(\mathbf{x}_2, \omega_2), \ldots$, and molecule $s$ at $(\mathbf{x}_s, \omega_s)$. More precisely, $F^{(s)}(\mathbf{x}_1, \ldots, \mathbf{x}_s, \omega_1, \ldots, \omega_s, t) \, d\mathbf{x}_1 \cdots d\mathbf{x}_s \, d\omega_1 \cdots d\omega_s$ is the probability that, at time $t$, molecule 1 is in the volume element[†] $d\mathbf{x}_1$ (located about $\mathbf{x}_1$) and has a momentum in $d\omega_1$ about $\omega_1$, that molecule 2 is in $d\mathbf{x}_2$ about $\mathbf{x}_2$ and has a momentum in $d\omega_2$ about $\omega_2$, etc.

The distribution function $F^{(s)}$ satisfies Liouville's equation (the statistical-mechanical analog of Newton's second law), so that $F^{(s)}(\mathbf{x}_1, \ldots, \omega_s, t)$

---

[†] $d\mathbf{x}_1, d\omega_1$ are, of course, not vectors; they are just a shorthand way of writing, e.g., $dx_{1x} \, dx_{1y} \, dx_{1z}$ and $d\omega_{1x} \, d\omega_{1y} \, d\omega_{1z}$.

can be determined from the initial value $F^{(s)}(\mathbf{x}_1, \ldots, \omega_s, 0)$. The value of any physical macroscopic property (e.g., pressure) can be determined from knowledge of $F^{(s)}$.

An important aspect of this level of description of the system is that it is now continuous in that certainties have been replaced by probabilities. We can now consider that each point of phase space is occupied by a certain fraction of molecules or, in other words, given an arbitrary point of phase space and any one of the $s$ molecules, there is a certain probability that the molecule will be at that point.

A less detailed description than the above is given by the one-particle distribution function $F^{(1)}(\mathbf{x}, \omega, t)$, which is the probability that a given molecule is at $(\mathbf{x}, \omega)$ at time $t$ regardless of the position and momenta of the other $s - 1$ molecules. If the given molecule is molecule 1, this probability is given by the integration of $F^{(s)}$ over all of phase space for all the other $s - 1$ molecules; since all the molecules are identical (we are considering a one-component system), the result of the integration would be the same for any of the $s$ molecules, and so

$$F^{(1)}(\mathbf{x}, \omega, t)$$
$$= \int \cdots \int F^{(s)}(\mathbf{x}_1, \mathbf{x}_2, \ldots, \mathbf{x}_s, \omega_1, \omega_2, \ldots, \omega_s, t) \, d\mathbf{x}_2 \cdots d\mathbf{x}_s \, d\omega_2 \cdots d\omega_s$$

The function $F^{(1)}$ is suitable for the description of systems with properties that do not depend on the relative position of two or more molecules, which means that it is suitable for the description of dilute gases. This is the level of approximation with which we will deal in developing the kinetic theory of gases (both the "simplified" and "rigorous" forms). For simplicity, we will from now on omit the superscript (1).

The distribution function $F(\mathbf{x}, \omega, t)$ can also be interpreted as the fraction $dn/n$ of molecules in the volume element $d\mathbf{x}\,d\omega$ at $(\mathbf{x}, \omega)$. Another function, $f(\mathbf{x}, \omega, t) \equiv nF(\mathbf{x}, \omega, t)$, is frequently used—we shall later use this function exclusively: $f(\mathbf{x}, \omega, t)\,d\mathbf{x}\,d\omega$ is the average number of molecules per unit volume in the phase-space element $d\mathbf{x}\,d\omega$.

A still simpler description than those described above is the thermodynamic one, which assumes that the system is near a local equilibrium state, i.e., a state which can be described by a set of differential volumes, each of which contains a large number of molecules and is in equilibrium. Each differential volume may have a different temperature, density, composition, and average velocity. We will follow this approach in Part A of Chapter 3.

In this chapter, we will (a) derive the equilibrium distribution function for the velocities (i.e., starting from the Boltzmann energy distribution law, which we will merely state but not derive); (b) use the velocity distribution function to calculate quantities such as the average speed of the molecules; (c) assume there is a perturbation of the equilibrium state and that the perturbation is small enough so that the distribution function is not modified, introduce *ad hoc* the concept of the mean free path, and then use a simplified-kinetic-theory derivation of the transport coefficients.

In the rigorous kinetic theory (to be taken up in Chapter 3) we will (a) derive the equilibrium velocity distribution function in a more rigorous way (starting from the integrodifferential Boltzmann equation); (b) perturb the equilibrium state; (c) derive transport coefficients; (d) derive the constitutive equations of diffusion.

# PART A.  GASES IN EQUILIBRIUM

## 2.1.  Kinetic Theory of Gases in Equilibrium

### 2.1.1.  The Maxwell–Boltzmann Distribution

In those gaseous systems that obey the laws of classical mechanics and are in equilibrium (i.e., all the molecular properties are uniform and constant), the Maxwell–Boltzmann distribution is valid. In this distribution, the fraction of molecules with energy $\Xi(\mathbf{x}, \omega)$, with spatial coordinates between $\mathbf{x}$ and $\mathbf{x} + d\mathbf{x}$ and with momenta between $\omega$ and $\omega + d\omega$, is given by[†]

$$F^{(0)}(\mathbf{x}, \omega)\, d\mathbf{x}\, d\omega = \frac{dn}{n} = \frac{\exp(-\Xi/kT)\, d\mathbf{x}\, d\omega}{\displaystyle\int_{-\infty}^{\infty} \cdots \int_{-\infty}^{\infty} \exp(-\Xi/kT)\, d\mathbf{x}\, d\omega} \qquad (2.1)$$

where the superscript (0) denotes the equilibrium distribution function (note that the time $t$ need not appear in $F^{(0)}$).

Knowledge of the distribution function for a system enables us to calculate the average value of any physical property $\Psi(\mathbf{x}, \omega)$ by

$$\overline{\Psi} = \int \Psi \frac{dn}{n} \qquad (2.2)$$

Equation (2.2) is the well-known definition of a statistical average.

---

[†]We should point out here that the Maxwell–Boltzmann distribution is often defined as the quantity $f^{(0)} = nF^{(0)} = dn/d\mathbf{x}\,d\omega$ rather than as $dn/n$.

## 2.1.2.  Ideal Gas

By definition, an ideal gas consists of point molecules with kinetic energy only, so the energy of a molecule of an ideal gas is

$$\Xi = (2m)^{-1}\left(\omega_x^2 + \omega_y^2 + \omega_z^2\right) = \tfrac{1}{2}m\left(v_x^2 + v_y^2 + v_z^2\right) \tag{2.3}$$

and the volume $V$ of the system is[†]

$$V = \int_{-\infty}^{\infty}\int_{-\infty}^{\infty}\int_{-\infty}^{\infty} dx\,dy\,dz \tag{2.4}$$

The volume $V$ is accessible to every one of the molecules (since the molecules are "points," the volume they occupy is zero).

Equations (2.1), (2.3), and (2.4) give the mathematical description of an ideal gas.

Since $\Xi$ is independent of $\mathbf{x}$ for an ideal gas, equation (2.1) can be integrated with respect to $\mathbf{x}$ to give

$$\frac{dn}{n} = \frac{\exp(-\Xi/kT)\,d\omega_x\,d\omega_y\,d\omega_z}{\int_{-\infty}^{\infty}\int_{-\infty}^{\infty}\int_{-\infty}^{\infty}\exp(-\Xi/kT)\,d\omega_x\,d\omega_y\,d\omega_z} \tag{2.5}$$

### 2.1.2.1.  Energy Distribution

To find the distribution of energies in an ideal gas, it is convenient to express the momenta differentials $d\omega_x, d\omega_y, d\omega_z$ in terms of the energy differential $d\Xi$. For this, we rewrite equation (2.3) as

$$2m\Xi = \omega_x^2 + \omega_y^2 + \omega_z^2 = K_0^2 \tag{2.6}$$

where $K_0$ can be interpreted as the radius of a sphere in momentum space (i.e., we use "spherical coordinates" $K_0, \Theta_\omega, \Phi_\omega$ instead of Cartesian coordinates $\omega_x, \omega_y, \omega_z$). Since the integrand in equation (2.5) depends only on $K_0$ and not on the direction of the momentum (i.e., not on $\Theta_\omega, \Phi_\omega$), we can write for the volume element in momentum space

$$d\omega_x\,d\omega_y\,d\omega_z = 4\pi K_0^2\,dK_0 \tag{2.7}$$

---

[†]For simplicity we assume the system is infinite in extent.

Using equation (2.6) we can write equation (2.7) as

$$d\omega_x \, d\omega_y \, d\omega_z = 2\pi (2m)^{3/2} \Xi^{1/2} \, d\Xi \qquad (2.8)$$

Introducing equation (2.8) into the denominator of the right-hand side of equation (2.5) and integrating, we obtain

$$\int_{-\infty}^{\infty} \int_{-\infty}^{\infty} \int_{-\infty}^{\infty} \left[ \exp(-\Xi/kT) \right] d\omega_x \, d\omega_y \, d\omega_z = (2\pi mkT)^{3/2} \qquad (2.9)$$

Carrying out the same substitution in the numerator of equation (2.5) and using equation (2.9), we finally arrive at

$$\frac{dn}{n} = \frac{2\Xi^{1/2}}{\pi^{1/2}(kT)^{3/2}} \exp\left( -\frac{\Xi}{kT} \right) d\Xi \qquad (2.10)$$

which is the Boltzmann distribution of energies in an ideal gas.

### 2.1.2.2.  Average Energy

The average energy $\bar{\Xi}$ of a molecule in an ideal gas can be obtained using equation (2.2),

$$\bar{\Xi} = \int_0^{\infty} \Xi \frac{dn}{n} \qquad (2.11)$$

Introducing equation (2.10) into equation (2.11) and integrating, we obtain

$$\bar{\Xi} = \tfrac{3}{2}kT \qquad (2.12)$$

It can be shown that each coordinate (or better, each degree of freedom) contributes $\tfrac{1}{2}kT$ to the translational energy (principle of equipartition of energy).

### 2.1.2.3.  Distribution of Molecular Speeds

Since [equation (2.3)]

$$\Xi = \tfrac{1}{2}mv^2 \qquad (2.13)$$

we have

$$d\Xi = mv \, dv \qquad (2.14)$$

Substitution of equation (2.13) and (2.14) into equation (2.10) yields

$$F^{(0)}(v)\,dv \equiv \frac{dn}{n} = \left(\frac{2}{\pi}\right)^{1/2}\left(\frac{m}{kT}\right)^{3/2}\exp\left(-\frac{mv^2}{2kT}\right)v^2\,dv \qquad (2.15)$$

Here $F^{(0)}(v)\,dv$ is the fraction of molecules in the gas with speeds between $v$ and $v + dv$.

Equation (2.15) is Maxwell's law for the distribution of molecular speeds in an ideal gas [again, the Maxwell distribution is often defined as $f^{(0)} = nF^{(0)}(v)$].

### 2.1.2.4.  Mean Speeds

Using equation (2.2) we have

$$\bar{v} = \int_0^\infty v\,\frac{dn}{n} \equiv \int_0^\infty vF^{(0)}(v)\,dv \qquad (2.16)$$

where $F^{(0)}(v)$ is the distribution of molecular speeds.

By substituting equation (2.15) into equation (2.16) and integrating, we obtain the mean molecular speed

$$\bar{v} = (8kT/\pi m)^{1/2} \qquad (2.17)$$

From equations (2.12) and (2.13) we have

$$\bar{\Xi} = \tfrac{1}{2}m\overline{v^2} = \tfrac{3}{2}kT \qquad (2.18)$$

from which we have for the root-mean-square speed of a molecule in an ideal gas

$$(\overline{v^2})^{1/2} = (3kT/m)^{1/2} \qquad (2.19)$$

The ratio of the mean speeds given in equations (2.17) and (2.19) is thus

$$\frac{(\overline{v^2})^{1/2}}{\bar{v}} = \left(\frac{3\pi}{8}\right)^{1/2} = 1.085 \qquad (2.20)$$

### 2.1.2.5.  Distribution of the Velocity Components

Using $\omega_x = mv_x$, $\omega_y = mv_y$, $\omega_z = mv_z$, and using the expression for $\Xi$ in equation (2.3), we obtain from equation (2.5)

$$F^{(0)}(\mathbf{v})\,d\mathbf{v} \equiv \frac{dn}{n} = \frac{m^3\left[\exp(-m/2kT)\left(v_x^2 + v_y^2 + v_z^2\right)\right]dv_x\,dv_y\,dv_z}{\int_{-\infty}^\infty\int_{-\infty}^\infty\int_{-\infty}^\infty\left[\exp(-\Xi/kT)\right]d\omega_x\,d\omega_y\,d\omega_z} \qquad (2.21)$$

By introducing equation (2.9) we obtain

$$F^{(0)}(\mathbf{v})\, d\mathbf{v} \equiv \frac{dn}{n} = \left(\frac{m}{2\pi kT}\right)^{3/2} \left\{ \exp\left[ -\frac{m}{2kT}\left(v_x^2 + v_y^2 + v_z^2\right)\right]\right\} dv_x\, dv_y\, dv_z$$

$$(2.22)$$

which represents the fraction of molecules per unit volume with velocities lying between $v_x$ and $v_x + dv_x$, $v_y$ and $v_y + dv_y$, $v_z$ and $v_z + dv_z$. Equation (2.22) is the Maxwell–Boltzmann velocity distribution in an ideal gas at rest.[†] Again, we should point out that the Maxwell–Boltzmann velocity distribution for a gas in equilibrium is usually given as $f^{(0)}(\mathbf{v}) = nF^{(0)}(\mathbf{v})$:

$$nF^{(0)}(\mathbf{v}) = \frac{dn}{dv_x\, dv_y\, dv_z} = n\left(\frac{m}{2\pi kT}\right)^{3/2} \exp\left[ -\frac{m}{2kT}\left(v_x^2 + v_y^2 + v_z^2\right)\right]$$

$$(2.23)$$

To obtain the fraction of molecules with velocities between $v_x$ and $v_x + dv_x$ for any value of $v_y$ and $v_z$, equation (2.22) must be integrated between $-\infty$ and $\infty$ with respect to $v_y$ and $v_z$,

$$\left.\frac{dn}{n}\right|_{v_x} = \left(\frac{m}{2\pi kT}\right)^{3/2} \exp\left(-\frac{mv_x^2}{2kT}\right) dv_x$$

$$\times \int_{-\infty}^{\infty}\left[\exp\left(-\frac{mv_y^2}{2kT}\right)\right] dv_y \int_{-\infty}^{\infty}\left[\exp\left(-\frac{mv_z^2}{2kT}\right)\right] dv_z \quad (2.24)$$

or

$$\left.\frac{dn}{n}\right|_{v_x} = \left(\frac{m}{2\pi kT}\right)^{1/2}\left[\exp\left(-\frac{mv_x^2}{2kT}\right)\right] dv_x \qquad (2.25)$$

[†] The calculation of

$$\bar{\mathbf{v}} = \int_{-\infty}^{\infty} \mathbf{v} F(\mathbf{v})\, d\mathbf{v} \int_{-\infty}^{\infty} F(\mathbf{v})\, d\mathbf{v} \equiv 1 \qquad (2.22a)$$

must verify that the system is at rest. When the system moves as a whole at velocity $\vartheta$ (observe that, since the system is in equilibrium this velocity must be the same at every point), the Maxwell–Boltzmann distribution function must be written in terms of velocities in the reference frame, so equation (2.22) becomes

$$nF^{(0)}(\mathbf{v}) = n(m/2\pi kT)^{3/2}\exp[(-m/2kT)(\mathbf{v} - \vartheta)^2] \qquad (2.22b)$$

and the application of equation (2.22a) will lead to $\bar{\mathbf{v}} = \vartheta$.

### 2.1.2.6.  Average Value of the Velocity Component

To calculate the average value of the velocity component $v_z$ in the direction of positive $z$, we can integrate equation (2.2) over $v_z$ from 0 to $\infty$,

$$\bar{v}_z = \int_0^\infty v_z \left( \frac{dn}{n} \Big|_{v_z} \right) = \left( \frac{kT}{2\pi m} \right)^{1/2} \tag{2.26}$$

A comparison of equations (2.17) and (2.26) leads to

$$\bar{v}_z = \tfrac{1}{4}\bar{v} \tag{2.27}$$

Equation (2.27) can be used to calculate the frequency $Z$ of molecular bombardment of a plane from one side (i.e., the number of molecules per unit area per unit time crossing the plane from one side)[†]:

$$Z = \tfrac{1}{4}n\bar{v} \tag{2.28}$$

This result will be used later to predict transport coefficients by simplified methods.

### 2.1.2.7.  Mean Free Path

In an ideal gas, the molecules are considered as point particles that do not interact, and in our work above no mention was made of intermolecular collisions. To study transport properties, we have to consider such collisions (which are, of course, assumed to be elastic). We will assume the molecules are effectively small hard spheres of radius $\sigma$, where $\sigma$ is much less than the average distance between the molecules, and we will use the concept of the *mean free path*— the average distance traveled by a molecule between two successive collisions. This quantity can be calculated in several ways. Using Maxwell's definition of the mean free path as the total distance traveled by the molecules divided by the total number of free paths, and using equation (2.5), we arrive at the following[‡]:

$$\lambda = 1/2^{1/2}\pi\sigma^2 n \tag{2.29}$$

---

[†] Equation (2.28) should really be obtained from

$$dZ \equiv v_z\, dn = nv_z\, dn/n \tag{2.28a}$$

which, by integration, leads to

$$Z \equiv n\bar{v}_z = \tfrac{1}{4}n\bar{v} \tag{2.28b}$$

[‡] The actual calculation of equation (2.29) from Maxwell's definition of the mean free path is very intricate and for details we refer the reader to, e.g., Loeb.[5]

Using the ideal gas law we can write equation (2.29) in terms of temperature and pressure:

$$\lambda = kT/2^{1/2}\pi\sigma^2 p \qquad (2.30)$$

## PART B.    FIRST-APPROXIMATION EXTENSION TO NONEQUILIBRIUM GASES (SIMPLIFIED KINETIC THEORY)

A nonequilibrium gas is characterized by a lack of uniformity of properties such as density, pressure, temperature, and/or velocity. This nonuniformity gives rise to the transport of momentum, heat, and/or mass.

In the simplified kinetic theory, a gas is in equilibrium except for a small perturbation, the perturbation being small enough so that the Maxwell–Boltzmann distribution is not altered. This method is only valid for systems near equilibrium and where the mean free paths are determined exclusively by collisions in the gaseous phase. In spite of the high degree of simplification of these working hypotheses, the theory predicts transport coefficients fairly well.

## 2.2. Transport Coefficients

Consider the plane represented by ABCD in Figure 2.1 and assume that a certain property $\psi(z)$ is distributed throughout the gas. If the distribution of $\psi$ is not symmetrical with respect to the plane ABCD there will be a net transport of $\psi$ through this plane.

To obtain expressions for the transport coefficients, we assume that the molecules that reach the plane have undergone their last collision at an average distance $\lambda$ from the plane ABCD and that therefore they carry the properties of the gas prevailing at the distances $\pm\lambda$ from that plane. Since we are considering systems near equilibrium, the values of $\psi$ at the distances $\pm\lambda$ from ABCD correspond to the first two terms of a Taylor series expansion.

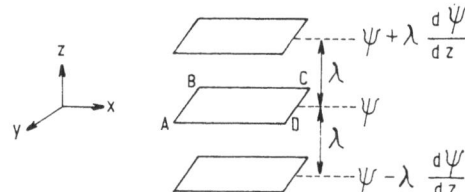

Figure 2.1. Schematic representation of a transport process at the molecular level.

The flux $\Gamma_{z(+)}$ of the property $\psi$ coming from positive $z$ is given by the product of the molecular flux in a given direction [equation (2.28)] multiplied by the value of $\psi$ in the plane from which the molecules are coming,

$$\Gamma_{z(+)} = \frac{n\bar{v}}{4}\left(\psi + \lambda\frac{d\psi}{dz}\right) \tag{2.31}$$

Similarly, the flux of $\psi$ coming from negative $z$ is given by

$$\Gamma_{z(-)} = \frac{n\bar{v}}{4}\left(\psi - \lambda\frac{d\psi}{dz}\right) \tag{2.32}$$

The net transport of $\psi$ along positive $z$ is then

$$\Gamma = \Gamma_{z(-)} - \Gamma_{z(+)} = -\tfrac{1}{2}n\bar{v}\lambda(d\psi/dz) \tag{2.33}$$

Observe that in obtaining equation (2.33) it is assumed that $n$ is not a function of $z$, and we assume that the value of the mean free path used is given by equation (2.29), i.e., the same as that obtained using the Maxwell–Boltzmann distribution for equilibrium gases. We now apply the general expression (2.33) to the transport of momentum, mass, and energy.

### 2.2.1.  Viscosity

Viscosity involves the transport of momentum. For the example described in Section 1.7.2, the corresponding component of the momentum transport tensor is

$$\tau_{zx} = m\vartheta_x(z) \tag{2.34}$$

where $\vartheta_x(z)$ is the $x$ component of the nonseparative velocity (on a mass basis) of the gas at height $z$. We have $\Gamma = \tau_{zx}$ and $\psi = m\vartheta_x(z)$, so equation (2.33) yields

$$\tau_{zx} = -\tfrac{1}{2}mn\bar{v}\lambda(d\vartheta_x/dz) \tag{2.35}$$

Comparing equation (2.35) and Newton's law of viscosity [equation (1.61)] we have

$$\mu = \tfrac{1}{2}mn\bar{v}\lambda \tag{2.36}$$

On introducing equations (2.17) and (2.29) into equation (2.36), we can write the viscosity as

$$\mu = \frac{1}{\pi\sigma^2}\left(\frac{kTm}{\pi}\right)^{1/2} \tag{2.37}$$

It is interesting to note that in Maxwell's original work on viscosity, he obtained equation (2.36) but with the factor $\frac{1}{3}$ instead of $\frac{1}{2}$. The factor $\frac{1}{3}$ rather than the correct factor $\frac{1}{2}$ still appears in many contemporary results. The reasons for this will be discussed in Chapter 6.

## 2.2.2. Slip Friction Coefficient

In Section 1.7.3 we analyzed the slip flux $N_z^S$ in a capillary. As we saw there, this flux is the second term of equation (1.70):

$$N_z^S = -\frac{R}{2\beta}\frac{p}{R_g T}\frac{dp}{dz} \tag{2.38}$$

Equation (2.38) does not correspond to a pure mechanism of flux; rather, it describes the transition regime flux. It is zero for the limiting case of the viscous regime ($\beta \to \infty$) and for the limiting case of the Knudsen regime.

Our aim now is to derive the slip friction coefficient $\beta$ using the simplified kinetic theory. Such a derivation was first carried out by Maxwell[†] and later by Millikan,[6] and is based on an analysis of the momentum transferred from the moving gas to the walls of the capillary; this momentum transport is given phenomenologically by Newton's law of viscosity as

$$\tau_{rz}|_R = -\mu\frac{d\vartheta_z}{dr}\bigg|_R \tag{2.39}$$

On the basis of kinetic theory, $\tau_{rz}|_R$ can be written as

$$\tau_{rz}|_R = F\zeta m\vartheta_z|_\lambda \tag{2.40}$$

where $F$ is the number of molecules striking the wall per unit time per unit area, $\zeta$ (usually known as the *coefficient of diffuse reflection*) is the fraction of the molecule–wall collisions that are diffuse, $m\vartheta_z|_\lambda$ is the momentum transferred to the wall by each diffuse collision (i.e., the average $z$ component of the momentum of a molecule when it strikes the wall), and $\vartheta_z|_\lambda$ is the gas velocity at a distance $\lambda$ from the wall, where the molecules, on the average, undergo their last collision.

[†]See Reference 5.

If we assume that the Maxwell–Boltzmann distribution is valid locally, the molecular flux $F$ is given by equation (2.28).

Since the system is not too far from equilibrium, the velocity $\vartheta_z|_\lambda$ can be approximated by the first two terms of a Taylor series,

$$\vartheta_z|_\lambda = \vartheta_z|_R - \lambda \frac{d\vartheta_z}{dr}\bigg|_R \tag{2.41}$$

By substituting equation (2.28) and (2.41) into equation (2.40) and using equation (2.39) we have

$$-\mu \frac{d\vartheta_z}{dr}\bigg|_R = \frac{mn\bar{v}}{4} \varsigma \left(\vartheta_z|_R - \lambda \frac{d\vartheta_z}{dr}\bigg|_R\right) \tag{2.42}$$

By substituting the simplified-kinetic-theory expression for the viscosity [equation (2.36)] into equation (2.42) and rearranging some, we obtain

$$-\frac{d\vartheta_z}{dr}\bigg|_R = \frac{\varsigma}{2-\varsigma}\frac{1}{\lambda}\vartheta_z\bigg|_R \tag{2.43}$$

Now comparing the definition of the slip friction coefficient [equation (1.69)] with equation (2.43), we find

$$\beta = \frac{\mu}{\lambda}\frac{\varsigma}{2-\varsigma} \tag{2.44}$$

By substituting equation (2.36) into equation (2.44), we obtain the following alternative expression:

$$\beta = \frac{1}{2}mn\bar{v}\frac{\varsigma}{2-\varsigma} \tag{2.45}$$

On introducing the ideal gas law to eliminate $n$ and using equation (2.17), we obtain

$$\beta = \left(\frac{2m}{\pi kT}\right)^{1/2} p\frac{\varsigma}{2-\varsigma} \tag{2.46}$$

This simplified-kinetic-theory expression for the slip friction coefficient enables us to obtain some interesting further results.

Using equations (1.67), (2.38), and (2.45), and writing $\lambda_p = 2R$, we find

$$\frac{N_z^S}{N_z^V} = 8\,\mathrm{Kn}\frac{2-\varsigma}{\varsigma} \tag{2.47}$$

Thus, for $\zeta \to 1$ and a Knudsen number $\text{Kn} < 1.25 \times 10^{-3}$, the slip flux is negligible in comparison with the viscous flux with an error margin of 1%. The reader should observe, however, that the viscous regime is approached asymptotically, as shown by equation (2.47).

In the related literature many of the experimental results for rarefied gases are expressed in terms of the Reynolds and Mach numbers (this reflects the influence of aeronautical engineering). We define the Mach number Ma and the Reynolds number Re as

$$\text{Ma} \equiv \vartheta_c / \vartheta_s \tag{2.48}$$

$$\text{Re} \equiv 2\rho \vartheta_c R / \mu \tag{2.49}$$

where $\vartheta_c$ is a characteristic velocity and $\vartheta_s$ is the velocity of sound. The latter velocity can be written as

$$\vartheta_s = (\pi \gamma / 8)^{1/2} \bar{v} \tag{2.50}$$

where $\gamma \equiv c_p / c_v$ is the ratio of the specific heats at constant pressure and constant volume, respectively.

Using equation (2.36), we can write the Knudsen number as

$$\text{Kn} = \mu / \rho \bar{v} R \tag{2.51}$$

By substituting equations (2.48)–(2.50) into equation (2.51) we obtain

$$\text{Kn} = \left(\frac{\pi \gamma}{2}\right)^{1/2} \frac{\text{Ma}}{\text{Re}} \tag{2.52}$$

It is readily seen that, for the regime obtained by extension from the viscous regime, it does not matter whether we express the results in terms of the Knudsen number or in terms of the group $\gamma^{1/2} \text{Ma} / \text{Re}$ (very frequently the experimental data are for a pure gas, so $\gamma = \text{const}$ and the group reduces to $\text{Ma}/\text{Re}$). For example, for the laminar flow of air in a tube (for this case, $\gamma = 1.4$), Tsien[9] gave an empirical relationship for the friction factor[†] as a function of Ma and Re. When $\text{Ma} \to 0$, the friction factor is the same as the one predicted by the Hagen–Poiseuille law.

It is also customary in the related literature to express the slip flux as

$$\mathbf{N}^S = -D^S (R_g T)^{-1} \nabla p \tag{2.53}$$

where $D^S$ is the slip self-diffusivity. From equations (2.38), (2.46), (2.17),

[†]The friction factor is a dimensionless parameter defined as the force (excluding buoyancy) that a fluid exerts on a surface due to its relative motion, per unit surface area and per unit characteristic kinetic energy per unit volume.

and the ideal gas law we find

$$D^S = \pi R \bar{v} \frac{2-\zeta}{8\zeta} \tag{2.54}$$

The fact that $D^S$ is independent of the pressure indicates that the magnitude of the slip flux does not depend on the pressure, thus justifying an intercept greater than zero when $-NR_g T/\nabla p$ is plotted as a function of $p$ (see Figure 1.14).

### 2.2.3.  Knudsen Diffusivity

The results obtained by Knudsen[4] in the region $p \to 0$ are described by equation (1.43) adapted to an ideal pure gas in a capillary, i.e., with $\nabla c_A = (R_g T)^{-1}(dp/dz)$ and $D^K_{AA, \text{eff}} = D^K_{AA}$.

Knudsen himself evaluated $D^K_{AA}$ analytically, assuming that, since the number of molecule–molecule collisions is negligible compared to the number of molecule–wall collisions, at every point there is an equilibrium state determined by the pressure. The molecular flux thus takes place between equilibrium regions at different pressures.

To evaluate $D^K_{AA}$ we write the principle of conservation of momentum for a capillary slice as follows:

$$\tau_{rz}|_R 2\pi R \, dz = -\pi R^2 \, dp \tag{2.55}$$

or

$$\tau_{rz}|_R = -\frac{R}{2}\frac{dp}{dz} \tag{2.56}$$

The momentum flux can also be written in a form similar to equation (2.40), but with the difference that now it is convenient to write it in a differential form, i.e., if $d\tau_{rz}|_R$ is the rate at which momentum is transferred to the walls by molecules colliding with the wall and that have velocities in the range $\mathbf{v}$ and $\mathbf{v} + d\mathbf{v}$, then

$$d\tau_{rz}|_R = m v_z \zeta \, dF \tag{2.57}$$

where $dF$ is the molecular flux on the wall for molecules with velocities between $\mathbf{v}$ and $\mathbf{v} + d\mathbf{v}$ and $v_z$ is the $z$ component of that velocity.

If we assume that, for an ideal gas in equilibrium, and therefore for our slightly perturbed gas (since the perturbation does not modify the velocity distribution),

$$\frac{v_z}{v} = \frac{\bar{v}_z}{\bar{v}} = \frac{1}{4} \tag{2.58}$$

where the last equality follows from equation (2.27), we have

$$dF = \tfrac{1}{4} v \, dn \qquad (2.59)$$

The fraction of molecules with speeds between $v$ and $v + dv$ is given by the Maxwell distribution (2.15). In consequence, introducing equations (2.58), (2.59), and (2.15) into equation (2.57), we obtain

$$\int d\tau_{rz}\big|_R = \frac{nm^{5/2}\zeta}{(8\pi)^{1/2}(kT)^{3/2}} \frac{\bar{v}_z}{\bar{v}} \int_0^\infty v^4 \left[ \exp\left( \frac{-mv^2}{2kT} \right) \right] dv \qquad (2.60)$$

Integration of equation (2.60) yields[†]

$$\tau_{rz}\big|_R = \tfrac{3}{4} nkT\zeta \bar{v}_z / \bar{v} \qquad (2.61)$$

The Knudsen molar flux is given by

$$N_z^K \equiv c\bar{v}_z = \frac{p}{R_g T} \bar{v}_z \qquad (2.62)$$

By substituting equation (2.61) into equation (2.62) we obtain

$$N_z^K = \frac{4}{3} \frac{\bar{v}}{\zeta} \frac{1}{R_g T} \tau_{rz}\big|_R \qquad (2.63)$$

Finally, from equations (2.56) and (2.63) we obtain

$$N_z^K = -\frac{2}{3} \frac{R\bar{v}}{R_g T\zeta} \frac{dp}{dz} \qquad (2.64)$$

From a comparison of equations (2.64) and (1.43) [adapted to this case it reads $N_{Az}^D = -D_{AA}^K (R_g T)^{-1}(dp/dz)$] it follows that[‡]

$$D_{AA}^K = \frac{2}{3} R\bar{v}_A \frac{1}{\zeta} \qquad (2.65)$$

---

[†] If $\tau_{rz}\big|_R$ had been evaluated by using average values in equation (2.57), namely, $\tau_{rz}\big|_R = m\bar{v}_z\zeta F$, the numerical coefficient obtained would have been different: $\tau_{rz}\big|_R = (2/\pi)nkT\zeta\bar{v}_z/\bar{v}$. Pollard and Present[8] have taken the coefficient $2/\pi$ as correct rather than Knudsen's coefficient $3/4$, despite the fact that the latter has been experimentally verified.

[‡] For other geometries and $\zeta = 1$, von Smoluchowski[10] obtained

$$D_{AA}^K = \frac{\bar{v}_A}{16A} \int_x \int \int_{-\pi/2}^{\pi/2} k^2 \cos\Theta \, d\Theta \, dx \qquad (2.65a)$$

where $A$ is the cross-sectional area, $dx$ is an element of the perimeter of the pore, and $k$ is the length of a chord that extends across the pore from the element $dx$ and is drawn at an angle $\Theta$ to the normal to $dx$.

We have thus verified the phenomenological prediction given by equation (1.46).

We interrupt our discussion here to make two remarks concerning equations (2.64) and (2.65). In his original development, Knudsen took the coefficient of diffuse reflection $\zeta$ to be 1, and so obtained $D_{AA}^K = \frac{2}{3} R\bar{v}_A$. While equation (2.65) is the correct generalization of Knudsen's result to the case $\zeta < 1$, in some studies[1, 3] the Knudsen diffusivity is defined as

$$D_{AA}^K = \frac{2}{3} R\bar{v}_A \frac{2-\zeta}{\zeta} \tag{2.66}$$

with Knudsen's original work, as well as a subsequent demonstration due to von Smoluchowsky,[11] being quoted. This incorrect expression was probably assumed to be the generalization of Knudsen's result, by analogy with the expression for the slip friction coefficient [equation (2.45)], which has the same $\zeta$ dependence as equation (2.66).

Knudsen provided an alternative demonstration of equation (2.64) (again with $\zeta = 1$). He introduced the inertial force term into equation (2.56),

$$(\tau_{rz}|_R)\frac{2}{R} + \rho\bar{v}_z\frac{d\bar{v}_z}{dz} = -\frac{dp}{dz} \tag{2.67}$$

and then demonstrated that that term was negligible for the working conditions of his experiments.

We can now continue the analysis, begun in Chapter 1, of the semiempirical expression (1.71) proposed by Knudsen for the flux of a pure gas in a capillary.

As we stated in Section 1.7.3., Knudsen found theoretical expressions for the coefficients $a^K$ and $b^K$ of his equation. We noted there that $a^K$ is the viscous flux coefficient $(R^2/8\mu)$ and that $b^K$ is the Knudsen diffusivity. Thus we have

$$a^K = R^2/8\mu; \qquad b^K = \frac{2}{3} R\bar{v} \tag{2.68}$$

We also saw in Section 1.7.3 that Knudsen's application of the method of least squares to his results led to $c_1^K/c_2^K = 0.81$, which agrees with the experimental results of other authors (see Table 1.5). We can now find theoretical expressions for $c_1^K/c_2^K$, starting from the expression for the slip friction coefficient derived above.

In the discussion following equation (1.71) we saw that $b^K c_1^K/c_2^K$ is the slip flux coefficient and so from equation (2.54) with $\zeta = 1$ we obtain

$$D^S = b^K c_1^K/c_2^K = \left(\frac{2}{3}R\bar{v}\right)(3\pi/16) = 0.59\left(\frac{2}{3}R\bar{v}\right) \tag{2.69}$$

and using equation (2.68) we have[†]

$$c_1^K / c_2^K = 0.59 \tag{2.70}$$

All the values of $c_1^K / c_2^K$ given in Table 1.6 are higher than the simplified-kinetic-theory theoretical value (2.70). The difference is usually ascribed[1] to the assumption that $\zeta = 1$ rather than $\zeta < 1$, since for $\zeta < 1$ the theoretical value would be $c_1^K / c_2^K = 0.59(2 - \zeta)$. But, as emphasized by Pollard and Present,[8] it is possible that the working hypotheses of the simplified kinetic theory themselves lead to the erroneous values for $c_1^K / c_2^K$. In Chapter 4 we will find a better expression for $c_1^K / c_2^K$ based on more refined arguments.

### 2.2.4. Graham's Law

If we introduce equation (2.17) into equation (2.65) and the latter into equation (1.53) we obtain

$$\sum_{i=1}^{\nu} \mathbf{F}_i^D m_i^{1/2} = 0 \tag{2.71}$$

where we have assumed uniform pressure. Equation (2.71) is known as *Graham's law of diffusion*—a law that was first accepted, later misinterpreted and so forgotten, then rediscovered, but still often misinterpreted (see Chapter 6 for a more detailed history of this law).

Equation (2.71) shows that, as a consequence of the molecule–wall collisions, the total molecular diffusive fluxes of the different species are related to each other and that relationship depends on the square root of the molecular masses of the species, no matter what diffusion regime (molecular or Knudsen) prevails.

---

[†]We must point out that Knudsen's derivation does not take account of inlet and exit effects and, therefore, it is rigorously valid only for capillaries for which $L/R \to \infty$. Pollard and Present[8] proposed the correction factor $1 - 3R/4L$. Clausing proposed the semiempirical correction factor[2]

$$\frac{3 + 15R/2L}{2 + 19R/L + 20(R/L)^2} \tag{2.70a}$$

which has an error margin of 1.5% for any value of $R/L$.

## 2.2.5.   Thermal Conductivity

We now consider the transport of thermal energy. The property $\psi$ is now identified with the average energy $\overline{\overline{\Xi}}$ of a molecule. In accordance with equation (2.12), we can write

$$\psi \equiv \overline{\overline{\Xi}} = \tfrac{3}{2}kT \tag{2.72}$$

If we call $q_z$ the heat flux along the $z$ coordinate and substitute equation (2.72) into equation (2.33), we have

$$q_z = -\tfrac{3}{4}n\bar{v}k\lambda(dT/dz) \tag{2.73}$$

Fourier's law of heat transfer (a phenomenological expression similar to Newton's law of viscosity and Fick's laws) is

$$q_z = -\lambda_K(dT/dz) \tag{2.74}$$

and a comparison of equations (2.73) and (2.74) gives for the thermal conductivity $\lambda_K$

$$\lambda_K = \tfrac{3}{4}n\bar{v}k\lambda \tag{2.75}$$

The specific heat per molecule for a monatomic gas can be calculated from its definition $(\partial \Xi / \partial T)_V$, and using equation (2.72) we find

$$c_V = \tfrac{3}{2}k \tag{2.76}$$

Combining equations (2.75) and (2.76), an alternative expression for the thermal conductivity is obtained:

$$\lambda_K = \tfrac{1}{2}n\bar{v}c_V\lambda \tag{2.77}$$

As in the case of viscosity, a common error in the related literature is the use of $\tfrac{1}{3}$ as the coefficient instead of $\tfrac{1}{2}$ in the expression for $\lambda_K$.

## 2.2.6.   Molecular Diffusivity

We shall now consider mass transport in a binary system containing two species, A and B, and in which the molecular concentrations $n_A$ and $n_B$ vary only along the $z$ coordinate. Since we are considering a system without walls, it is necessary to assume that the total molecular concentration (and therefore the pressure) remains uniform throughout the system,

$$\frac{dn}{dz} = \frac{dn_A}{dz} + \frac{dn_B}{dz} = 0 \tag{2.78}$$

Let us first analyze the transport of A along $z$ assuming it is independent of the transport of B.

Using Figure 2.1 and equation (2.31) written for component A of the gas, the molecular flux of A coming from positive $z$ is

$$F_{Az(+)}^D = \frac{\bar{v}_A}{4}\left(n_A + \lambda_A \frac{dn_A}{dz}\right) \tag{2.79}$$

Similarly, the molecular flux of A coming from negative $z$ is

$$F_{Az(-)}^D = \frac{\bar{v}_A}{4}\left(n_A - \lambda_A \frac{dn_A}{dz}\right) \tag{2.80}$$

Consequently, the net molecular flux of A in the direction of positive $z$ is

$$F_{zA}^D = -\frac{1}{2}\bar{v}_A \lambda_A \frac{dn_A}{dz} \tag{2.81}$$

Similarly, the molecular flux of B is

$$F_{zB}^D = -\frac{1}{2}\bar{v}_B \lambda_B \frac{dn_B}{dz} \tag{2.82}$$

Equations (2.81) and (2.82) describe the molecular fluxes of the two components of a binary, isobaric mixture, assuming that both fluxes are independent. However, the fluxes cannot in general be independent, as can be seen from the following. Unless $m_A = m_B$ and $\sigma_A = \sigma_B$, in which case $\bar{v}_A \lambda_A = \bar{v}_B \lambda_B$, the total molecular flux $F_A^D + F_B^D$ predicted by equations (2.81) and (2.82) will be different from zero, a result incompatible with equation (2.78). Consequently, in the most general case, it is necessary to assume the existence of a nonseparative mass flux that makes it possible to keep the pressure uniform.

In such a case, the molecular fluxes of A and B must be written as

$$F_{zA}^D = n_A \vartheta_z - \frac{1}{2}\bar{v}_A \lambda_A \frac{dn_A}{dz} \tag{2.83}$$

$$F_{zB}^D = n_B \vartheta_z - \frac{1}{2}\bar{v}_B \lambda_B \frac{dn_B}{dz} \tag{2.84}$$

As previously discussed, in order to keep the pressure uniform it is necessary that

$$F_z^D = F_{zA}^D + F_{zB}^D = 0 \qquad (2.85)$$

Eliminating $\vartheta_z$ from equations (2.83)–(2.85) and using equation (2.78) we have

$$F_{zA}^D = -\frac{n_A \bar{v}_B \lambda_B + n_B \bar{v}_A \lambda_A}{2(n_A + n_B)} \frac{dn_A}{dz} \qquad (2.86)$$

$$F_{zB}^D = -\frac{n_A \bar{v}_B \lambda_B + n_B \bar{v}_A \lambda_A}{2(n_A + n_B)} \frac{dn_B}{dz} \qquad (2.87)$$

By comparing equations (2.86), (2.87), and (1.39), we find that the binary molecular diffusivity in terms of the simplified kinetic theory is given by

$$D_{AB} = D_{BA} = \frac{n_A \bar{v}_B \lambda_B + n_B \bar{v}_A \lambda_A}{2n} \qquad (2.88)$$

In order to apply equation (2.88) it is necessary to explicitly write the mean free path for both components. In the system under consideration, $\lambda_A$ and $\lambda_B$ are determined by collisions between like molecules as well as by collisions between unlike molecules. The expression for the mean free path under these conditions is given by[7]

$$\lambda_A = \left[ 2^{1/2} \pi n_A \sigma_A^2 + \pi n_B \sigma_{AB}^2 (1 + m_A/m_B)^{1/2} \right]^{-1} \qquad (2.89)$$

where $\sigma_{AB}$ is the collision diameter for unlike molecules: $\sigma_{AB} = \frac{1}{2}(\sigma_A + \sigma_B)$.

Equation (2.89) leads to erroneous values for the binary molecular diffusivities, so it has been proposed (as we saw in the phenomenological description in Sections 1.1 and 1.2) that collisions between like molecules do not influence the binary molecular diffusivity or, in other words, that the resistance to diffusion is due only to collisions between unlike molecules. Thus the correct mean free paths to be used for diffusion are

$$\lambda_A = \left[ \pi n_B \sigma_{AB}^2 (1 + m_A/m_B)^{1/2} \right]^{-1} \qquad (2.90)$$

$$\lambda_B = \left[ \pi n_A \sigma_{AB}^2 (1 + m_B/m_A)^{1/2} \right]^{-1} \qquad (2.91)$$

Introducing equations (2.90), (2.91), and (2.17) into equation (2.88), we obtain

$$D_{AB} = \frac{1}{\pi \sigma_{AB}^2 n} \left[ \frac{2kT}{\pi} \left( \frac{1}{m_A} + \frac{1}{m_B} \right) \right]^{1/2} \qquad (2.92)$$

Since this equation can be obtained from Stefan–Maxwell equations such as equation (1.37), it is called the Stefan–Maxwell equation for the binary molecular diffusivity.

If the ideal gas law is used to eliminate $n$, equation (2.92) shows that the binary molecular diffusivity is directly proportional to $T^{3/2}$ and inversely proportional to $p$. The dependence of $D_{AB}$ on pressure agrees with the experimental results for many gas pairs up to 10 atm. However, the experimentally found temperature dependence is steeper than is predicted by equation (2.92).

The simplified kinetic theory has (a) enabled us to obtain approximate expressions for the transport coefficients in terms of basic parameters, and (b) provided a physical picture of the phenomenon, a picture that is lost in other, more refined, developments. For example, the physical picture of diffusion in a binary mixture led us to an expression for the binary molecular diffusivity that takes account of two effects: (1) the molecular flux $\frac{1}{2}\bar{v}_A\lambda_A(dn_A/dz)$; (2) the mass transport that restores the uniform pressure.

# 3

## *Constitutive Equations of Diffusion in Multicomponent Systems without Walls*

In Chapter 1 we proposed, on a phenomenological basis, constitutive equations to describe transport properties of a gas. In Chapter 2 we used a simplified kinetic theory method to derive the coefficients in these equations—the transport coefficients—but the equations themselves were still just assumed to be valid, i.e., we did not derive them from the theory.

In this chapter our goal is to determine the morphology and structure of the constitutive equations of diffusion, i.e., (1) to derive the constitutive equations from theory and (2) to derive expressions for the transport coefficients in terms of the molecular parameters of the system. In Part A we will approach this using the *thermodynamics of irreversible processes*. With this method, only objective (1) can be attained, but the result is independent of the state of aggregation of the fluid under study. We will attain both objectives (morphology and structure) by our second approach —the application of the rigorous kinetic theory to sufficiently dilute gases.

In both cases we will consider only the equations for mass transport, although these equations contain transport coefficients for other properties.

There are excellent references [14, 19, 37, 78] giving the detailed development of the theory, so we will avoid cumbersome mathematical developments, restricting ourselves to analysis of the beginning equations, the scheme of the solution method employed, and the results obtained.

The only simplification we will make is to consider systems without walls. The complete problem, systems with walls, will be analyzed in Chapter 4.

# PART A.   THERMODYNAMICS OF IRREVERSIBLE PROCESSES

From the thermodynamics of irreversible processes we can obtain correct expressions for the flux of a given property in terms of different "driving forces" for a fluid, independently of its state of aggregation. The coefficients that appear in these expressions are, however, phenomenological, and we can obtain (using the Onsager reciprocal relations) only a few relationships between the transport coefficients.

## 3.1.   Postulates

In every transport phenomenon there are fluxes $\Gamma_i$ of $i$ different properties associated with driving forces or affinities $X_i$. In general, we say that the flux of a given property can be produced by driving forces of different kinds.

The fundamental postulate of the thermodynamics of irreversible processes states that, for systems which are not too far from equilibrium, the fluxes can be expressed as

$$\Gamma_i = \sum_j \alpha_{ij} X_j \tag{3.1}$$

where the coefficients $\alpha_{ij}$ are phenomenological. Equation (3.1) states that the fluxes are linear functions of the driving forces. The diagonal elements $\alpha_{ii}$ are the coefficients for the direct effects, while the off-diagonal elements $\alpha_{ij}$, $i \neq j$, are the coefficients for the crossed effects.

It is important to note that the assumption of a linear relationship between $\Gamma_i$ and $X_i$ restricts the applicability of the thermodynamics of irreversible processes to states that are near equilibrium.

The fundamental theorem of the thermodynamics of irreversible processes, due to Onsager,[58, 59] states that, for a suitable selection of the $\Gamma_i$ and $X_j$, the phenomenological coefficients $\alpha_{ij}$ are symmetric:

$$\alpha_{ij} = \alpha_{ji} \tag{3.2}$$

Equations (3.2), usually known as the *reciprocal relations*, are based on arguments of statistical mechanics and on the concept of microscopic reversibility (invariance of the equations of motion under time reversal).

The equilibrium state corresponds to maximum entropy, and for a system near equilibrium it can be shown[†] that the rate of irreversible

[†]Reference 37, pp. 706–707.

production of entropy is given by

$$\frac{d(\Delta S)}{dt} = T^{-1} \sum_{i=1}^{\nu} X_i \Gamma_i \qquad (3.3)$$

where $\Delta S$ is the difference between the entropies of the perturbed and equilibrium states.

## 3.2.  Application to Transport Phenomena

Onsager[60] analyzed transport phenomena using the equation of change of entropy in a multicomponent system.

At equilibrium, the entropy depends only on the state of the system, and the differential change of entropy per unit mass for a differential volume element of a gas is given by

$$T d\hat{S} = d\hat{U} + p \, d(\rho^{-1}) - \sum_{i=1}^{\nu} \tilde{G}_{fi} d(n_i/\rho) \qquad (3.4)$$

where $\hat{S}$ and $\hat{U}$ denote the entropy per unit mass and the internal energy per unit mass, respectively, and $\tilde{G}_{fi}$ is the partial molar Gibbs free energy (also called the free enthalpy or chemical potential).

For a system in a nonequilibrium state, it is postulated that equation (3.4) is valid and that $p$ can be expressed in terms of an equation of state. Then the substantive derivative of equation (3.4) is written as follows:

$$T \frac{D\hat{S}}{Dt} = \frac{D\hat{U}}{Dt} + p \frac{D(\rho^{-1})}{Dt} - \sum_{i=1}^{\nu} \tilde{G}_{fi} \frac{D(n_i/\rho)}{Dt} \qquad (3.5)$$

The three terms on the right-hand side of equation (3.5) can be evaluated using the momentum, energy, and mass conservation equations developed in Section 1.5. The equation obtained for a differential volume element of a fluid system without chemical reaction[†] is [37]

$$\rho \frac{D\hat{S}}{Dt} = - \nabla \cdot \left[ T^{-1} \left( \mathbf{q} - \sum_{i=1}^{\nu} n_i \tilde{G}_{fi} \mathbf{v}_i^d \right) \right]$$

$$- \left\{ T^{-1} \sum_{i=1}^{\nu} (\mathbf{j}_i \cdot \Lambda_i) + T^{-1} (\tau : \nabla \vartheta) + T^{-2} (\Xi \cdot \nabla T) \right\} \qquad (3.6)$$

---

[†] It is not necessary to use the chemical reaction term to obtain an expression for the diffusive flux.

where

$$\Lambda_i \equiv m_i^{-1} \sum_{\substack{j=1 \\ j \neq i}}^{\nu} \left( \frac{\partial \tilde{G}_{fi}}{\partial x_j} \right)_{\substack{T,p,x_k \\ k \neq i,j}} \nabla x_j + \frac{\tilde{V}_i}{m_i} \nabla p - m_i^{-1} \mathbf{f}_i \qquad (3.7)$$

$$\tau \equiv \mathbf{T} - p\mathbf{I} \qquad (3.8)$$

$$\Xi \equiv \mathbf{q} - \sum_{i=1}^{\nu} n_i \tilde{H}_i \mathbf{v}_i^d \qquad (3.9)$$

and where $\mathbf{T}$ is the shear tensor (or pressure tensor) of the fluid; $\mathbf{I}$ is the unit tensor; $\tilde{H}_i$ and $\tilde{V}_i$ are the partial molal enthalpy and partial molar volume, respectively; and $\mathbf{v}_i^d$ is the diffusion velocity [defined by equation (1.6)].

The first term on the right-hand side of equation (3.6) is the rate of reversible production of entropy and the term in the curly brackets is the rate of irreversible production of entropy, both per unit volume in a differential volume element moving downstream with the fluid.

A comparison of equations (3.3) and (3.6), both of which give the rate of production of entropy, allows us to pick out the flux-driving force pairs. These are given in Table 3.1.

In general, using equation (3.1) any flux can be expressed as a linear combination of all the driving forces. For the particular case of an isotropic system, however, it can be shown[19, 37] that there can be no terms corresponding to the coupling of tensors whose orders differ by an odd number (this circumstance ensures the validity of the reciprocal relations). For example, the tensor $\nabla \vartheta$ cannot couple with the vector $\mathbf{j}_i$, but $\mathbf{j}_i$ can couple with $\nabla \ln T$, giving rise to thermal diffusion (called the *Soret effect* in liquids).[†] The driving force of a chemical reaction is a scalar and so, in the most general case, cannot couple with the mass transport flux vector.

We can now write the diffusive flux $\mathbf{j}_i$ as follows:

$$\mathbf{j}_i = -\boldsymbol{\alpha}_{i0} \nabla \ln T - \sum_{j=1}^{\nu} \boldsymbol{\alpha}_{ij} \Lambda_j \qquad (3.10)$$

where the suffix 0 corresponds to the temperature variable and the other suffixes refer to the components of the mixture.

The fluxes $\mathbf{j}_i$ are not independent since $\Sigma_i \mathbf{j}_i = 0$ [see equation (1.7)]. Then [reverting to the form of the right-hand side of equation (3.1) for

---

[†] The coupling between $\Xi$ and $\Lambda_i$ gives rise to the so-called *Dufour effect*.

Table 3.1. The Various Fluxes and the Corresponding Driving Forces

| Flux $\Gamma_i$ | Driving force $X_i$ |
|---|---|
| $\mathbf{j}_i$ | $-\Lambda_i$ |
| $\tau$ | $-\nabla \vartheta$ |
| $\Xi$ | $-\nabla \ln T$ |

simplicity],

$$\mathbf{j}_k = -\sum_{\substack{i=1 \\ i\neq k}}^{\nu} \mathbf{j}_i = -\sum_{\substack{i=1 \\ i\neq k}}^{\nu}\sum_{j=1}^{\nu} \alpha_{ij}\mathbf{X}_j = -\sum_{j=1}^{\nu} \mathbf{X}_j \sum_{\substack{i=1 \\ i\neq k}}^{\nu} \alpha_{ij}$$

But since $\mathbf{j}_k = \sum_j \alpha_{kj}\mathbf{X}_j$, we must have

$$\alpha_{kj} = -\sum_{\substack{i=1 \\ i\neq k}}^{\nu} \alpha_{ij}$$

It then follows that $\sum_{i=1}^{\nu}\alpha_{ij} = 0$.

Let us define the vector

$$\mathbf{d}_i = \frac{n_i m_i}{nkT}\left[\Lambda_i - \rho^{-1}\nabla p + \rho^{-1}\sum_{j=1}^{\nu} n_j \mathbf{f}_j\right] \tag{3.11}$$

Equation (3.10) can be rewritten in terms of $\mathbf{d}_i$, and, using the fact that $\sum_i \alpha_{ij} = 0$, we find

$$\mathbf{j}_i = -\frac{\alpha_{i0}}{T}\nabla T - nkT\frac{\rho}{\rho_i}\sum_{j=1}^{\nu} \frac{\alpha_{ij}}{n_j m_j}\mathbf{d}_j$$

$$= -\frac{\alpha_{i0}}{T}\nabla T - \frac{\rho}{\rho_i}\sum_{j=1}^{\nu} \frac{\alpha_{ij}}{m_j}\left[\sum_{\substack{k=1 \\ k\neq j}}^{\nu}\left(\frac{\partial \tilde{G}_{fj}}{\partial x_k}\right)_{\substack{T,p,x_g \\ g\neq j,k}} \nabla x_k\right.$$

$$\left. + \left(\tilde{V}_j - \frac{m_j}{\rho}\right)\nabla p - \mathbf{f}_j + \frac{m_j}{\rho}\sum_{k=1}^{\nu} n_k \mathbf{f}_k\right] \tag{3.12}$$

The first expression of $\mathbf{j}_i$ given in equation (3.12) is similar to equation (3.10) and, as we will see, it has a morphology similar to that of the equations arising from the rigorous kinetic theory.

This is as far as we can go using the thermodynamics of irreversible processes. The coefficients $\alpha_{ij}$ remain, in this theory, quantities to be determined experimentally. However, the positive contribution of this approach lies in the fact that equation (3.12) is independent of the kind of fluid, i.e., a dilute or a dense gas or even a liquid.

Even though the theory does not tell us anything about the coefficients in equations (3.10)–(3.12), these equations show that the diffusion rate of a given species depends on composition gradients (*ordinary diffusion*), on pressure gradients (*pressure diffusion*), on temperature gradients (*thermal diffusion*), and on external forces (*forced diffusion*). Further, these equations permit some other interesting conclusions. For example, we observe that, since $\mathbf{j}_i$ (and thus $\mathbf{v}_i^d$) depends on the $\nabla x_j$, a given component can diffuse because of the composition gradients of the other species even though its own composition is uniform throughout the system; this phenomenon was studied by Hellund[36] and is known as *osmotic diffusion*. It is also possible for a given component to diffuse in a direction opposite to its gradient, as observed by Duncan and Toor[21]; this phenomenon is known as *inverted (or reverse) diffusion*. Further, there may be a *diffusive barrier*,[76] i.e., the diffusive flux of a component may be zero even though its composition gradient is not zero. These phenomena were foreseen in Section 1.6.2.2.


## PART B.   RIGOROUS KINETIC THEORY

The starting point of this theory is the Boltzmann distribution function $f(\mathbf{x}, \omega, t)$ of a gaseous system removed from equilibrium. In Section 3.3 we will derive the integrodifferential equation (called the Boltzmann equation) that $f$ must satisfy, noting the hypotheses involved in the derivation of this equation, and we will determine the equilibrium solution and the transport and change equations which it provides. In Part C we will study approximate solutions to the Boltzmann equation.


## 3.3.   The Boltzmann Equation

Consider a monatomic gaseous mixture that is very dilute and removed from equilibrium. Only binary collisions will be important, so the behavior of the system (cf. p. 64) is described by one-particle distribution functions[†] (one for each species) $f_i(\mathbf{x}, \omega_i, t)$, where $f_i(\mathbf{x}, \omega_i, t)\,d\mathbf{x}\,d\omega_i$ is the

---

[†]As in Chapter 2, we omit the superscript (1) on $f_i$.

average number of molecules of species $i$ that at a time between $t$ and $t + dt$ have position coordinates between $\mathbf{x}$ and $\mathbf{x} + d\mathbf{x}$ and momenta between $\omega_i$ and $\omega_i + d\omega_i$. For generality, we will assume that the molecules of $i$ are acted upon by an external force $\mathbf{f}_i(\mathbf{x}, t)$.

The equation of change of the distribution function (the Boltzmann equation) can be obtained from the Liouville equation,[37, 78] but this approach is quite abstract and we find it more suitable, from a didactic point of view, to base the derivation on a physical picture of the system.[†]

We begin by considering a volume element $d\mathbf{x}\,d\omega_i$ about the point $(\mathbf{x}, \omega_i)$ of phase space and the group of molecules of $i$ within this differential volume at time $t$. The number of such molecules in $d\mathbf{x}\,d\omega_i$ at time $t$ is

$$f_i(\mathbf{x}, \omega_i, t)\, d\mathbf{x}\, d\omega_i \tag{3.13}$$

The molecules in this group will, as time goes on, move to different points of phase space. After a time $dt$, a given molecule of the group will have the position vector (using Cartesian coordinates) $\mathbf{x} + (\omega_i/m_i)dt$ and momentum vector $\omega_i + \mathbf{f}_i dt$. If there are no collisions, the original group of molecules will occupy a region $d\mathbf{x}'d\omega_i'$ around the point $(\mathbf{x} + (\omega_i/m_i)\,dt, \omega_i + \mathbf{f}_i\,dt)$ in phase space, and so the number of molecules in the original volume element [given by equation (3.13)] must equal the number in $d\mathbf{x}'\,d\omega_i'$. It can be shown[‡] that $d\mathbf{x}'d\omega_i' = d\mathbf{x}\,d\omega_i$ (i.e., these volume elements are the same size, although they may differ in shape), so we have

$$f_i(\mathbf{x}, \omega_i, t)\, d\mathbf{x}\, d\omega_i =$$

$$f_i\left(\mathbf{x} + \frac{\omega_i}{m_i}dt,\ \omega_i + \mathbf{f}_i\,dt, t + dt\right) d\mathbf{x}\,d\omega_i \quad \text{(no collisions)}$$

$$\tag{3.14}$$

Equation (3.14) is an expression for the conservation of molecules in the absence of collisions.

The effect of collisions[§] of a molecule of $i$ with a molecule of the same or another species is of two types. First, not all the molecules of $i$ that leave the point $(\mathbf{x}, \omega_i)$ arrive at the point $\left(\mathbf{x} + (\omega_i/m_i)\,dt, \omega_i + \mathbf{f}_i\,dt\right)$, because the collisions will generally produce a change in their momenta. Second, there are molecules which were not at the point $(\mathbf{x}, \omega_i)$ at time $t$, but at a time $dt$

---

[†]The reader will find both derivations in Reference 37 (pp. 442–452).

[‡]The phase-space volume occupied by a given group of molecules throughout the dynamic evolution of the group is one of the "integral invariants of Poincaré" (see Reference 29).

[§]Here and below we will only consider collisions between the molecules in the gas phase, i.e., we will not consider molecule–wall collisions or other surface effects.

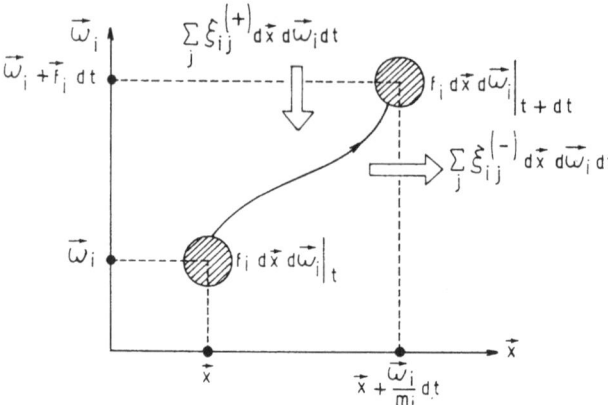

Figure 3.1. In the absence of molecular collisions, the group of molecules in the shaded element of phase space centered on $(x, \omega_i)$ moves in time $dt$ to the other shaded element. The large arrows indicate the gain and loss of molecules from this group as a result of collisions.

later have arrived at the point $(x + (\omega_i/m_i)\, dt;\ \omega_i + f_i\, dt)$ (see Figure 3.1). We define functions $\xi_{ij}^{(+)}$ and $\xi_{ij}^{(-)}$ for all $i, j$ so that $\xi_{ij}^{(+)}\, dxd\omega_i dt$ is the number of molecules of $i$ that join the original group as a result of collisions with species $j$ and $\xi_{ij}^{(-)}\, dxd\omega_i\, dt$ is the number of molecules of $i$ lost from the original group as a result of collisions with molecules of $j$. We then have

$$f_i\!\left(x + \frac{\omega_i}{m_i}\, dt, \omega_i + f_i\, dt, t + dt\right) dx\, d\omega_i$$

$$= f_i(x, \omega_i, t)\, dx\, d\omega_i + \sum_{j=1}^{\nu} \left(\xi_{ij}^{(+)} - \xi_{ij}^{(-)}\right) dx\, d\omega_i\, dt \qquad (3.15)$$

The left-hand side of this equation can be expanded in a Taylor series about the point $(x, \omega_i, t)$:

$$f_i\!\left(x + \frac{\omega_i}{m_i}\, dt, \omega_i + f_i\, dt, t + dt\right) dx\, d\omega_i$$

$$= \left[ f_i(x, \omega_i, t) + \frac{1}{m_i}\, \omega_i \cdot \frac{\partial f_i}{\partial x}\, dt + f_i \cdot \frac{\partial f_i}{\partial \omega_i}\, dt + \frac{\partial f_i}{\partial t}\, dt + \cdots \right] dx\, d\omega_i \qquad (3.16)$$

Combining equations (3.15) and (3.16) and dropping the $dx\, d\omega_i$ leads to the *Boltzmann equation*,

$$\frac{\partial f_i}{\partial t} + \frac{1}{m_i}\, \omega_i \cdot \frac{\partial f_i}{\partial x} + f_i \cdot \frac{\partial f_i}{\partial \omega_i} = \sum_{j=1}^{\nu} \left(\xi_{ij}^{(+)} - \xi_{ij}^{(-)}\right) \qquad (3.17)$$

which is a differential expression for the conservation of molecules in a dilute gaseous system.

The Boltzmann equation is the starting point for the derivation of the transport equations in a dilute gas since, as we shall see, knowledge of the distribution functions $f_i$ makes possible the calculation of the various fluxes that develop in a system removed from equilibrium.

It is important to point out that *the functions* $\xi_{ij}$ *do not include the contributions of molecule–wall collisions. The influence of these collisions can be included by imposing boundary conditions on solutions of equation* (3.17) *or through some artifice such as a special model* (e.g., our use of particles to represent walls in Section 1.3.2 and later in Chapter 4).

From analysis of the dynamics of a binary (simple) molecular collision we can obtain an expression for the right-hand side of equation (3.17) in terms of the parameters characterizing such a collision[†]:

$$\sum_{j=1}^{\nu}\left(\xi_{ij}^{(+)}-\xi_{ij}^{(-)}\right)=\sum_{j=1}^{\nu}\int\int\int\left(f_i^a f_j^a - f_i f_j\right)g_{ij}\,b\,db\,d\Theta d\omega_j \quad (3.18)$$

Here

$$f_i = f_i(\mathbf{x},\omega_i,t), \qquad f_i^a = f_i^a(\mathbf{x},\omega_i^a,t)$$
$$f_j = f_j(\mathbf{x},\omega_j,t), \qquad f_j^a = f_j^a(\mathbf{x},\omega_j^a,t)$$

where $\omega_i$ and $\omega_i^a$ are, respectively, the momenta of a molecule of $i$ before and after a collision with a molecule of $j$ (the collision occurs at $\mathbf{x}$ at time $t$), and $\omega_j$ and $\omega_j^a$ are, respectively, the momenta of a molecule of $j$ before and after the collision; $g_{ij}$ is the initial relative velocity of the colliding molecules, $g_{ij} \equiv |\omega_i/m_i - \omega_j/m_j|$; $b$ is an impact parameter and is equal to the distance between the initial lines of travel (in the center-of-mass frame) of the colliding pair of molecules $(0 \leqslant b \leqslant \infty)$; $\Theta$ is the polar angular coordinate $(0 \leqslant \Theta \leqslant 2\pi)$ in the center-of-mass frame.

The integrals on the right-hand side of equation (3.18) are usually called *collision integrals*; they can be evaluated from the geometric parameters characterizing the impact $(b,g_{ij})$ and from the potential energy of interaction, which makes the calculation of $f_i^a f_j^a$ from $f_i f_j$ possible.

Substituting equation (3.18) into equation (3.17), we arrive at

$$\frac{\partial f_i}{\partial t} + \frac{1}{m_i}\omega_i \cdot \frac{\partial f_i}{\partial \mathbf{x}} + \mathbf{f}_i \cdot \frac{\partial f_i}{\partial \omega_i} = \sum_{j=1}^{\nu}\int\int\int\left(f_i^a f_j^a - f_i f_j\right)g_{ij}\,b\,db\,d\Theta d\omega_j$$

$$(3.19)$$

which is Boltzmann's well-known integrodifferential equation for the distribution function of a dilute gas removed from equilibrium.

---

[†] The details of this analysis are given in References 37 and 78.

We can write an equation such as equation (3.19) for every one of the components of the mixture. Since each of these equations contains the distribution functions of the other components in the collision integrals and, further, these integrals depend on the interaction potential between the molecules, the distribution function of a species will in general depend upon the distribution functions of all the other species.

For our purposes it is now convenient to write equation (3.19) in terms of the distribution function $f_i(\mathbf{x}, \mathbf{v}_i, t)$ in a position–velocity space. The function $f_i d\mathbf{v}_i$ is defined as the number of molecules per unit volume with velocities in the range between $\mathbf{v}_i$ and $\mathbf{v}_i + d\mathbf{v}_i$, i.e., $dn_i = f_i d\mathbf{v}_i$.

A derivation similar to that leading to equation (3.19) gives

$$\frac{\partial f_i}{\partial t} + \mathbf{v}_i \cdot \frac{\partial f_i}{\partial \mathbf{x}} + \frac{1}{m_i}\mathbf{f}_i \cdot \frac{\partial f_i}{\partial \mathbf{v}_i} = \sum_{j=1}^{\nu} \int \int \int (f_i{}^{\mathrm{a}} f_j{}^{\mathrm{a}} - f_i f_j) g_{ij} \, b \, db \, d\Theta d\mathbf{v}_j$$

(3.20)

Equation (3.20) is our starting point for the study of transport properties in dilute gaseous systems.

### 3.3.1   Working Hypotheses of the Boltzmann Equation

The Boltzmann integrodifferential equation can be obtained in a rigorous way from the Liouville theorem[37] or by the method used by Grad.[78] Both of these derivations involve cumbersome mathematical developments, but they permit a precise analysis of the working hypotheses involved in the derivation of the Boltzmann equation. These hypotheses are:

1.   The continuum hypothesis.
2.   All collisions are binary and are "complete."
3.   The distribution function varies slowly in space and time.
4.   The assumption of molecular chaos (there is no correlation between the states of any two randomly selected molecules).
5.   Quantum effects are negligible.

The continuum hypothesis asserts that the molecular density is high enough to insure the validity of statistical laws, i.e., we can write the distribution function as a continuous function of $\mathbf{x}$, $\omega$, and $t$.

The hypothesis of binary collisions simply requires that the average duration $t_m$ of a collision be small compared to the average time $t_b$ between collisions, so that the possible interference of a third molecule can be ignored. The more severe condition of completeness requires the existence of an interval[†] $dt$ such that $dt$ is large compared to $t_m$ but small compared to $t_b$,

$$t_m \ll dt \ll t_b \tag{3.21}$$

This makes it possible to ignore incomplete collisions during $dt$ ($dt \gg t_m$).

The condition $t_m \ll t_b$ is equivalent to

$$\sigma \ll \lambda \tag{3.22}$$

For a gas in equilibrium, the mean free path is given by equation (2.30), so equation (3.22) can be written as

$$\sigma^3 n \ll 1 \tag{3.23}$$

which means that the volume occupied by the molecules must be a small fraction of the total gas volume.

The third hypothesis asserts that the distribution function does not vary significantly over the distance traveled by a molecule during the time $dt$.[‡] Consequently, $f$ must remain constant over a distance of the order of the size of a molecule, but it may vary over distances of the order of the mean free paths. The Boltzmann equation is thus not suitable for flows around bodies of molecular size, but it can be used for sizes of the order of the mean free paths.

The hypothesis of molecular chaos asserts that, given any two molecules of species $i$, the probability of finding the first with velocity $v_{i1}$ and the other with velocity $v_{i2}$ in the volume element $dx$ at $x$ at time $t$ is simply the product $f_i(x, v_{i1}, t) f_i(x, v_{i2}, t) \, dx \, dv_{i1} \, dx \, dv_{i2}$, i.e., the probabilities $f_i(x, v_{i1}, t) \, dx \, dv_{i1}$ and $f_i(x, v_{i2}, t) \, dx \, dv_{i2}$ are not correlated. This restriction leads, through the Boltzmann equation, to the condition of irreversibility of the system, e.g., with the assumption of molecular chaos it can be demonstrated[78] that the function

$$H_1(t) \equiv \int f \ln f \, dv \, dx \tag{3.24}$$

which is related to the entropy of an isolated system removed from equilibrium, is a decreasing function of time. In contrast, the Liouville equation of classical statistical mechanics describes a dynamically reversible system.

---

[†] This interval $dt$ corresponds to the time element we used earlier (Section 3.3) as the time for a group of molecules to move from the point $(x, \omega_i)$ to $(x + (\omega_i/m_i) \, dt, \omega_i + f_i dt)$.

[‡] We made this assumption on truncating the Taylor expansion of equation (3.16).

Hypothesis 5 is unnecessary because we use classical mechanics to arrive at the Boltzmann equation, and this means that the study of low-temperature phenomena is excluded from the analysis, since quantum effects may be important there. According to the quantum theory there is a wavelength, the de Broglie wavelength $\lambda = h/mv$, associated with every molecule. The de Broglie wavelength corresponding to the mean molecular speed is $27.4(MT)^{-1/2}$ Å, which at low temperatures is large. When the wavelength is of the order of the molecule's size, quantum diffraction effects are observed, and when the wavelength is of the order of the distance between molecules we observe statistical effects related to Pauli's exclusion principle. For temperatures higher than 200°K, quantum effects are smaller than 1%, even for hydrogen and helium. Below this temperature, diffraction effects become important for the light gases. Statistical effects are only important below 2°K and appear in the peculiar behavior of liquid helium.

There are, of course, kinetic theories for systems outside the range of validity of the Boltzmann equation, e.g., there are kinetic theory models for dense gases and liquids, and the quantum statistical mechanics of gases can be used to study transport phenomena at low temperatures.

### 3.3.2 Equilibrium Solution

The distribution functions that describe the behavior of a gaseous mixture are the solutions of a set of Boltzmann equations with one equation for every component.

The properties of a system in equilibrium are constant in time, so the equilibrium function will be independent of $t$, i.e., $\partial f_i/\partial t = 0$. If we further assume that there is no external force acting on the system, the system will be uniform and $f_i$ will be independent of x. Thus for a uniform gas in equilibrium, the Boltzmann equation (3.20) becomes

$$\sum_{j=1}^{v} \int\int\int \left[ f_i^{a(0)}(v_i^a) f_j^{a(0)}(v_j^a) - f_i^{(0)}(v_i) f_j^{(0)}(v_j) \right] g_{ij} \, b \, db \, d\Theta dv_j = 0$$

$$(3.25)$$

The fact that equation (3.25) is zero means (see Section 3.3) that there is no gain or loss of molecules of $i$ from a given volume element $dxdv$ as it evolves in time, i.e., the number of molecules lost from the group as a result of collisions is exactly compensated by the number gained by the group through other collisions.

To make the collision integrals zero, it is sufficient (and, it can be shown, necessary) that $f_i^{a(0)} f_j^{a(0)} = f_i^{(0)} f_j^{(0)}$. We can then write

$$\ln f_i^{a(0)}(v_i^a) + \ln f_j^{a(0)}(v_j^a) = \ln f_i^{(0)}(v_i) + \ln f_j^{(0)}(v_j) \qquad (3.26)$$

i.e., at equilibrium the quantity $\ln f_i^{(0)} + \ln f_j^{(0)}$ is an invariant of a molecular collision. Comparing equation (3.26) with the conservation laws for a molecular collision (e.g., conservation of mass, momentum, and energy), we see that equation (3.26) will hold if $\ln f_i^{(0)}(\mathbf{v}_i)$ is equal to $m_i, m_i v_{ix}, m_i v_{iy}, m_i v_{iz}$ (i.e., the components of the momentum of $m_i$), $\frac{1}{2} m_i v_i^2$, or to a linear combination of these quantities. We can thus write

$$\ln f_i^{(0)} = C_1 m_i + \mathbf{C}_2 \cdot (m_i \mathbf{v}_i) + C_3 \left( \tfrac{1}{2} m_i v_i^2 \right) \tag{3.27}$$

where the constants $C_1$, $C_2$, and $C_3$ must be the same for all species. Equation (3.27) can also be written as

$$f_i^{(0)} = C_4 \exp \left[ -C_5 (\mathbf{v}_i - \mathbf{C}_6)^2 \right] \tag{3.28}$$

which generates exponentials in $v_i^0$, $\mathbf{v}_i$, and $v_i^2$; the constants $C_4$, $C_5$, and $\mathbf{C}_6$ are in general different for each species (i.e., we should really use a subscript $i$ on them).

To determine the constants $C_4$, $C_5$, and $\mathbf{C}_6$, we use equation (3.28) to evaluate the following properties, which we define here for the general (equilibrium and nonequilibrium) case.

1. *The number density*:

$$n_i = \int f_i d\mathbf{v}_i \tag{3.29}$$

where the integration is to be carried out over all of $\mathbf{v}_i$ space, e.g., if Cartesian coordinates are used for $\mathbf{v}_i$, then equation (3.29) becomes

$$n_i = \int_{-\infty}^{\infty} \int_{-\infty}^{\infty} \int_{-\infty}^{\infty} f_i \, dv_{ix} \, dv_{iy} \, dv_{iz}$$

Equation (3.29) is consistent with the definition of $f_i$.

2. *The average molecular velocity of species $i$* [*equation (2.17)*]:

$$\bar{\mathbf{v}}_i(\mathbf{x}, t) = n_i^{-1} \int \mathbf{v}_i f_i d\mathbf{v}_i \tag{3.30}$$

Here again the integration is to be carried out over all of $\mathbf{v}_i$ space. For a system in equilibrium, $\bar{\mathbf{v}}_i$ will be independent of position and time and will be equal to $\vartheta$, the average velocity on mass basis [defined by equation (1.5)].[†]

---

[†] In the equilibrium case, there is no segregation of species, so $\vartheta = \vartheta_M$.

3a. *The average kinetic energy of a molecule of species i in the center-of-mass frame*:

$$\langle \tfrac{1}{2} m v_i^2 \rangle = (2n_i)^{-1} \int m_i (\mathbf{v}_i - \boldsymbol{\vartheta})^2 f_i d\mathbf{v}_i \tag{3.31}$$

with the integration to be carried out over all of $\mathbf{v}_i$ space.

3b. *The generalized kinetic theory definition of the temperature of a gas*[†]:

$$\langle \tfrac{1}{2} m_i v_i^2 \rangle = \tfrac{3}{2} kT \tag{3.32}$$

For the equilibrium case this agrees with the result obtained in Chapter 2. Since this average is the same for all species, the *total* kinetic energy of the gas per unit volume is $\tfrac{3}{2} nkT$.

If we carry out the integrations in equations (3.29)–(3.31) using equation (3.28) for $f_i$ and solve for the constants $C_4$, $C_5$, and $\mathbf{C}_6$, we obtain the equilibrium values:

$$C_4 \equiv n_i (m_i/2\pi kT)^{3/2} \tag{3.33}$$

$$C_5 \equiv m_i/2kT \tag{3.34}$$

$$\mathbf{C}_6 \equiv \boldsymbol{\vartheta} \tag{3.35}$$

Substituting equations (3.33)–(3.35) into equation (3.28), we obtain the Maxwell–Boltzmann distribution:

$$f_i^{(0)} = n_i \left( \frac{m_i}{2\pi kT} \right)^{3/2} \exp\left[ -\frac{m_i(\mathbf{v}_i - \boldsymbol{\vartheta})^2}{2kT} \right] \tag{3.36}$$

which we obtained earlier [equation (2.22b)] by a different method.

## 3.4  Transport Equations

In a gas removed from equilibrium there are gradients of one or more of the macroscopic properties of the system: composition, average velocity on mass basis, pressure, temperature, etc. The gradients of these properties are the "cause" of the molecular transport of momentum, kinetic energy, and mass through the gas. We will continue to use the generalized expressions $\Gamma_i$ and $\psi_i$ for these fluxes and properties as we did in Chapter 2.

---

[†]This result can be arrived at for the equilibrium case by calculating the pressure $p$ of a gas using equation (3.28) for $f_i$ and using the equation of state $p = nkT$ (see Reference 40).

Our goal in this section is to find expressions for the various fluxes in the system in terms of the distribution function.

We begin by defining the *peculiar velocity* $v_i^P$ of the molecules of $i$ by

$$v_i^P(v_i, x, t) \equiv v_i - \vartheta \tag{3.37}$$

i.e., the peculiar velocity of a molecule of $i$ is its velocity relative to coordinates moving at the average velocity on mass basis $\vartheta(x, t)$.

The diffusion velocity of species $i$ is (see Table 1.3)

$$v_i^d \equiv \bar{v}_i^P(x, t) = \bar{v}_i - \vartheta \tag{3.38}$$

or

$$v_i^d = n_i^{-1} \int (v_i - \vartheta) f_i(x, v_i, t) \, dv_i \tag{3.39}$$

Consider a differential surface $dS$ moving at velocity $\vartheta$ and with a normal $\hat{n}$ as shown in Figure 3.2. According to equation (3.37), the velocity of a molecule of species $i$ with respect to the surface $dS$ is $v_i^P$. All molecules of $i$ with velocity $v_i^P$ in the cylinder with $dS$ as base and $v_i^P \, dt$ as generator will cross $dS$ in time $dt$; the number of such molecules is equal to the volume of the cylinder, $v_i^P \, dt \cdot \hat{n} \, dS$, multiplied by $f_i(x_i, v_i, t) \, dv_i = f_i(x_i, v_i^P + \vartheta, t) \, dv_i^P$, which is the number of molecules of $i$ per unit volume with velocity $v_i^P$. Consequently, the number of molecules of $i$ with velocity $v_i^P$ crossing the surface $dS$ during time $dt$ is given by

$$(\hat{n} \cdot v_i^P) f_i(x_i, v_i^P + \vartheta, t) \, dv_i^P dS dt \tag{3.40}$$

This flux of molecules in general contributes to the flux $\Gamma_i$ of a property $\psi_i$ across the surface $dS$ in the direction $\hat{n}$. The component $\Gamma_{in}$ of the flux $\Gamma_i$ along $\hat{n}$ is

$$\Gamma_{in} = \int \psi_i (\hat{n} \cdot v_i^P) f_i \, dv_i^P \tag{3.41}$$

The quantity

$$\Gamma_i = \int \psi_i f_i v_i^P \, dv_i^P \tag{3.42}$$

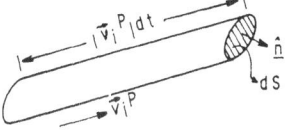

Figure 3.2. The number of molecules with velocity $v_i^P$ that cross the surface $dS$ in time $dt$ are all those contained in the cylinder of length $|v_i^P| \, dt$ and with generator parallel to $v_i^P$.

is the flux vector[†] corresponding to the property $\psi_i$, and its component along $\hat{\mathbf{n}}$ is

$$\Gamma_{in} = \hat{\mathbf{n}} \cdot \Gamma_i \tag{3.43}$$

We now write equation (3.42) for the transport of mass, momentum, and energy.

1.   *Mass transport.* In this case, $\psi_i = m_i$ and $\Gamma_i = \mathbf{j}_i$, so we have

$$\mathbf{j}_i = m_i \int f_i \mathbf{v}_i^P \, d\mathbf{v}_i^P \tag{3.44}$$

Using the definition of $\mathbf{v}_i^d$ given in equation (3.39), we can write equation (3.44) as

$$\mathbf{j}_i = n_i m_i \mathbf{v}_i^d = \rho_i \mathbf{v}_i^d \tag{3.45}$$

2. *Momentum transport.* In this case, $\psi_i$ is the vector[‡] $m_i \mathbf{v}_i^P$ and $\Gamma_i$ is the tensor $T_i$, the shear tensor associated with component $i$. We have

$$T_i = m_i \int f_i \mathbf{v}_i^P \mathbf{v}_i^P \, d\mathbf{v}_i^P = n_i m_i \overline{\mathbf{v}_i^P \mathbf{v}_i^P} \tag{3.46}$$

The sum of $T_i$ over all $i$ gives the shear tensor of the gas,

$$T = \sum_{i=1}^{\nu} T_i = \sum_{i=1}^{\nu} n_i m_i \overline{\mathbf{v}_i^P \mathbf{v}_i^P} \tag{3.47}$$

which is the momentum flux through the gas. The diagonal elements $T_{xx}, T_{yy}, T_{zz}$ of $T$ are normal shears, while the off-diagonal elements are tangential shears, in both cases as measured in a coordinate frame moving at the velocity $\vartheta$.

For a gas in equilibrium, the tangential shears are zero and the normal shears are all equal. The shear stress on every elementary gas surface is thus constant, normal to the surface, and independent of its orientation:

$$T_{xx} = T_{yy} = T_{zz} = p \tag{3.48}$$

---

[†]Cf. our discussion in Section 1.4.2. leading to equation (1.1).

[‡]We could, of course, take each component of the momentum as a property $\psi_i$ and determine the corresponding $\Gamma_i$. Since each $\Gamma_i$ has three components and there is a $\Gamma_i$ for each component of the momentum, we would obtain in this way the nine components of $T_i$. Cf. our discussion of the momentum flux in Sections 1.7.2 and 2.2.1.

3. *Kinetic energy transport.* In this case, $\psi_i = \frac{1}{2} m_i (v_i^P)^2, \Gamma_i = \mathbf{q}_i$, and we have

$$\mathbf{q}_i = \frac{1}{2} m_i \int (v_i^P)^2 \mathbf{v}_i^P f_i \, d\mathbf{v}_i^P = \frac{1}{2} n_i m_i \overline{(v_i^P)^2 \mathbf{v}_i^P} \qquad (3.49)$$

The sum of $\mathbf{q}_i$ over all components $i$ gives the total heat flux vector:

$$\mathbf{q} = \sum_{i=1}^{\nu} \mathbf{q}_i = \sum_{i=1}^{\nu} \frac{1}{2} n_i m_i \overline{(v_i^P)^2 \mathbf{v}_i^P} \qquad (3.50)$$

We see that knowledge of the distribution function of a gas not in equilibrium makes possible the calculation of all the transport phenomena that occur in a gas. In particular, the constitutive equations for diffusion, our main objective, are obtained from equation (3.44). The possibility, then, of writing the constitutive equation of diffusion explicitly depends on the possiblity of solving the Boltzmann equation. Before we deal with the existing attempts to solve the Boltzmann equation, we will show that the equations of change given in Section 1.5 can be obtained from the Boltzmann equation.

## 3.5   Generalized Equation of Change

### 3.5.1   Derivation

The differential equations for the conservation of mass, momentum, and energy can be obtained from the Boltzmann equation without explicitly determining the form of the distribution function. Although these equations are derived for dilute gases, their form is valid for dense gases and liquids.

If we multiply equation (3.20) by the property $\psi_i$ and integrate over $\mathbf{v}_i$, we obtain

$$\int \psi_i \left[ \frac{\partial f_i}{\partial t} + \mathbf{v}_i \cdot \frac{\partial f_i}{\partial \mathbf{x}} + \frac{1}{m_i} \left( \mathbf{f}_i \cdot \frac{\partial f_i}{\partial \mathbf{v}_i} \right) \right] d\mathbf{v}_i$$

$$= \sum_{j=1}^{\nu} \int \int \int \int \psi_i (f_i^a f_j^a - f_i f_j) g_{ij} \, b \, db \, d\Theta \, d\mathbf{v}_i d\mathbf{v}_j \qquad (3.51)$$

We can write equation (3.51) in a different form using the following: (1) the variables $\mathbf{x}$, $\mathbf{v}_i$, and $t$ are independent; (2) $\psi_i$ is in general a function of $\mathbf{x}$

and $t$, e.g., if $\psi_i = m_i v_i^P = m_i(v_i - \vartheta)$ it depends in general on $\mathbf{x}$, $t$ through $\vartheta$; (3) the force $\mathbf{f}_i(\mathbf{x}, t)$ is assumed to depend only on position (i.e., on $\mathbf{x}$) and not on the velocity $\mathbf{v}_i$ of the molecules; (4) the average value of $\psi_i$ is

$$\overline{\psi}_i = n_i^{-1} \int \psi_i f_i d\mathbf{v}_i \tag{3.52}$$

(5) the product $\psi_i f_i$ is assumed to drop rapidly to zero for large $|\mathbf{v}_i|$, i.e., $f_i$ must drop off rapidly enough so that, e.g., $v_i^2 f_i \to 0$ for large $|\mathbf{v}_i|$.

On the basis of these arguments, the terms on the left-hand side of equation (3.51) can be rewritten as follows:

$$\int \psi_i \frac{\partial f_i}{\partial t} d\mathbf{v}_i = \frac{\partial}{\partial t} \int \psi_i f_i d\mathbf{v}_i - \int f_i \frac{\partial \psi_i}{\partial t} d\mathbf{v}_i$$

$$= \frac{\partial(n_i \overline{\psi}_i)}{\partial t} - n_i \frac{\overline{\partial \psi_i}}{\partial t} \tag{3.53}$$

$$\int \psi_i \left( \mathbf{v}_i \cdot \frac{\partial f_i}{\partial \mathbf{x}} \right) d\mathbf{v}_i = \int \psi_i (\mathbf{v}_i \cdot \nabla f_i) d\mathbf{v}_i$$

$$= \nabla \cdot \int \psi_i f_i \mathbf{v}_i d\mathbf{v}_i - \int f_i (\mathbf{v}_i \cdot \nabla \psi_i) d\mathbf{v}_i$$

$$= \nabla \cdot (n_i \overline{\psi_i \mathbf{v}_i}) - n_i \overline{(\mathbf{v}_i \cdot \nabla \psi_i)} \tag{3.54}$$

$$\int \frac{\psi_i}{m_i} \left( \mathbf{f}_i \cdot \frac{\partial f_i}{\partial \mathbf{v}_i} \right) d\mathbf{v}_i = \frac{\mathbf{f}_i}{m_i} \cdot \int \psi_i \frac{\partial f_i}{\partial \mathbf{v}_i} d\mathbf{v}_i$$

$$= \frac{\mathbf{f}_i}{m_i} \cdot \left[ \int \frac{\partial}{\partial \mathbf{v}_i} (\psi_i f_i) d\mathbf{v}_i - \int f_i \frac{\partial \psi_i}{\partial \mathbf{v}_i} d\mathbf{v}_i \right]$$

$$= -\frac{\mathbf{f}_i}{m_i} \cdot \int f_i \frac{\partial \psi_i}{\partial \mathbf{v}_i} d\mathbf{v}_i = -\frac{\mathbf{f}_i}{m_i} \cdot n_i \frac{\overline{\partial \psi_i}}{\partial \mathbf{v}_i} \tag{3.55}$$

where the integral containing $\partial(\psi_i f_i)/\partial \mathbf{v}_i$ is zero by virtue of the hypothesis 5.

Substituting equations (3.53)–(3.55) into equation (3.51), we have

$$\frac{\partial(n_i \overline{\psi}_i)}{\partial t} + \nabla \cdot (n_i \overline{\psi_i \mathbf{v}_i}) - n_i \left[ \frac{\overline{\partial \psi_i}}{\partial t} + \overline{(\mathbf{v}_i \cdot \nabla \psi_i)} + \frac{\mathbf{f}_i}{m_i} \cdot \frac{\overline{\partial \psi_i}}{\partial \mathbf{v}_i} \right]$$

$$= \sum_{j=1}^{\nu} \int \int \int \int \psi_i (f_i^a f_j^a - f_i f_j) g_{ij} b \, db \, d\Theta d\mathbf{v}_i d\mathbf{v}_j \tag{3.56}$$

Equation (3.56) is known as *Enskog's equation* or the *generalized equation of change* for the physical property $\psi_i$ associated with the molecules of species *i*. The sum of equation (3.56) over *i* gives the equation of change of $\psi$ for the entire gas.

## 3.5.2  Particular Cases

The three particular cases of interest for equation (3.56) are: $\psi_i = m_i$, $\psi_i = m_i \mathbf{v}_i^P$, and $\psi_i = \frac{1}{2} m_i (v_i^P)^2$.

Since mass, momentum, and kinetic energy are invariants of a molecular collision, it can be shown[37] that the following holds:

$$\int \int \int \int m_i ( f_i^a f_j^a - f_i f_j ) g_{ij} b \, db \, d\Theta \, d\mathbf{v}_i \, d\mathbf{v}_j = 0 \tag{3.57}$$

and, for $\psi_i$ equal to $m_i \mathbf{v}_i^P$ or $\frac{1}{2} m_i (v_i^P)^2$,

$$\sum_{i=1}^{\nu} \sum_{j=1}^{\nu} \int \int \int \int \psi_i ( f_i^a f_j^a - f_i f_j ) g_{ij} b \, db \, d\Theta \, d\mathbf{v}_i \, d\mathbf{v}_j = 0 \tag{3.58}$$

In equation (3.58) the integrand is summed over *i* and *j*, while in equation (3.57) it is not; this is because the mass of each individual molecule is conserved in any molecular collision between molecules of *i* and *j* [equation (3.57)], while momentum and kinetic energy are conserved only if we consider all the components of the system (in a collision of molecules of *i* and *j* it is the total momentum and total kinetic energy of the molecules that are conserved and not the momentum and kinetic energy of each molecule by itself).

### 3.5.2.1  Equation of Change of Mass

Putting $\psi_i = m_i$ in equation (3.56), using equation (3.57), and taking into account that $\bar{m}_i = m_i$ and that $m_i$ does not depend on $\mathbf{x}$, $\mathbf{v}_i$, or $t$, we find

$$\frac{\partial \rho_i}{\partial t} + \nabla \cdot \rho_i \bar{\mathbf{v}}_i = 0 \tag{3.59}$$

Using equation (3.38) for the velocity $\bar{\mathbf{v}}_i$, equation (3.59) reads

$$\frac{\partial \rho_i}{\partial t} + \nabla \cdot \rho_i ( \boldsymbol{\vartheta} + \mathbf{v}_i^d ) = 0 \tag{3.60}$$

Using the definition of $\mathbf{j}_i$ given in Table 1.3, we can write equation (3.60) as

$$\frac{\partial \rho_i}{\partial t} + \nabla \cdot \rho_i \vartheta + \nabla \cdot \mathbf{j}_i = 0 \tag{3.61}$$

The sum of equation (3.61) over all species leads to the equation of change of the total mass of the system:

$$\frac{\partial \rho}{\partial t} + \nabla \cdot \rho \vartheta = 0 \tag{3.62}$$

### 3.5.2.2  Equation of Change of Momentum

Introducing $\psi_i = m_i \mathbf{v}_i^P$ into equation (3.56), we obtain

$$\frac{\partial \mathbf{j}_i}{\partial t} + \nabla \cdot (\rho_i \overline{\mathbf{v}_i \mathbf{v}_i^P}) - \rho_i \left[ \overline{\frac{\partial v_i^P}{\partial t}} + (\overline{\mathbf{v}_i \cdot \nabla \mathbf{v}_i^P}) + \frac{\mathbf{f}_i}{m_i} \cdot \overline{\frac{\partial \mathbf{v}_i^P}{\partial \mathbf{v}_i}} \right]$$

$$= \sum_{j=1}^{\nu} \int \int \int \int m_i \mathbf{v}_i^P (f_i^a f_j^a - f_i f_j) g_{ij} b \, db \, d\Theta \, d\mathbf{v}_i d\mathbf{v}_j \tag{3.63}$$

Note that *this equation contains diffusive fluxes and diffusion velocities; it is therefore plausible to assume that the equation of change of momentum for a given species is to be used to obtain diffusion equations.* The appearance of diffusive fluxes in the equation of change of momentum for a given species was pointed out in Section 1.6.1 in connection with equation (1.38), and we will see this again (Section 3.7.1) when we consider the Grad–Zhdanov theory.

The sum of equation (3.63) over all species leads to the equation of change of momentum of the system,

$$\rho \left( \frac{\partial \vartheta}{\partial t} + \vartheta \cdot \nabla \vartheta \right) = -\nabla \cdot T + \sum_{i=1}^{\nu} n_i \mathbf{f}_i \tag{3.64}$$

The shear tensor $T$ may be written as

$$T = pl + \tau \tag{3.65}$$

where $l$ is the unit tensor. The term in parentheses in equation (3.64) is the substantive derivative of the average velocity on mass basis, i.e., it is the acceleration of the fluid volume element:

$$\frac{D\vartheta}{Dt} = \frac{\partial \vartheta}{\partial t} + \vartheta \cdot \nabla \vartheta \tag{3.66}$$

If the only force acting on the molecules is the gravitational force, the last term of equation (3.64) becomes

$$\sum_{i=1}^{\nu} n_i \mathbf{f}_i = \sum_{i=1}^{\nu} n_i m_i \frac{\mathbf{f}_i}{m_i} = \sum_{i=1}^{\nu} \rho_i \mathbf{g} = \rho \mathbf{g} \qquad (3.67)$$

For this case, it is customary to define a variable $P$ such that

$$\nabla P = \nabla p - \rho \mathbf{g} \qquad (3.68)$$

Using equations (3.65)–(3.68) we can write equation (3.64) as

$$\rho \frac{D\boldsymbol{\vartheta}}{Dt} = -\nabla P - \nabla \cdot \boldsymbol{\tau} \qquad (3.69)$$

### 3.5.2.3.  Equation of Change of Energy

Introducing $\psi_i = \frac{1}{2} m_i (v_i^p)^2$ into equation (3.56), summing over all species, and using a little algebra, we obtain

$$\frac{\partial}{\partial t} \left( \rho \hat{U}^{(\mathrm{tr})} \right) + \nabla \cdot \rho \hat{U}^{(\mathrm{tr})} \boldsymbol{\vartheta} + \nabla \cdot \mathbf{q} + (T : \nabla \boldsymbol{\vartheta}) - \sum_{i=1}^{\nu} n_i \left( \mathbf{f}_i \cdot \mathbf{v}_i^d \right) = 0 \quad (3.70)$$

where

$$\hat{U}^{(\mathrm{tr})} = \rho^{-1} \sum_{i=1}^{\nu} \tfrac{1}{2} n_i m_i \overline{(v_i^p)^2} \qquad (3.71)$$

is the translational contribution to the internal energy.

## PART C.    APPROXIMATE SOLUTIONS TO THE BOLTZMANN EQUATION

The aim of this part is to provide a brief outline of the Chapman–Enskog,[10–12, 22–24] Grad–Zhdanov,[31, 83] and BGK[6, 81] theories and other methods of solution of the Boltzmann equation. We shall emphasize the results obtained rather than the method of solution employed.

## 3.6.    The Chapman–Enskog Theory

In the technique developed by Enskog to solve Boltzmann's integro-differential equation, the distribution function of every component in the mixture is written as the sum of the equilibrium distribution function and a perturbation term:

$$f_i = f_i^{(0)}(1 + \psi_{pi})$$

(3.72)

where $f_i^{(0)}$ is the Maxwell–Boltzmann distribution [equation (2.23), (2.22b), and (3.36)], and $\psi_{pi}(\mathbf{x}, \mathbf{v}_i, t)$ is the perturbation function (this mathematical approach is usually called the *perturbation technique*).

Substituting equation (3.72) into the Boltzmann equation with the assumption that the perturbation is small (i.e., terms of second order in $\psi_{pi}$ are ignored), and making certain simplifications, we can write a linearized differential equation for $\psi_{pi}$. The form of the resulting equation suggests the following solution for the perturbation function (the solution is somewhat forced to achieve the desired form):

$$\psi_{pi} = -(\mathbf{A}_i \cdot \nabla \ln T) - (\mathbf{B}_i : \nabla \boldsymbol{\vartheta}) + n \sum_{j=1}^{\nu} (\mathbf{C}_i^{(j)} \cdot \mathbf{d}_j)$$

(3.73)

where $\mathbf{A}_i$, $\mathbf{B}_i$, and $\mathbf{C}_i^{(j)}$ are functions to be determined and which depend on the peculiar velocity $\mathbf{v}_i^P$, and $\mathbf{d}_j$ is given by

$$\mathbf{d}_j = \nabla x_j + \left( x_j - \frac{n_j m_j}{\rho} \right) \nabla \ln p - \frac{n_j m_j}{\rho p} \left( \frac{\rho}{m_j} \mathbf{f}_j - \sum_{i=1}^{\nu} n_i \mathbf{f}_i \right)$$

(3.74)

where $x_j = n_j/n$, as defined in Section 1.4.4. It can be verified from equation (3.74) that

$$\sum_{j=1}^{\nu} \mathbf{d}_j = 0$$

(3.75)

The substitution of equation (3.73) into the Boltzmann equation shows that the coefficients $\mathbf{A}_i$, $\mathbf{B}_i$, and $\mathbf{C}_i$ can be written in terms of the velocities $\mathbf{v}_i$ as follows:

$$\mathbf{A}_i = \mathbf{v}_i^* A_i(v_i^*), \qquad \mathbf{B}_i = \left( \mathbf{v}_i^* \mathbf{v}_i^* - \tfrac{1}{3} v_i^{*2} \mathbf{I} \right) B_i(v_i^*),$$
$$\mathbf{C}_i^{(j)} = \mathbf{v}_i^* C_i^{(j)}(v_i^*)$$

(3.76)

where $\mathbf{v}_i^*$ is the dimensionless form of the velocity $\mathbf{v}_i$,

$$\mathbf{v}_i^* \equiv (m_i/2kT)^{1/2} \mathbf{v}_i^P$$

(3.77)

and $\mathbf{I}$ is the unit tensor. Note that each of the right-hand sides in equations

(3.76) is the product of a vector (or tensor) and a scalar function of the magnitude of $v_i^*$. Using these general forms for the coefficients we can determine the dependence of the mass flux vectors on the variables of the system. From equation (3.39) and (3.45) we obtain

$$\mathbf{v}_i^d = \frac{\mathbf{j}_i}{n_i m_i} = \frac{1}{n_i} \int \mathbf{v}_i^P f_i \, d\mathbf{v}_i^P = \frac{1}{n_i} \int \mathbf{v}_i^P f_i^{(0)} \psi_{pi} \, d\mathbf{v}_i^P \qquad (3.78)$$

since $\int \mathbf{v}_i^P f_i^{(0)} \, d\mathbf{v}_i^P = 0$. Now substituting equations (3.73) and (3.76) into equation (3.78) and noting that the integral over **B** is zero (the integrand is "odd" in $v_i^*$), we obtain

$$\mathbf{v}_i^d = \frac{n^2}{n_i \rho} \sum_{j=1}^{\nu} m_j D_{ij}^m \mathbf{d}_j - \frac{1}{n_i m_i} D_i^T \nabla \ln T \qquad (3.79)$$

where the coefficients $D_{ij}^m$ and $D_i^T$ are the *multicomponent diffusivity*[†] and the *multicomponent thermal diffusivity*, respectively:

$$D_{ij}^m = \frac{\rho}{3 n m_j} \left( \frac{2kT}{m_i} \right)^{1/2} \int C_i^{(j)}(v_i^*) v_i^{*2} f_i^{(0)} \, d\mathbf{v}_i^P \qquad (3.80)$$

$$D_i^T = \frac{m_i}{3} \left( \frac{2kT}{m_i} \right)^{1/2} \int A_i(v_i^*) v_i^{*2} f_i^{(0)} \, d\mathbf{v}_i^P \qquad (3.81)$$

Before the work of Chapman and Enskog, thermal diffusion was unknown, both theoretically and experimentally. Experiments by Chapman and Dootson[15] verified the Chapman–Enskog theoretical prediction of this phenomenon. This is an example of an infrequent occurrence in science, since a phenomenon is usually observed first and then explained.

To evaluate the coefficients $D_{ij}^m$ and $D_i^T$ it is first necessary to evaluate the coefficients $A_i(v_i^*)$ and $C_i^{(j)}(v_i^*)$. The procedure is as follows. The perturbation function of equation (3.73) is introduced into the linearized Boltzmann equation obtained by assuming that $\psi_{pi}$ is small. Then, equating the coefficients of like gradients, we obtain integral equations for each of the coefficients $A_i(v_i^*)$ and $C_i^{(j)}(v_i^*)$. The solution of these integral equations has been obtained by two equivalent methods, that of Chapman and Cowling[14] and a variational one developed by Curtiss and Hirschfelder.[18] In both methods the coefficients are expanded in a series of Sonine

---

[†] $D_{ij}^m$ is the multicomponent diffusivity, while $D_{ij}$ is the binary diffusivity.

polynomials, which are defined by

$$S_n^{(m)}(x) = \sum_j \frac{(-1)^j(m+n)!x^j}{(n+j)!(m-j)!j!} \tag{3.82}$$

The number of terms of the series that are used will be called the *degree of approximation*[†] to which the diffusion coefficients are calculated. For the multicomponent diffusivity, the use of the first term of the series gives very good results with the second term providing only a small correction. However, if only the first term is used, the thermal multicomponent diffusivity is zero and so it is necessary to take at least the first two terms of the series.

The diffusion coefficients can thus be written as a function of a set of collision integrals characterizing the molecular collision. These integrals have the dimensions of area and depend on the temperature and the intermolecular potential. In their most general form they can be written as[(51)]

$$\Omega_{i,j}^{(L_r,s)}(T) = \frac{1}{(s+1)!(kT)^{s+2}} \int_0^\infty \left[ \exp\left(-\frac{E_r}{kT}\right) \right] E_r^{s+1} s_{i,j}^{(L_r)}(E_r)\,dE_r \tag{3.83}$$

where $E_r$ is the relative initial kinetic energy of a pair of colliding molecules and $s_{i,j}^{(L_r)}(E_r)$ is the transverse section for diffusion, given by

$$s_{i,j}^{(L_r)}(E_r) = 2\pi \left[ 1 - \frac{1+(-1)^{L_r}}{2(1+L_r)} \right]^{-1} \int_0^\infty \left(1 - \cos^{(L_r)}\check{\alpha}_{ij}\right)b\,db \tag{3.84}$$

Where $\check{\alpha}_{ij}$ is the deflection angle for a collision with an impact parameter $b$ and depends explicitly on the interaction potential.[‡] It can be shown that $\Omega_{ij} = \Omega_{ji}$. It is convenient to deal with dimensionless collision integrals defined by

$$\Omega_{i,j}^{(L_r,s)*} = \Omega_{i,j}^{(L_r,s)}/\pi\sigma_{ij}^2 \tag{3.85}$$

---

[†]We will reserve the term "*order* of approximation" to indicate the order to which the perturbation expansion is taken. The approximations used throughout this section are first-order in the perturbation $\psi_{pi}$.

[‡]The mathematical foundation of the developments leading to $\Omega_{i,j}^{(L_r,s)}$ and $s_{i,j}^{(L_r)}$ may be found elsewhere[(14, 37)]; $L_r$ and $s$ are summation indexes corresponding to the expansion of different functions (see Section 9.32 of Reference 14).

The numerical values of $\Omega_{i,j}^{(L_r,s)*}$ lie near one. Deviations from this value reflect the "softness" of the potential in comparison with an ideal rigid sphere of diameter $\sigma$. The collision integrals can be calculated for plausible intermolecular potentials, but the integrations are cumbersome.

### 3.6.1. *The Constitutive Equations of Diffusion Expressed in Terms of the Binary Coefficients*

Hirschfelder *et al.*[37] have demonstrated that there is a relationship between the multicomponent diffusivity $D_{ij}^m$ and the binary diffusivity $D_{ij}$ when both are calculated in the first-degree approximation. Using that relationship, the constitutive equation of diffusion [equation (3.79)] can be written in terms of the binary diffusivities, a transformation called *inversion*. For a multicomponent gaseous mixture of $\nu$ components, the following $\nu - 1$ independent equations result:

$$\sum_{\substack{j=1 \\ j \neq i}}^{\nu} \frac{n_i n_j}{n^2 [D_{ij}]_1} \left( \mathbf{v}_j^d - \mathbf{v}_i^d \right) = \mathbf{d}_i - \nabla \ln T \sum_{\substack{j=1 \\ j \neq i}}^{\nu} \frac{n_i n_j}{n^2 [D_{ij}]_1} \left( \frac{D_j^T}{n_j m_j} - \frac{D_i^T}{n_i m_i} \right) \quad (3.86)$$

where $[D_{ij}]_1$ is the first-degree approximation[†] (see p. 108) to the binary diffusivity. Equation (3.86) is usually known as the Stefan–Maxwell equation since when reduced to a binary, isobaric, isothermal system without external forces it is the expression originally proposed independently by Stefan[74, 75] and Maxwell[54] [cf. equation (1.37)].

Equation (3.86) involves a mathematical inconsistency not usually mentioned. In the development of equation (3.86), the first-degree diffusivities are used, while for the thermal diffusivities the second-degree expressions are used. However, this inconsistency does not lead to incorrect results, as shown by Muckenfuss,[56] who found a method for inverting equation (3.79) for a multicomponent mixture to any degree of approximation. The equation obtained is identical to equation (3.86) but with the diffusivities in the desired approximation.

Since the multicomponent diffusivity is very difficult to calculate and it depends on composition, the $\nu$ constitutive equations in which these coefficients appear are usually replaced by $\nu - 1$ equations involving the binary diffusivities.[‡]

---

[†] The term $n_i n_j / n^2 [D_{ij}]_1$ is called the *impedance*.

[‡] The remaining $\nu$th equation is

$$\sum_i \mathbf{j}_i = 0 \quad \text{or} \quad \sum_i n_i m_i \mathbf{v}_i^d = 0 \quad (3.86a)$$

### 3.6.2.  Binary Diffusivity

We shall now determine the functional dependence of the diffusion transport coefficients. According to the Chapman–Enskog theory, the binary diffusivity in the first degree is given by

$$[D_{AB}]_1 = [D_{BA}]_1 \equiv \frac{3}{16}\left(\frac{2\pi kT}{m_{AB}^+}\right)^{1/2}\frac{1}{n\Omega_{A,B}^{(1,1)}} \qquad (3.87)$$

where $m_{AB}^+ \equiv m_A m_B/(m_A + m_B)$ is the reduced mass. According to equation (3.85), the collision integral $\Omega_{A,B}^{(1,1)}$ may be replaced by $\pi\sigma_{AB}^2\Omega_{A,B}^{(1,1)*}$. Using the ideal gas law, we arrive at the rigorous kinetic theory prediction for the dependence of $[D_{AB}]_1$ on the variables of the system:

$$[D_{AB}]_1 \propto \begin{cases} (x_A, x_B)^0 & (3.88a) \\ p^{-1} & (3.88b) \\ T^{3/2}/\Omega_{A,B}^{(1,1)}(T) & (3.88c) \end{cases}$$

Although the diffusivity in the first-degree approximation is independent of composition [equation (3.88a)], the experimental results show a slight dependence of the diffusivity on composition; this dependence appears in the second-degree coefficient.

According to equation (3.88b), the rigorous kinetic theory predicts that the binary diffusivity is inversely proportional to the molecular density $n$. For the functional dependence of the diffusivity for dense gases, see References 17 and 25.

To find the dependence of the binary diffusivity on temperature [equation (3.88c)], it is necessary to determine the temperature dependence of the collision integral, which depends on the intermolecular potential. Calculations for some plausible potentials show that $(\partial \ln[D_{AB}]_1/\partial \ln T)_p$ ranges between $\frac{3}{2}$ and 2, which agrees with experimental observation. A semiquantitative estimation of this dependence has been made by Mason and Marrero[51] and is shown in Figure 3.3, where $\Xi$ is the energy corresponding to the intermolecular potential well. The shape of the curve arises by assuming that the important part of the molecular interaction at low temperatures are the London dispersion forces, the potential of which is proportional to (distance)$^{-6}$, and that at high temperatures the important part of the interaction is a short-range repulsion that varies exponentially with distance.

Marrero and Mason[48] have made an excellent compilation of all available measurements up to 1970 of binary diffusivities in the gaseous phase; they discuss the ranges of validity and give a critical comparison

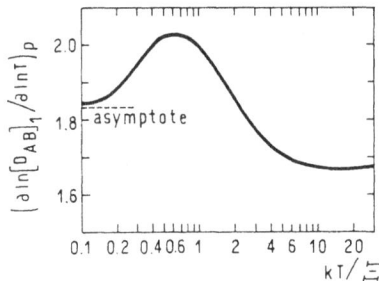

Figure 3.3. Variation of the molecular diffusivity with temperature. Reprinted from Mason and Marrero[51] by permission.

between theory and experiment. They also give a very precise analysis of the different methods for measuring diffusivities.

If experimental values do not exist, the binary diffusivity can be calculated using an approximation to the molecular interaction potential. This approximation can be drawn from various sources such as quantum mechanics, spectroscopic data, virial coefficients, data on molecular diffraction, the interaction potential of the pure species, etc. Once the interaction potential has been selected, we can refer to the tabulated data[48, 51] to calculate the collision integral and the binary diffusivity.

The binary diffusivity can also be obtained from other transport properties such as the viscosity of the mixture, thermal conductivity, and thermal diffusivity.[48]

Quite reliable methods for estimating binary diffusion coefficients are available as semiempirical correlations of experimental data. These include propositions by Arnold,[3] Gilliland,[28] Andrussow,[2] Wilke and Lee,[82] Slattery and Bird,[72] Chen and Othmer,[16] Othmer and Chen,[61] Fuller et al.,[27] and Bailey and Chen.[4] Most of them are based on equation (2.92) and add empirical modifications.

Marrero and Mason[49] have recently found an expression for the binary diffusivity for $x_A \to 0$ (i.e., $M_A \gg M_B$); it is a function of two adjustable parameters and is valid for temperatures ranging from those at which quantum effects become significant up to $10,000°K$.

For $x_B = 0$, we have $D_{AB} = D_{AA}$, the *self-diffusivity*; $D_{AA}$ can be calculated from empirical relationships.[53]

### 3.6.2.1.  The Second-Degree Approximation to the Binary Diffusivity

To obtain transport coefficients in higher approximations, two mathematical methods are usually employed, one due to Chapman and Cowling[14] and the other to Kihara[41]; the latter is valid for a Lorentzian mixture $(M_A \ll M_B, x_A \ll x_B)$ or a Maxwellian gas [potential energy $\propto$ (distance)$^{-5}$].

The second-degree binary diffusivity is usually written as

$$[D_{AB}]_2 \equiv \frac{[D_{AB}]_1}{1 - \Delta_{AB}} \simeq [D_{AB}]_1(1 + \Delta_{AB}) \qquad (3.89)$$

and is used in place of the first-degree binary diffusivity in equation (3.86); $\Delta_{AB}$ is a correction factor. It can be verified that $[D_{AB}]_2 = [D_{BA}]_2$.

The correction factor $\Delta_{AB}$ according to Chapman and Cowling is

$$\Delta_{AB} = \frac{(6C_{AB}^* - 5)^2}{60(X_K + Y_K)} Z_K \qquad (3.90)$$

where $C_{AB}^*$ is a value near 1,

$$C_{AB}^* \equiv \Omega_{A,B}^{(1,2)*} / \Omega_{A,B}^{(1,1)*}$$

and

$$X_K = X_K(x_A, x_B, [\lambda_{KAA}]_1, [\lambda_{KBB}]_1, [\lambda_{KAB}]_1)$$

$$Y_K = Y_K(x_A, x_B, [\lambda_{KAA}]_1, [\lambda_{KBB}]_1, [\lambda_{KAB}]_1, \Omega_{A,B}^{(1,1)*}, \Omega_{A,B}^{(1,2)*}, \Omega_{A,B}^{(1,3)*}, \Omega_{A,B}^{(2,2)*}, M_A, M_B) \qquad (3.91)$$

$$Z_K = Z_K(x_A, x_B, [\lambda_{KAA}]_1, [\lambda_{KBB}]_1, [\lambda_{KAB}]_1, \Omega_{A,B}^{(1,1)*}, \Omega_{A,B}^{(2,2)*}, M_A, M_B)$$

where the parameter $[\lambda_{Kii}]_1$ is the first-degree thermal conductivity of the pure gas, and $[\lambda_{Kij}]_1$ is the first-degree thermal conductivity of the binary mixture $i$–$j$. Explicit expressions for the functions $X_K$, $Y_K$, and $Z_K$ can be found in related literature.[14, 37]

Equation (3.90) shows that the second-degree binary diffusivity is a function of composition and temperature, but for the majority of actual systems the composition dependence may be neglected.

As the functional dependence of $\Delta_{AB}$ on its variables is complicated, simplified procedures have been proposed for its straightforward estimation.[1, 48, 51, 52, 82]

The third-degree approximation, which is not necessary because of the rapid convergence of the series, has been calculated by Mason.[50] Simultaneously, Muckenfuss[56] has shown that all correction factors to $[D_{ij}]_1$ are equal for every degree of approximation.

### 3.6.2.2. Convergence of Approximations

The closeness of approximation of $[D_{AB}]_1$ to $\lim_{j \to \infty} [D_{AB}]_j$ depends on the composition, molecular masses, and the intermolecular potential. When the molecular masses are nearly the same, the error may be as large as 2%, irrespective of the composition and intermolecular potential. If the masses are very different and the concentration of the heavier component is small (quasi-Lorentzian gas) the error is less than 1%; but if the concentration of the heavier component is high (Lorentzian gas), the error can be larger; in this case, the Kihara approximation[41] can be used.

### 3.6.3.  Quantum Effects and Inelastic Collisions

The original Chapman–Enskog theory assumes the working hypotheses of the Boltzmann integrodifferential equation plus the hypothesis of a small perturbation. Quantum effects, which can be important for light gases at very low temperatures, require modification of the integral that gives the transverse section to diffusion, $s_{ij}^{(L,)}(E_r)$.

To extend the results to polyatomic and polar molecules, it is necessary to consider inelastic collisions. Wang Chang et al.[80] have developed a theoretical framework to modify the Chapman–Enskog theory for inelastic collisions; it is well known, however, that the diffusivities (as well as viscosities) of polyatomic gases are well predicted by potential models that ignore inelastic collisions. This is a result of the fact that mass and momentum, but not kinetic energy, are conserved in the inelastic collisions between polyatomic molecules; consequently, the only modified transport coefficient is that for the thermal conductivity.

### 3.6.4.  Multicomponent Diffusivity

To use equation (3.79) to calculate the diffusive flux of a given species in a multicomponent mixture, the multicomponent diffusivity $D_{ij}^m$ must be known.

The coefficients $D_{ij}^m$ have the following properties:

$$D_{ii}^m = 0; \quad \text{for } \nu > 2,\ D_{ij}^m \neq D_{ji}^m \tag{3.92}$$

The Chapman–Enskog theory gives the following expression for $D_{ij}^m$:

$$D_{ij}^m = \frac{1}{M_j}\left(\sum_{k=1}^{\nu} M_k x_k\right)\frac{K^{ji} - K^{ii}}{|K|} \tag{3.93}$$

where $K^{ji}$ and $K^{ii}$ are the minors of the determinant $|K| = |K_{ij}|$, with $K_{ii} = 0$ and

$$K_{ij} \equiv \frac{x_i}{[D_{ij}]_1} + \frac{M_j}{M_i}\sum_{\substack{k=1 \\ k \neq i}}^{\nu}\frac{x_k}{[D_{ik}]_1}, \quad i \neq j \tag{3.94}$$

i.e.,

$$
K^{ji} = (-1)^{i+j} \begin{vmatrix}
0 & \cdots & K_{1,i-1} & K_{1,i+1} & \cdots & K_{1,\nu} \\
\vdots & & \vdots & \vdots & & \vdots \\
K_{j-1,1} & \cdots & K_{j-1,i-1} & K_{j-1,i+1} & \cdots & K_{j-1,\nu} \\
K_{j+1,1} & \cdots & K_{j+1,i-1} & K_{j+1,i+1} & \cdots & K_{j+1,\nu} \\
\vdots & & \vdots & \vdots & & \vdots \\
K_{\nu,1} & \cdots & K_{\nu,i-1} & K_{\nu,i+1} & \cdots & K_{\nu,\nu}
\end{vmatrix}
$$

$$(3.95)$$

### 3.6.5. Thermal Diffusivity

The thermal diffusivity of a multicomponent mixture can be written as

$$
[D_i^T]_2 = [D_i^T]_2 (n_j, m_j, kT, \Omega_{ij}^{(1,1)}, \Omega_{ij}^{(1,2)}, \Omega_{ij}^{(1,3)}, \Omega_{ij}^{(2,2)}) \tag{3.96}
$$

where $j = 1, \ldots, \nu$, i.e., it runs over all components of the gaseous mixture. The exact expression for $[D_i^T]_2$ can be found elsewhere.[37] The suffix 2 means it is necessary to take at least two terms of the Sonine series, i.e., to take the second-degree approximation. It can be verified that

$$
\sum_{i=1}^{\nu} D_i^T = 0 \tag{3.97}
$$

In binary systems it is customary to use other properties related to the thermal diffusivity. Thus, by writing equation (3.86) for a binary system and taking into account equation (3.97), it follows that

$$
v_A^d - v_B^d = -[D_{AB}]_1 \left( \frac{d_A}{x_A x_B} + \alpha_{AB} \nabla \ln T \right) \tag{3.98}
$$

where $\alpha_{AB}$ is the *generalized thermal diffusivity*:

$$
\alpha_{AB} \equiv \frac{k_T}{x_A x_B} = \left( \frac{1}{n_A m_A} + \frac{1}{n_B m_B} \right) \frac{D_A^T}{[D_{AB}]_1} \tag{3.99}
$$

The coefficient $k_T$ is a measure of the relative importance of thermal versus ordinary diffusivity and is known as the *thermal diffusion ratio*. A value of $k_T > 0$ means that A moves towards the cold region and vice versa.

The Chapman–Enskog theory predicts a very complex dependence of the thermal diffusion ratio on temperature, composition, molecular masses, and molecular interaction potential[37] (through the collision integrals). Values of $k_T$ can be positive or negative and the temperature at which the sign inverts is known as the inversion temperature. For example, for the 6–12 potential of Lennard-Jones, two inversion temperatures are predicted: one for $T^* = 0.4$ and the other for $T^* = 0.95$, $T^*$ being a dimensionless temperature defined by

$$T^* \equiv \frac{kT}{\Xi_{AB}} \tag{3.100}$$

where $\Xi_{AB} \equiv (\Xi_A \Xi_B)^{1/2}$ and $\Xi$ is the energy of the potential well in the Lennard-Jones potential. Inversion at the higher temperature has been observed in hydrogen–deuterium mixtures.[20]

### 3.6.6.  Binary Systems

In a $\nu$-component system the Chapman–Enskog theory provides the necessary equations to calculate the diffusive flux $\mathbf{j}_i$ of each component [$\nu$ equations like equation (3.79) can be used, as well as $\nu - 1$ equations like equation (3.86), which is more convenient].

To illustrate the calculation of $\mathbf{j}_i$, consider a binary mixture of the gases A and B. Equation (3.79) [or equation (3.86)] then reduces to

$$\mathbf{j}_A = -\frac{n^2}{\rho} m_A m_B [D_{AB}]_1 \left\{ \nabla x_A + \left( \frac{n_B m_B}{\rho} - x_B \right) \nabla \ln p \right.$$

$$\left. + \frac{n_B m_B}{\rho p} \left[ \frac{\rho}{m_B} \mathbf{f}_B - (n_A \mathbf{f}_A + n_B \mathbf{f}_B) \right] \right\} - D_A^T \nabla \ln T$$

$$\tag{3.101}$$

and

$$\mathbf{j}_A = -\mathbf{j}_B \tag{3.102}$$

We can now obtain Fick's law of diffusion [equations (1.32) and (1.35)] if we assume: (1) uniform temperature; (2) uniform pressure; (3) that the external forces are such that they produce the same acceleration of both components (the gravitational force is just such a force). Under these conditions equation (3.101) becomes

$$\mathbf{j}_A = -\frac{n^2}{\rho} m_A m_B [D_{AB}]_1 \nabla x_A = -\rho [D_{AB}]_1 \nabla w_A \tag{3.103}$$

### 3.6.7. Application to Adsorbed Gases

Popielawski et al.,[62-64,70,71] starting from the Chapman–Enskog theory, developed general expressions that can be used for the calculation of surface self-diffusivities, binary surface diffusivities, and thermal surface diffusivities for binary gaseous mixtures physically adsorbed on an energetically homogeneous surface.

The equations developed assume that:

1. The adsorbed molecules are located on a plane that is parallel to the solid surface and situated at a distance corresponding to the minimum of the total potential energy; the adsorbed molecules can move randomly over this surface.

2. The adsorbed molecules form a two-dimensional rarefied gas in which only binary collisions are important.

3. The adsorbed molecules interact with the molecules of the solid, which are oscillating thermally around their equilibrium positions.

4. The solid is at thermodynamic equilibrium.

5. The phenomena of adsorption, desorption, and absorption have a negligible effect on the transport properties of the adsorbed gas.

The surface diffusion constitutive equation derived from this theory is given by

$$v_A^{d,s} - v_B^{d,s} = -\frac{(n^s)^2}{n_A^s n_B^s}\left(D_{AB}^s d_{AB}^s + D_A^{sT}\nabla \ln T\right) \qquad (3.104)$$

where the superscript s indicates a surface property; the vector $d_{AB}^s$ includes both the surface-pressure gradient and the effect of external forces.

To calculate the collision integrals that appear in expressions for the surface diffusivities it is necessary to determine the interaction potential between adsorbed molecules. For this, we can apply equations developed by Sinagoglu and Pitzer[68] and by Barker and Everett,[5] whose coefficients can be estimated theoretically or can be calculated from experimental values of the second virial coefficient for adsorbed gases.

The binary and thermal surface diffusivities have been calculated for the isotopic mixture of argon $^{36}Ar-^{40}Ar$ adsorbed on a graphite surface, in the temperature range 100–300°K and for surface concentrations ranging from $10^{12}$ to $10^{14}$ g mole/cm$^2$. The reason for the selection of this system is that it is one of the few (if not the only one) in which all the variables and parameters necessary for the calculation are known. There has been, unfortunately, no experimental verification of this calculation.

### 3.6.8.  Higher Approximations

So far we have considered only the first-order perturbation of the distribution function. For second-order perturbations, the equations obtained are much more complex—they involve second-order derivatives and products of first derivatives, and terms corresponding to the viscous transport of momentum appear in the constitutive equation of diffusion. Chapman and Cowling[13, 14] have shown that all such new terms can be neglected in comparison with the first-order terms except the one proportional to the shear tensor gradient, $\nabla \cdot \mathbf{T}$. This new term can be handled in such a way that only the pressure-diffusion coefficient in equation (3.86) is modified. This modified coefficient was evaluated by Chapman and Cowling for the particular case of a gaseous mixture of similar molecular masses and the same intermolecular potential for all species. For the most general case, the Grad–Zhdanov molecular theory of transport (Section 3.7) can be used. While this theory uses a different mathematical method, the constitutive equations derived from it are equivalent to those of Chapman–Enskog for a second-order perturbation.

### 3.6.9.  Comments

Before we take up the Grad–Zhdanov theory it is of interest to determine the order of approximation of the constitutive equation of diffusion obtained by the method of the thermodynamics of irreversible processes (Sections 3.1 and 3.2). To do this, it is convenient to write equation (3.12) for a binary mixture and use the following values for an ideal gas mixture:

$$\tilde{V}_i = n^{-1}, \qquad \left( \frac{\partial \tilde{G}_{fi}}{\partial x_j} \right)_{\substack{T,p,x_k \\ k \neq i,j}} = -\frac{kT}{x_i} \tag{3.105}$$

After some manipulation we obtain[37] an expression exactly equal to the Chapman–Enskog equation (3.101), with

$$\alpha_{i0} = D_i^{\mathrm{T}} \tag{3.106}$$

$$\alpha_{ij} = -\frac{n}{\rho^2 kT} n_i n_j m_i^2 m_j^2 D_{ij} \tag{3.107}$$

For an ideal $\nu$-component mixture, equation (3.106) is still valid, but equation (3.107) must be replaced by

$$\alpha_{ij} = \frac{nn_j m_i m_j}{\rho^2 kT} \left[ -\rho m_j D_{ij}^m + \sum_{\substack{k=1 \\ k \neq i}}^{\nu} \rho_k m_k D_{ik}^m \right] \qquad (3.108)$$

We thus conclude that the thermodynamics of irreversible processes provides constitutive equations of diffusion to the same order of approximation as those arising from the Chapman–Enskog theory, and to a lower order (as we shall see) than the Grad–Zhdanov theory: it does not, of course, give the functional dependence of the transport coefficients.

## 3.7. The Grad–Zhdanov Theory

### 3.7.1. The 13-Moment Approximation

In 1949, Grad[31] developed a mathematical technique that makes it possible to solve the Boltzmann integrodifferential equation for one component with an approximation similar to the Chapman–Enskog second-order perturbation.

In 1962, Zhdanov et al.[83] applied this technique to obtain the constitutive equations of diffusion for a multicomponent mixture. For the sake of convenience and to follow tradition, it is common to write the Grad–Zhdanov equations in a form similar to the first order Chapman–Enskog equations. When this is done, new terms (such as the correction factor $\Delta_{ij}$ for the second-degree approximation) appear, and the coefficient in the pressure-diffusion and external force terms are modified.

Schematically, the Grad–Zhdanov theory involves the following steps:

(1) The distribution function for component $i$ is written as a product of the equilibrium distribution function $f_i^{(0)}$ and a series of Hermite polynomials, $H_{i,k_1,\ldots,k_s}^{(s)}$ (for a definition of these, see Reference 31):

$$f_i(\mathbf{x}, \mathbf{v}_i, t) \equiv f_i^{(0)} \sum_{s=0}^{\infty} \frac{1}{s!} \left( \frac{m_i}{kT} \right)^s A_{i,k_1,\ldots,k_s}^{(s)}(\mathbf{x}, t) H_{i,k_1,\ldots,k_s}^{(s)}(\mathbf{v}_i^p) \quad (3.109)$$

where the coefficients $A^{(s)}$ are tensors of order $s$ and are equal to the average values of the corresponding Hermite polynomials,

$$A_{i,k_1,\ldots,k_s}^{(s)}(\mathbf{x}, t) \equiv \int H_{i,k_1,\ldots,k_s}^{(s)}(\mathbf{v}_i^p) f_i \, d\mathbf{v}_i^p \qquad (3.110)$$

(2) "Moments" of the Boltzmann equation are generated for component $i$ by multiplying the Boltzmann equation by the tensor velocity of order $s$,

$$v^{(s)}_{i,k_1,\ldots,k_s} \equiv v_{i,k_1} v_{i,k_2} \cdots v_{i,k_s} \tag{3.111}$$

where the $k_j$ indicate the components of the velocities, and integrating over the velocities. Corresponding to each velocity tensor, a "momentum" tensor of the distribution function may be defined as follows:

$$P^{(s)}_{i,k_1,\ldots,k_s} \equiv \int v^{(s)}_{i,k_1,\ldots,k_s} f_i(\mathbf{x}, \mathbf{v}_i, t)\, d\mathbf{v}_i \tag{3.112}$$

There are five "momenta" $P^{(s)}_i$ and five corresponding moments of the Boltzmann equation with physical meaning: The moment of order 0 generates the equation of change of mass for component $i$; the moment of order 1 generates the equation of change for the partial momentum of $i$; the contracted moment of order 2 (sum of the diagonal elements of the tensor) generates the equation of change of the partial energy of component $i$; the complete moment of order 2 generates the equation of change for the components of the pressure or shear tensor; and the contracted moment of order 3 leads to the equation of change for the components of the heat flux vector of $i$. In each case, the sum over all species leads to the corresponding equation of change for the entire gas.

(3) To solve the set of equations using the distribution function defined by equation (3.109), the 13-moment approximation is employed. In this approximation

$$A^{(s)}_{i,k_1,\ldots,k_s}(\mathbf{x}, t) = 0 \qquad \text{for } s \geqslant 4 \tag{3.113}$$

and

$$\left.\begin{array}{l} A^{(s)}_{i,k_i k_j k_j}(\mathbf{x}, t) \neq 0 \\[2mm] A^{(s)}_{i,k_i k_j k_k}(\mathbf{x}, t) = 0 \end{array}\right\} \quad \text{for } s = 3 \tag{3.114}$$

Equation (3.113) indicates that the 13-moment technique is a third-degree approximation, and with it and equation (3.114) we can express the distribution function in terms of 13 scalars that represent the unknown quantities in the problem [if we do not use the restriction given by equation (3.114), 20 scalars would have appeared, giving rise to the 20-moment method; it can be demonstrated[83] that both the 13- and 20-moment approximations are equivalent].

In the 13-moment approximation the distribution function is given by

$$f_i = f_i^{(0)}\left\{ n_i + \frac{j_{i,k_i}v_{i,k_i}^P}{kT} + \frac{\rho_i}{2kTp_i}T_{i,k_i,k_j}\left(v_{i,k_i}^P v_{i,k_k}^P - \frac{kT}{m_i}\delta_{k_i,k_k}\right)\right.$$

$$\left. + \frac{1}{5}\frac{\rho i}{kTp_i}\left(q_{i,k_i} - \frac{5kTj_{i,k_i}}{2m_i}\right)v_{i,k_i}^P\left[\left(\frac{m_i(v_i^P)^2}{kT}\right) - 5\right]\right\}$$

$$(3.115)$$

where the suffixes $k_i$, $k_j$, and $k_k$ indicate the components of the vectors and tensors $\mathbf{j}_i$, $\mathbf{T}_i$, $\mathbf{q}_i$, and $\mathbf{v}_i^P$; $\delta_{k_i,k_k}$ is the Kronecker delta.

### 3.7.2. Constitutive Equations of Diffusion

#### 3.7.2.1. Multicomponent Systems

By introducing equation (3.115) into the equation of balance of partial momentum for component $i$ (first-order moment of the Boltzmann equation) and rearranging,[83] we obtain

$$\sum_{j=1}^{\nu} \frac{n_i n_j kT}{n[D_{ij}]_1}(\mathbf{v}_i^d - \mathbf{v}_j^d) = -\left(\nabla p_i - \frac{\rho_i}{\rho}\nabla p\right) + \left(n_i \mathbf{f}_i - \frac{\rho_i}{\rho}\sum_{j=1}^{\nu} n_j \mathbf{f}_j\right)$$

$$- \left(\nabla\cdot\mathbf{T}_i - \frac{\rho_i}{\rho}\nabla\cdot\mathbf{T}\right) + \sum_{j=1}^{\nu} \beta_{ij}\left(\frac{\mathbf{q}_j - 5kT\mathbf{j}_j/2m_j}{n_j m_j} - \frac{\mathbf{q}_i - 5kT\mathbf{j}_i/2m_i}{n_i m_i}\right)$$

$$(3.116)$$

where $[D_{ij}]_1$ is the first-degree Chapman–Enskog binary diffusivity and

$$\beta_{ij} \equiv \frac{n_i n_j}{n[D_{ij}]_1}m_{ij}\left(\tfrac{6}{5}C_{ij}^* - 1\right), \qquad m_{ij} \equiv \frac{m_i}{m_j} \qquad (3.117)$$

$$C_{ij}^* \equiv \Omega_{i,j}^{(1,2)}/\Omega_{i,j}^{(1,1)} \qquad (3.118)$$

It is important to observe that the Grad–Zhdanov constitutive equations of diffusion [equations (3.116)] are obtained by application of a partial momentum balance for a given species; this supports other derivations that obtain simplified equations of diffusion in this way.

A comparison of equations (3.116) and (3.86) (i.e., of the Grad–Zhdanov and Chapman–Enskog constitutive equations) shows that these expressions differ in three ways:

(1) The thermal diffusion term does not appear explicitly in equation (3.116).

(2) The last two terms of equation (3.116) do not appear in equation (3.86). However, as we shall see immediately, the last term of equation (3.116) can be written as the sum of two terms: one is the thermal diffusion term identical to that in equation (3.86) and the other becomes the Chapman–Enskog second-degree correction factor $\Delta_{ij}$.

(3) There is an additional term in the Grad–Zhdanov equation for the mass segregation due to the viscous transport of momentum through the moving gas.

By using the differential equations for the complete moment of order 2 and the contracted moment of order 3 of the Boltzmann equation, the tensor $\mathbf{T}_i$ and the vector $\mathbf{q}_i - 5kT\mathbf{j}_i/2m_i$ can be eliminated from equation (3.116). The vector $\nabla \cdot \mathbf{T}$ can also be eliminated by using the equation of change of momentum for the entire gas. With these substitutions and the assumption of creeping flow,[†] the following equation results:

$$
\sum_{j=1}^{\nu} \frac{n_i n_j kT}{n[D_{ij}]_1}(\mathbf{v}_i^d - \mathbf{v}_j^d) + k\left(\frac{T}{p}\right)^2 \sum_{j=1}^{\nu} \sum_{k'=1}^{\nu} \sum_{l=1}^{\nu} \frac{kT}{m_{k'}}
$$

$$
\times \beta_{ij}\beta_{k'l}\left(\frac{|b|_{k'j}}{m_j|b|} - \frac{|b|_{k'i}}{m_i|b|}\right)(\mathbf{v}_{k'}^d - \mathbf{v}_l^d) = -\left(\nabla p_i - \frac{\eta_i + W_i}{\eta}\nabla p\right)
$$

$$
+ \left(n_i\mathbf{f}_i - \frac{\eta_i + W_i}{\eta}\sum_{j=1}^{\nu} n_j\mathbf{f}_j\right) + \left[T\sum_{j=1}^{\nu} \beta_{ij}\left(\frac{\lambda_{Ki}}{n_i m_i} - \frac{\lambda_{Kj}}{n_j m_j}\right)\right]\nabla \ln T
$$

$$\tag{3.119}$$

where

$$
W_i \equiv \frac{2}{5}k\left(\frac{T}{p}\right)^2 \sum_{j=1}^{\nu} \sum_{k'=1}^{\nu} \beta_{ij}\eta_{k'}\left(\frac{|b|_{k'j}}{m_j|b|} - \frac{|b|_{k'i}}{m_i|b|}\right) \tag{3.120}
$$

$$
\lambda_{Ki} \equiv x_i \sum_{k=1}^{\nu} x_k \frac{|b|_{ki}}{|b|}, \qquad \eta_i \equiv x_i \sum_{j=1}^{\nu} x_j \frac{|a|_{ji}}{|a|}, \qquad \eta = \sum_{i=1}^{\nu} \eta_i \tag{3.121}
$$

[†]Creeping flow around spheres prevails at low Reynolds numbers, i.e., Re < 0.1; this means the inertial force terms are considered negligible. See the discussion preceding equation (1.64).

The $|a|, |b|$ are the determinants of the coefficients $a_{ij}$ and $b_{ij}$ (and $|a|_{kj}, |b|_{kj}$ are the cofactors of the element $kj$ of these determinants), with

$$a_{ii} = \frac{x_i^2}{[\mu_{ii}]_1} + 2 \sum_{\substack{j=1 \\ j \neq i}}^{\nu} \frac{x_i x_j}{n(m_i + m_j)[D_{ij}]_1} \left(1 + \frac{3}{5} \frac{m_j}{m_i} A_{ij}^*\right) \tag{3.122}$$

$$a_{ij} = -\frac{2 x_i x_j}{n(m_i + m_j)[D_{ij}]_1}\left(1 - \tfrac{3}{5} A_{ij}^*\right), \qquad i \neq j \tag{3.123}$$

$$b_{ii} = \frac{x_i^2}{[\lambda_{Kii}]_1} + \frac{4}{25} \frac{T}{P} \sum_{\substack{j=1 \\ j \neq i}}^{\nu} \frac{\left(\tfrac{15}{2}m_i^2 + \tfrac{25}{4}m_j^2 - 3m_j^2 B_{ij}^* + 4 m_i m_j A_{ij}^*\right) x_i x_j}{(m_i + m_j)^2 [D_{ij}]_1}$$

$$\tag{3.124}$$

$$b_{ij} = -\frac{4}{25} \frac{T}{P} \frac{m_i m_j x_i x_j}{(m_i + m_j)^2 [D_{ij}]_1}\left(\frac{55}{4} - 3 B_{ij}^* - 4 A_{ij}^*\right), \qquad i \neq j \tag{3.125}$$

Here $[\mu_{ii}]_1$ and $[\lambda_{Kii}]_1$ are the coefficient of viscosity and the thermal conductivity of the pure gas $i$ calculated using the first-degree approximation of the Chapman–Enskog theory, and

$$A_{ij}^* \equiv \Omega_{i,j}^{(2,2)}/2\Omega_{i,j}^{(1,1)} \tag{3.126}$$

$$B_{ij}^* \equiv \left(5\Omega_{i,j}^{(1,2)} - \Omega_{i,j}^{(1,3)}\right)/3\Omega_{i,j}^{(1,1)} \tag{3.127}$$

The $\nu - 1$ independent equations (3.119) are the most rigorous constitutive equations of diffusion that the kinetic theory of gases has provided. It is convenient to write these equations in a form similar to those of the Chapman–Enskog theory. We write

$$\frac{\text{left-hand side of equation (3.119)}}{nkT} = \sum_{j=1}^{\nu} \frac{(1 - \Delta_{ij}) n_i n_j}{n^2 [D_{ij}]_1}\left(v_i^d - v_j^d\right)$$

$$\tag{3.128}$$

and

$$\alpha_{ij} = (nk)^{-1} \frac{\beta_{ij}}{x_i x_j}\left(\frac{\lambda_{Kj}}{n_j m_j} - \frac{\lambda_{Ki}}{n_i m_i}\right) = -\alpha_{ji} \tag{3.129}$$

Equation (3.128) provides the definition of the correction factor $\Delta_{ij}$ and equation (3.129) is the definition of the generalized thermal diffusivity.

Using equations (3.128) and (3.129) we can write equation (3.119) as

$$\sum_{j=1}^{\nu} \frac{(1 - \Delta_{ij})}{n^2 [D_{ij}]_1} n_i n_j \left( v_i^d - v_j^d \right) = -p^{-1} (\nabla p_i - A_{Gi} \nabla p)$$

$$\text{I} \qquad\qquad \text{II}$$

$$+ p^{-1} \left( n_i \mathbf{f}_i - A_{Gi} \sum_{j=1}^{\nu} n_j \mathbf{f}_j \right) - \sum_{j=1}^{\nu} x_i x_j \alpha_{ij} \nabla \ln T \qquad (3.130)$$

$$\text{III} \qquad\qquad\qquad\qquad \text{IV}$$

where

$$A_{Gi} \equiv (\eta_i + W_i)/\eta \qquad (3.131)$$

Terms I–IV in equation (3.130) describe the four mechanisms of diffusion we referred to at the end of Section 3.2:

I   describes *ordinary diffusion*
II   describes *pressure diffusion*
III   describes *forced diffusion*
IV   describes *thermal diffusion*

The most common and important of these mechanisms is ordinary diffusion, the others being negligible. Forced diffusion becomes important in centrifugation (an example of this for a system without walls would be a vortex in open space) or when an electric field acts upon ions. Pressure gradients in a system without walls may arise from a force field (e.g., the pressure variation in the atmosphere is a gravitational effect).

### 3.7.2.2.  Binary Systems

For a binary system, the values of $\Delta_{AB}$ and $\alpha_{AB}$ are the same as those of the Chapman–Enskog theory. If we use $v_i^d = \mathbf{j}_i / m_i n_i$, $\Sigma_i \mathbf{j}_i = 0$, equation (3.130) becomes

$$\mathbf{j}_A = -\frac{n^2}{\rho} m_A m_B [D_{AB}]_2 \{ \nabla x_A + (x_A - A_{GA}) \nabla \ln p$$

$$- p^{-1} [n_A \mathbf{f}_A - A_{GA} (n_A \mathbf{f}_A + n_B \mathbf{f}_B)] + x_A x_B \alpha_{AB} \nabla \ln T \} \quad (3.132)$$

A comparison of equations (3.132) and (3.101) shows that the coefficients of pressure and forced diffusion differ. The Chapman–Enskog $\rho_A / \rho$ is replaced by the Grad–Zhdanov $A_{GA}$, and, in general, $\rho_i / \rho$ is replaced by

$A_{Gi}$. It is important to note that these coefficients can differ considerably, as was shown by Zhdanov et al.[83]

Note that if the acting force per unit mass is the same for all species (as is the case, for example, for the gravitational acceleration) the forced-diffusion term becomes zero in the Chapman–Enskog theory, but not in the Grad–Zhdanov theory. Gravity can consequently produce segregation of species.

### 3.7.2.3.  Expression in Terms of Total Diffusive Fluxes

It is convenient to write equation (3.130) in terms of $F_i^D$ [the total diffusive molecular flux relative to stationary coordinates, as given by equation (1.15) written on molecular basis]. With the aid of equation (1.22) and Table 1.4 we get

$$\sum_{j=1}^{\nu} \frac{(1 - \Delta_{ij})}{n^2 [D_{ij}]_1} (n_i F_j^D - n_j F_i^D) = p^{-1} \left[ (\nabla p_i - A_{Gi} \nabla p) \right.$$

$$\left. - \left( n_i f_i - A_{Gi} \sum_{j=1}^{\nu} n_j f_j \right) \right] + \sum_{j=1}^{\nu} x_i x_j \alpha_{ij} \nabla \ln T \quad (3.133)$$

where, since the system is devoid of walls, $\sum_{i=1}^{\nu} F_i^D = F^N = 0$. Equation (3.133) will be the starting point of Chapter 4.

One of the failures of the equations developed so far is that they do not correctly take account of the influence of the molecular collision cross sections. This will be analyzed in more detail in Chapter 4.

## 3.8.  Simplified Kinetic Models

### 3.8.1.  The BGK Model

In 1954, Bhatnagar, Gross, and Krook[6] and, independently, Welander[81] presented an approximate solution to the Boltzmann equation. Their method involves replacing the complicated collision integrals by relaxation terms which are mathematically simpler; these terms are selected so that mass, momentum, and kinetic energy are conserved.

The model, known as the BGK model, was constructed by analogy with the *local thermodynamic equilibrium model* (which was applied to the study of the continuous spectra of stellar atmospheres) and was developed for one-component systems. The extension of the model to binary gaseous mixtures was discussed between 1956 and 1967 by Gross and Krook,[32] Sirovich,[69] Morse,[55] Hamel,[34] and Oguchi.[57] In this model, the collision integrals are replaced by

$$\nu_{ii}\left(f_i^{(0)} - f_i\right) + \nu_{ij}\left(\mathring{f}_i^{(0)} - f_i\right) \tag{3.134}$$

where $\nu_{ii}$ is the frequency of collisions of molecules of kind $i$ among themselves, $\nu_{ij}$ is the frequency of collisions between molecules of $i$ and $j$, $f_i^{(0)}$ is the Maxwellian local distribution corresponding to the molecular concentration $n_i$, average velocity $\bar{\mathbf{v}}_i$, and temperature $T_i$,

$$f_i^{(0)} = n_i \left(\frac{m_i}{2\pi k T_i}\right)^{3/2} \exp\left[-\frac{m_i\left(\mathbf{v}_i - \bar{\mathbf{v}}_i\right)^2}{2kT_i}\right] \tag{3.135}$$

and $\mathring{f}_i^{(0)}$ is the Maxwellian distribution corresponding to the velocity $\mathring{\bar{\mathbf{v}}}_i$ and temperature $\mathring{T}_i$, where $\mathring{\bar{\mathbf{v}}}_i$ and $\mathring{T}_i$ are parameters to be determined,

$$\mathring{f}_i^{(0)} = n_i \left(\frac{m_i}{2\pi k \mathring{T}_i}\right)^{3/2} \exp\left[-\frac{m_i\left(\mathbf{v}_i - \mathring{\bar{\mathbf{v}}}_i\right)^2}{2k\mathring{T}_i}\right] \tag{3.136}$$

Analogous expressions are obtained for component $j$ by replacing the subscript $i$ by $j$. From the definition of the collision frequencies, we have

$$n_i \nu_{ij} = n_j \nu_{ji} \tag{3.137}$$

Equation (3.134) is a simple way of expressing the idea that collisions tend to relax the distribution function to the equilibrium value—it assumes that each collision produces a change in the distribution function $f_i$ proportional to the difference between $f_i$ and the equilibrium distribution.

To complete the statement of the model it is necessary to provide equations to compute the parameters $\nu_{ij}$, $\nu_{ii}$, $\nu_{jj}$, $\mathring{\bar{\mathbf{v}}}_i$, $\mathring{\bar{\mathbf{v}}}_j$, $\mathring{T}_i$, and $\mathring{T}_j$. These equations arise from the application of conservation principles, from the comparison of the equations of the model with the momenta of the Boltzmann integrodifferential equation (which makes it possible to calculate the relaxation of the velocity differences), and by comparing the transport coefficients predicted by the model with those of the Chapman–Enskog theory.

The conservation principles for momentum and energy are found to be[6, 81]

$$m_i \mathring{\bar{\mathbf{v}}}_i + m_j \mathring{\bar{\mathbf{v}}}_j = m_i \bar{\mathbf{v}}_i + m_j \bar{\mathbf{v}}_j \tag{3.138}$$

$$\tfrac{3}{2}k\left(\mathring{T}_i + \mathring{T}_j\right) + \tfrac{1}{2}m_i \mathring{\bar{v}}_i^2 + \tfrac{1}{2}m_j \mathring{\bar{v}}_j^2 = \tfrac{3}{2}k\left(T_i + T_j\right) + \tfrac{1}{2}m_i \bar{v}_i^2 + \tfrac{1}{2}m_j \bar{v}_j^2 \tag{3.139}$$

The model gives the following expressions for the relaxation of the velocities and temperatures:

$$\frac{\partial}{\partial t}\left(\bar{\mathbf{v}}_j - \bar{\mathbf{v}}_i\right) = \nu_{ji}\left(\mathring{\mathbf{v}}_j - \bar{\mathbf{v}}_j\right) - \nu_{ij}\left(\mathring{\mathbf{v}}_i - \bar{\mathbf{v}}_i\right)$$ (3.140)

$$\frac{\partial}{\partial t}(T_j - T_i) = n_i\nu_{ij}\left\{\frac{\mathring{T}_j}{n_j} - \frac{\mathring{T}_i}{n_i} + \frac{1}{3k}\left[\frac{m_j}{n_j}\left(\mathring{\mathbf{v}}_j - \bar{\mathbf{v}}_j\right)^2 - \frac{m_i}{n_i}\left(\mathring{\mathbf{v}}_i - \bar{\mathbf{v}}_i\right)^2\right]\right\}$$

(3.141)

Equations (3.140) and (3.141) must have the same form as the corresponding equations obtained from the complete Boltzmann equation for a particular interaction potential. Using equations (3.138) and (3.139), relationships are obtained which make it possible to calculate $\mathring{\mathbf{v}}_i$, $\mathring{\mathbf{v}}_j$, $\mathring{T}_i$, and $\mathring{T}_j$.

Hamel[34] obtained expressions for the transport coefficients by a comparison of his development with the results of Sirovich.[69] He found for the viscosity of the pure gas $i$

$$\mu_{ii} = \frac{n_i}{\nu_{ii}}kT$$ (3.142)

with a similar expression for component $j$; for the viscosity of the binary mixture $i$–$j$ he found

$$\mu_{ij} = \left(\frac{n_i}{\nu_{ii} + \nu_{ij}} + \frac{n_j}{\nu_{jj} + \nu_{ji}}\right)kT$$ (3.143)

and for the binary diffusivity

$$D_{ij} = \frac{n_j(m_i + m_j)kT}{(n_i + n_j)m_im_j\nu_{ij}}$$ (3.144)

A comparison of equations (3.142) and (3.143), or equation (3.142) and (3.144), with the corresponding Chapman–Enskog equations makes it possible to determine the coefficients $\nu_{ii}$, $\nu_{jj}$, and $\nu_{ij}$. It should be noted that not all coefficients of this model can be set equal to coefficients in the Chapman–Enskog equations, i.e., the two theories are not equivalent.

The application of the BGK model must be done with care. For example, when the BGK is applied to a flow problem, the collision frequency $\nu_{ij}$ is adjusted to give the correct value of the viscosity; if, with this adjusted value, the BGK is used for a self-diffusion problem, the resulting Schmidt number Sc is of the order of 1, which differs from the value $Sc \simeq 0.7$ that arises from the Chapman–Enskog theory; this means that the BGK model cannot describe problems of flow and self-diffusion simultaneously.

### 3.8.2. Other Models

In 1971, van Eekelen and Smit[77] proposed a new kinetic model, called the $\lambda$-*model* for a one-component system. This model does not lead to undetermined quantities in the transport coefficients as does the BGK model. In the $\lambda$-model, the collision frequency $\nu_{ij}$ is replaced by $v/\lambda$, where the mean free path $\lambda$ is independent of velocity. A consistent description of the problems of flow and self-diffusion is thus obtained in contrast to the situation with the BGK model.

Holway[38] proposed an *ellipsoidal statistical model* which contains the BGK model as a particular case and which can be used with success in the study of shock wave structures.[39] Walker and Tanenbaum[79] give a comparison of both models.

Numerous other kinetic models have been proposed. In general, the collision integrals are linearized and the Boltzmann equation is solved by a variational or an iterative method.

### 3.8.3. Particular Solutions. Wall Effects

The BGK model[†] has been successfully applied to the analysis of shock wave structures,[39,57] to the analysis of fluid-dynamic models of two fluids,[35] and to the description of surface diffusion.[33] In general, it can be used to solve transport problems for simple, well-defined geometries. For such problems, use is made of a boundary condition that specifies that the molecules, after colliding with the wall, are rebounded diffusely with a Maxwellian distribution corresponding to the wall temperature. Thus, while the BGK and the other simplified kinetic models do not contribute to our goal in this chapter—to obtain the structure and morphology of the constitutive equations of diffusion (the Chapman–Enskog or Grad–Zhdanov equations represent this goal)—they give some theoretical support for our phenomenological discussion of wall effects in Chapter 1. We shall thus briefly consider here some of the results obtained with the BGK model.

By applying the BGK model with the above boundary condition, Cercignani *et al.*[7–9] numerically solved the problem of the flow of a pure gas as a function of pressure (or Knudsen number) for several geometries.

---

[†]We include here all the kinetic models that linearize the collision integrals.

The results obtained for a cylindrical tube agree very well with Knudsen's experimental results, including the minimum in the curve of flux versus pressure. Subsequent studies by Simons,[66, 67] Ferziger,[26] and Sone and Yamamoto[73] give results in agreement with experiment.

Shendalman[65] analyzed diffusive phenomena in the flow of binary mixtures in long cylindrical tubes with the aid of the BGK model. The results obtained show the validity of Graham's law [equation (2.71)] in the limits of low and high pressures.

The solution of the linearized Boltzmann equation with variational techniques[42-47] has made it possible to obtain precise solutions of various diffusive phenomena in well-defined geometries. The treatment of these cases is beyond the scope of this book.

The validity of Graham's law leads to a very important conclusion (cf. Section 1.3.2.3): the *diffusive fluxes* $j_i$ *in the constitutive equations of diffusion are now, by virtue of the existence of walls, superimposed on a nonsegregative diffusive flux—the nonequimolar one* (usually called diffusion slip flux). In other words, if we sum the $F_i^D$ over all species, the sum is not zero—as was stated after equation (3.133); the sum now gives the non-equimolar flux. The mere introduction of a boundary condition to solve the Boltzmann equation has thus modified the physics of the diffusion phenomenon substantially. Theory has verified our phenomenological insight of Chapter 1.

# 4

## Constitutive Equations of Diffusion and Wall Effects

The equations developed in Chapter 3 took into account only molecule–molecule collisions and wall effects were only briefly mentioned at the end of that chapter. We shall now deal specifically with wall effects.

First let us note that it is important to know how to determine the effect of walls on diffusion since there are walls in every real system, from the simple case of a capillary to the complex case of a porous medium.

As we saw in Chapter 1, walls are the source of what we have called the "Knudsen" and "viscous" collisions, as well as the source of the nonequimolar flux.[†] "To determine the effect of walls" will for us be synonymous with obtaining the constitutive equations to be used to predict the molecular and Knudsen contributions to the diffusive flux and to predict the viscous flux.

In this connection, in Sections 1.3.3.4 and 1.4.4 we proposed that the total flux of a given component could be calculated by adding the diffusive and viscous contributions, i.e., by adding their respective constitutive equations. Is this proposition correct? In other words, is it possible that the total flux of a given component under composition and pressure gradients is given by only one constitutive equation? Much confusion and many errors have arisen in connection with this subject and the reader is referred to Chapter 6 for a detailed analysis of them.

Another point to be clarified is the following: According to the equations obtained in Chapter 3, the diffusion velocity remains constant over the cross section transverse to the direction of flow unless the gradients (pressure, composition, etc) driving the diffusion vary on that cross section. But when there are walls there is a velocity profile in the

---

[†] Walls also make the existence of a surface flux possible, but we will not deal with that here.

viscous flow, and we might well ask if walls also produce a diffusion-velocity profile.

These questions can only be properly answered when the problem is rigorously stated. And, generally speaking, the only rigorous way is to solve the Boltzmann equation with boundary conditions that determine the wall-induced perturbation of the velocity distribution of the molecules that collide with the walls. This is obviously an extremely complex problem and has been solved only for walls with very simple geometry. Such solutions help to answer the above questions, but are useless in real cases.

For example, Kucherov and Rikenglaz[60] studied the two-dimensional flow of a binary gas mixture close to a plane surface in the molecular diffusion regime. The velocity distribution function is separated into two parts, one for the incident molecules and the other for the reflected ones. The boundary conditions specify that a fraction of the molecules collides specularly with the wall while the remaining molecules collide diffusively; the velocity distribution of the diffusely reflected molecules is assumed to be the Maxwellian distribution corresponding to room temperature. *The solution obtained shows that the gas velocity at the wall can be written as the sum of two terms: a viscous one and a diffusive one.*

Breton[11] analyzed the one-dimensional flow of a binary gas mixture between parallel plates in the molecular diffusion regime and solved the Boltzmann equation with the aid of what he calls the "half-range moment." The results obtained indicate that there are three contributions to the flux of a given component: (a) a sum of two exponentials that go rapidly to zero in a distance of a few mean free paths from the plates; these terms, which describe Knudsen boundary layers, do not contribute to the flow and can be neglected; (b) a diffusive term; (c) a viscous term. This study was extended to a swarm of spherical particles for different flow regimes[12]: Knudsen, viscous, and a pseudo-Knudsen regime for which the addition of the viscous and diffusion contributions remains valid.

Shendalman[97] applied the BGK model (Section 3.8.1) to a binary gas mixture flowing in a capillary over a wide pressure range. The results were applied to a capillary connected to two bulbs initially containing two different gases at uniform pressure (System 2 of Table 1.1, a system to be analyzed in detail in Chapter 5); the variation of the pressure gradient as a function of axial position in the capillary was determined for the Knudsen and viscous regimes. The results are similar to those obtained by starting from constitutive equations in which the viscous flux is added to the diffusive flux.

*In conclusion, from the rigorous statement of the problem it seems plausible to assume that the addition of viscous and diffusive fluxes can also be performed in the transition regime of diffusion and that the walls do not generate diffusion-velocity profiles.*

The severe mathematical difficulties encountered in the application of the rigorous statement of the problem to multicomponent mixtures and/or porous media has led other authors to attempt different, more simplified, methods of solution. The method commonly used by chemical engineers is based on the application of the principle of conservation of momentum.[86, 96, 99, 103] The results obtained show the following general characteristics: (a) the contributions to diffusion due to molecule–molecule and molecule–wall collisions are clearly distinguished; (b) the Knudsen regime appears as a particular case; (c) in the molecular diffusion regime, the equations can be considered as a particular case of the Chapman–Enskog equations; (d) for a pure gas, the slip mechanism is not included (the minimum in the specific flux as a function of pressure cannot be predicted); (e) Graham's law of diffusion is obtained; (f) the Knudsen diffusivity appears as a phenomenological coefficient.

Another method begins with the constitutive equations of diffusion in the absence of walls (Chapman–Enskog or Grad–Zhdanov theory) and, with the aid of a model, "adds" the influence of walls. This is the approach used in the so-called dusty gas model (DGM) which Evans, Watson, and Mason[25] introduced using the Chapman–Enskog equations and which was further improved by Mason, Malinauskas, and Evans[73] by using the Grad–Zhdanov equations.

Unlike the method based on a momentum balance, the equations obtained by the DGM explain the "Kramers–Kistemaker effect" (which we will discuss in Chapter 5) as well as the deviations from Graham's law of diffusion. The Knudsen diffusivity for a molecule–particle collision can be predicted but this result cannot be applied directly to a molecule–wall collision inside a pore. However, in view of their origin, the equations have the same generality and scope as those analyzed in Chapter 3. Further, they have been experimentally verified for several porous media.

Since the model is based on the Grad–Zhdanov equations, it does not correctly predict the influence of collision cross sections, which play an important role in the deviations from Graham's law of diffusion (as is the case with the inversions of the law experimentally observed; see Section 4.2.2.5). But this is a minor omission if we consider the advantages of the model, one of them being its usefulness. We will devote our attention to it.

## 4.1. The Dusty Gas Model

Mason and colleagues went back to the old and simple manner[59,60,76] of representing a dispersed medium by means of spheres at rest suspended in a medium, making some modifications so that the model can be used for any degree of concentration of the solid.

The applicability of the model was suggested to Evans *et al.*[25] by a study of Waldmann's[106] which analyzed the forces acting upon small spheres distributed in an inhomogeneous gas; the Evans *et al.* DGM was suggested independently of the dusty gas model proposed by Derjaguin and Bakanov[23] for the study of the flow of a gas in a porous medium near the Knudsen regime.

The basic working hypotheses of the DGM are:

1. The suspended particles (a) are spherical; (b) can be treated as a component of the gas mixture; (c) are motionless and uniformly distributed,

$$v_p = 0; \qquad \nabla n_p = 0 \qquad\qquad (4.1)$$

where the subscript p indicates particle; (d) are very much larger and heavier than the gas molecules; (e) are acted upon by an external force that keeps them at rest even though a pressure gradient may exist in the gas mixture; (f) may have any packing distribution (the obstruction the particles present to the diffusing molecules will thus vary with the packing geometry); (g) are at rest, so they do not contribute to the viscosity $\mu$ and the thermal conductivity $\lambda_K$ of the gas mixture.

2. The gas molecules are assumed to have no external forces acting on them.

Hypothesis 1c needs further comment. For a nonporous plug with a cross section $S$, a pressure $p$ on one face, and a pressure $p + dp$ on the other face, the net force exerted on the plug is $S\,dp$. If we now drill some holes through the plug in such a way that their total cross section is $S_1$, the force exerted on the plug will decrease to $(S - S_1)\,dp$, but a new force appears—the drag force of the gas, which is now flowing through the holes. If the gas is not accelerated, i.e., if the inertial forces are neglected (which, as we have seen in Sections 1.7.3 and 3.7.1.1. is a sound assumption), the drag force the flowing gas exerts on the walls of the holes must be equal to the net fluid force $S_1\,dp$ exerted on the gas and causing it to flow. The net force exerted by the gas on the plug is thus still equal to $S\,dp$. This result can be immediately carried over to a system such as the DGM. In this case, the balance of forces per unit volume is

$$n_p f_p = \nabla p \qquad\qquad (4.2)$$

where $f_p$ is the (average) external force exerted on a particle to keep it at rest.

In applying the constitutive equation [equation (3.133)] to the DGM, we will need the following results. Because of hypothesis 2,

$$\sum_{k=1}^{\nu} n_k \mathbf{f}_k = n_p \mathbf{f}_p \tag{4.3}$$

where p is one of the subscripts $k$ in the summand. The total flux of the particles is zero, so using equation (1.16) we have $\mathbf{N}_p^D + \mathbf{N}_p^V = 0$; since by equation (4.2) the particles remain at rest even in the presence of a pressure gradient, $\mathbf{N}_p^V \equiv c_p \mathbf{v}_p^V = 0$, and so $\mathbf{N}_p^D = c_p \mathbf{v}_{pM}^D = 0$. We thus have $\mathbf{v}_p^V = \mathbf{v}_{pM}^D = 0.$[†]

We will follow Mason *et al.*'s notation and use a prime for the properties of the mixture when particles are taken into account:

$$n' = n + n_p; \qquad p' = p + n_p kT; \qquad \rho' = \rho + n_p m_p \tag{4.4}$$

### 4.1.1. Constitutive Equations of Diffusion

The gas molecules cannot, obviously, enter the space occupied by the particles. This means that the properties of the gas mixture (state properties, transport properties) are "interrupted" within some space regions. It thus appears necessary to introduce *local mean values* of such properties using the space occupied by both the gas and by the particles. This procedure was introduced by Slattery,[100] who defined local volume averages by

$$\bar{B} \equiv \frac{1}{V} \int_{V_g} B \, dV \tag{4.5}$$

where $B$ is any scalar, vector, or tensor property associated with the fluid, $V_g$ is the volume occupied by the gas, and $V$ is the total volume ($V_g$ plus the volume occupied by the particles). When $B = \mathbf{F}_i^D$,[‡] equation (4.5) leads to

---

[†] While this result appears trivial, we have "derived" it since in general we consider that the average velocity of a component of a system can be considered as the (vector) sum of a "viscous velocity" $\mathbf{v}_i^V$ and a diffusion velocity $\mathbf{v}_{iM}^D$, and it is in principle possible that, e.g., $\mathbf{v}_i^V = -\mathbf{v}_{iM}^D \neq 0$ so that $\bar{\mathbf{v}}_i = 0$.

[‡] To use the concept of the *local mean volumetric flux*, $V$ must be large enough to be compatible with the continuum hypothesis on the scale of the porous medium. Slattery[102] indicates that the surface surrounding $V$ must be large enough so that the addend containing the tortuous effect of the pores (which arises on applying the theorem of the volumetric mean value of the divergence) is not an explicit function of position, although it can be an implicit function of it through the dependence on other variables.

functions of the type $\overline{\mathbf{F}}_i^D = f(\overline{\nabla \Phi})$, where

$$\overline{\nabla \Phi} \equiv \frac{1}{V} \int_{V_g} \nabla \Phi \, dV$$

and $\Phi$ stands for any of the fluid properties under the operator $\nabla$ in equation (3.133); such quantities are difficult to calculate, so the definition (4.5) proves to be of restricted utility. Nevertheless, as shown by Slattery,[102] it is possible to define other functions of the type $\overline{\mathbf{F}}_i^D = f(\nabla \overline{\Phi})$ by introducing transport parameters other than those used up to now. If we then equate both functions, $f(\overline{\nabla \Phi}) = f(\nabla \overline{\Phi})$, we can obtain the relationship between the new (usually called *effective*) transport parameters and the original ones. Such a calculation would be extremely complex if some simplifying working hypotheses were not introduced. For example, Slattery[102] gives the calculation for an isobaric, isothermal binary gas mixture without external forces in a porous medium in which the influence of the walls is not taken into account.

In another procedure, due to Neale and Nader,[79] the constitutive equation valid in the absence of walls is applied to the volume occupied by the gas in a microscopic region of the porous medium. This equation is coupled with another equation that takes into account wall effects and which is applied macroscopically in the porous medium. In this case the fluxes involved in the second equation are averaged over the local cross section (over the gas *and* particles) while the state properties of the fluid are averaged over the local volume (interstitially). This type of averaging is the one most usually employed in practice. With this procedure, Neale and Nader studied an isobaric, isothermal gas mixture with no external forces and equimolar counterdiffusion $(\overline{\mathbf{F}}_i^D = \overline{\mathbf{J}}_{iM}^m)$ in a porous medium represented by a set of spheres of different diameters; they found that $D_{ij,\mathrm{eff}}/D_{ij} = 2\varepsilon/(3-\varepsilon)$, $\varepsilon$ being the void fraction (or *porosity*) of the porous solid. This result was arrived at by writing the constitutive equation for $\mathbf{F}_i^D$ for the gas surrounding one sphere and equating this to the constitutive equation for $\overline{\mathbf{F}}_i^D$ in the porous medium.

The application of the principle of the minimum generation of entropy led[85] to $D_{ij,\mathrm{eff}}/D_{ij} = \varepsilon(1+\varepsilon)/2$ for the same system as Neale and Nader's; this is numerically consistent with his results. Brugeman[18] studied the same system and found $D_{ij,\mathrm{eff}}/D_{ij} = \varepsilon^{2/3}$.

These results suggest that we can continue to use equations (1.32a) and (1.45a) in the general form

$$D_{ij,\mathrm{eff}} = D_{ij}Q \tag{4.6}$$

where

$$Q = Q_{\mathrm{m}} = \text{const} \qquad \text{for all } i, j, j \neq \mathrm{p} \qquad (4.7)$$

$$Q = Q_{\mathrm{p}} = \text{const} \qquad \text{for all } i \text{ and for } j = \mathrm{p} \qquad (4.8)$$

However, for the moment we are not concerned with the nature of the effective diffusivities (which we will analyze in Section 4.3) but rather with obtaining the constitutive equation of diffusion with wall effects; the analysis of the parameters in this equation will follow later.

We can now apply the Grad–Zhdanov constitutive equations of diffusion to the DGM. In the following, we will omit the bar indicating a local mean value, although from now on when we are dealing with porous media, all the variables will be locally averaged ones.

Taking into account hypotheses 1b and 2 and equations (4.3) and (4.4), we can write equation (3.133) as

$$\sum_{\substack{j=1 \\ j \neq i}}^{\nu} \frac{1 - \Delta'_{ij}}{n'^2 [D'_{ij}]_{1,\,\mathrm{eff}}} \left( n_i \mathbf{F}_j^{\mathrm{D}} - n_j \mathbf{F}_i^{\mathrm{D}} \right) = p'^{-1} \left[ (\nabla p_i - A_{\mathrm{G}i} \nabla p') + A_{\mathrm{G}i} n_{\mathrm{p}} \mathbf{f}_{\mathrm{p}} \right]$$

$$+ \sum_{j=1}^{\nu} x_i x_j \alpha'_{ij} \nabla \ln T \qquad (4.9)$$

where the particles are considered to be one of the $\nu$ components and $\Delta'_{ij}$, $A_{\mathrm{G}i}$, and $\alpha'_{ij}$ are to be regarded as effective parameters since they depend on the effective diffusivity (see Section 3.7.1).

Taking into account equation (4.2) written in terms of $p'$, equation (4.9) reduces to[†]

$$\sum_{\substack{j=1 \\ j \neq i}}^{\nu} \frac{1 - \Delta'_{ij}}{n'^2 [D'_{ij}]_{1,\,\mathrm{eff}}} \left( n_i \mathbf{F}_j^{\mathrm{D}} - n_j \mathbf{F}_i^{\mathrm{D}} \right) = p'^{-1} \nabla p_i + \sum_{j=1}^{\nu} x_i x_j \alpha'_{ij} \nabla \ln T$$

$$= \frac{1}{n'kT} \nabla p_i + \sum_{j=1}^{\nu} x_i x_j \alpha'_{ij} \nabla \ln T \qquad (4.10)$$

where, to write the last equality, we have used equation (4.4).

---

[†]Note that, in fact, even in a nonisothermal system, $\nabla p' = \nabla p$ since the particles are at rest. The generalized use of primes is just a mathematical aid in solving the problem.

Now using equation (1.49), we can write equation (4.10) as follows:

$$\sum_{\substack{j=1 \\ j \neq i}}^{\nu} \frac{1 - \Delta'_{ij}}{n \left[ D_{ij} \right]_{1,\,\text{eff}}} \left( n_i \mathbf{F}_j^{\text{D}} - n_j \mathbf{F}_i^{\text{D}} \right) = \frac{1}{kT} \nabla p_i + n' \sum_{j=1}^{\nu} x_i x_j \alpha'_{ij} \nabla \ln T \quad (4.11)$$

Observe that *the pressure contribution to diffusion is different from systems with* [equation (4.11)] *and without* [equation (3.133)] *walls;* for systems with walls, the term $A_{Gi}$ is cancelled because of equation (4.2). This fact must be taken into account whenever trying to go from equation (4.11) back to equation (3.133).

### 4.1.2.  Contribution of the Molecule–Particle Collisions

If we separate the terms corresponding to molecule–molecule collisions from those for molecule–particle collisions, equation (4.11) becomes

$$\sum_{\substack{j=1 \\ j \neq i,\,\text{p}}}^{\nu} \frac{x_i \mathbf{F}_j^{\text{D}} - x_j \mathbf{F}_i^{\text{D}}}{\left[ D_{ij} \right]_{2,\,\text{eff}}} - \frac{\mathbf{F}_i^{\text{D}}}{\left[ D_{ii}^{\text{K}} \right]_{2,\,\text{eff}}} = \frac{1}{kT} \nabla p_i + n' \sum_{j=1}^{\nu} x_i x_j \alpha'_{ij} \nabla \ln T$$

$$(4.12)$$

where we have used $\mathbf{F}_{\text{p}}^{\text{D}} = 0$ and

$$\left( 1 - \Delta'_{ij} \right) / \left[ D_{ij} \right]_{1,\,\text{eff}} \equiv 1 / \left[ D_{ij} \right]_{2,\,\text{eff}} \quad (4.13)$$

$$\left( n_{\text{p}}/n \right)\left( 1 - \Delta'_{i\text{p}} \right) / \left[ D_{i\text{p}} \right]_{1,\,\text{eff}} \equiv n_{\text{p}}/n \left[ D_{i\text{p}} \right]_{2,\,\text{eff}} \equiv 1 / \left[ D_{ii}^{\text{K}} \right]_{2,\,\text{eff}} \quad (4.14)$$

The first term (the summation) on the left-hand side of equation (4.12) is [cf. our discussion following equation (1.38)] the momentum lost by species $i$ through molecule–molecule collisions with species other than $i$ (but not with species p—the particles); the second term is the momentum lost by species $i$ through molecule–particle collisions (i.e., collisions with species p). Both phenomena are additive. The introduction of $[D_{ii}^{\text{K}}]_{2,\,\text{eff}}$ for $n_{\text{p}}/n[D_{i\text{p}}]_{2,\,\text{eff}}$ in equation (4.14) is for convenience and reflects the fact that this quantity is the diffusivity in the reduced form of equation (4.12) for the Knudsen regime.

Much confusion appears in related literature regarding the meaning of the flux terms (the F's) in equation (4.12). Possibly because some of the first steps in the derivation of this equation (in which, as we have seen in Chapter 3, the fluxes are only diffusive) were overlooked, such fluxes have often been taken to be the total fluxes, namely diffusive plus viscous, i.e., it was believed that the total flux of a given component was described by only one constitutive equation. Another important reason for confusion[†] is that, in the terms in equation (4.12) representing the molecule–molecule collisions, either diffusive *or* total fluxes can be used, since these terms can be written using relative velocities [see equations (1.38), (3.38), (3.86), (3.130)] and we have $v_{iM}^D - v_{jM}^D = \bar{v}_i - \bar{v}_j = v_i^d - v_j^d$; but in the term corresponding to molecule–particle collisions, there is only one velocity (and not a difference of velocities), so it must be exactly specified as either $v_{iM}^D$, $v_{iM}^d$, or $\bar{v}_i$ (see Section 1.6.2.2.)

It is also important to emphasize that, *from the point of view of the constitutive equations of diffusion, a system without walls is not the same as a system in which wall effects are negligible,[‡]* since equation (3.133) applies for the first case while equation (4.12), with $F_i^D/[D_{ii}^K]_{2,\,eff}$ regarded as negligible, applies for the second case.

We can also observe that *equation (4.12) is written in the same form regardless of whether the system is isobaric or nonisobaric; this is a consequence of wall effects since this assertion is not valid for equation (3.133).*

### 4.1.3.  Transport Coefficients

The coefficient $[D_{ip}]_1$, contained in $[D_{ip}]_{1,\,eff}$ [equation (4.11)], can be calculated from the expression given for $[D_{ij}]_1$ [equation (3.87)]. In using equation (3.87) we observe that, since $m_i \ll m_p$ by virtue of hypothesis 1d, the expression for $m_{ij}^+$ (which in this case reads $m_{ip}^+$), reduces to

$$m_{ip}^+ = m_i \tag{4.15}$$

Since the particles are spherical,

$$\sigma_{ij} = r_p \tag{4.16}$$

$$\Omega_{i,j}^{(1,1)*} = \Omega_{i,p}^{(1,1)*} = 1 + s_{ip} \tag{4.17}$$

where $s_{ip}$ is a collision parameter that depends on the angular scattering pattern of the gas molecules rebounding from the particles.[(73)] For specular reflection, $s_{ip} = 0$, while for diffuse reflection (cosine-law scattering pattern), $s_{ip} = \frac{4}{9}$; inelastic collisions obviously modify $s_{ip}$. The collision integral $\Omega_{i,p}^{(1,1)}$, as we have seen in Chapter 3, is in general a function of temperature.

---

[†]See Chapter 6 for more details.
[‡]Cf. our discussion of this in Section 1.3.3.4.

Equation (3.87) can thus be written as

$$n[D_{ip}]_1 = \frac{3}{8} \frac{(\pi kT/2m_i)^{1/2}}{\pi r_p^2 \Omega_{i,p}^{(1,1)*}} \tag{4.18}$$

Thus in the first approximation $n[D_{ip}]_1$ is equal to a constant times a function of $T$. The second-approximation correction to $D_{ip}$ involves the correction factor $\Delta'_{ip}$ [cf. equation (4.14)], and $\Delta'_{ip}$ depends in a complex fashion on the gas composition, as can be seen from equation (3.90) for $\Delta_{AB}$ or from equation (3.128) for $\Delta_{ij}$.

The expression for $\alpha'_{ip}$ is obtained from equation (3.129) and, for a multicomponent gas mixture, it is also very complex.

### 4.1.4. Relationship between Total Diffusive Fluxes

If we sum the first term in equation (4.12) over all the species, we find since $[D_{ij}]_{2,\text{eff}} = [D_{ji}]_{2,\text{eff}}$ [cf. Section 1.6.2.3 and our discussion following equation (3.89)],

$$\sum_{i=1}^{\nu} \sum_{\substack{j=1 \\ j \neq i,p}}^{\nu} \frac{F_j^D x_i - F_i^D x_j}{[D_{ij}]_{2,\text{eff}}} = 0 \tag{4.19}$$

Equation (4.19) is an expression of the conservation of momentum for the molecule–molecule collisions. Using equation (4.19), we find for the sum of equation (4.12) over all species

$$-\sum_{i=1}^{\nu} \frac{n_p}{n} \frac{1 - \Delta'_{ip}}{[D_{ip}]_{1,\text{eff}}} F_i^D = \frac{1}{kT} \nabla p + n' \sum_{i=1}^{\nu} \sum_{j=1}^{\nu} x_i x_j \alpha'_{ij} \nabla \ln T \tag{4.20}$$

According to equation (4.18), we should have an equation such as (4.20) for every particle size, since the DGM does not necessarily require uniform particle size. A sum over all the $r_p$ can be performed for a set of equations such as (4.20) and the result written in terms of a mean value of the particle radius. This is consistent with Neale and Nader's results.[79]

Taking into account equations (4.6), (4.8), and (4.18), and assuming that the scattering pattern (and thus $\Omega_{i,p}^{(1,1)*}$) is independent of $i$, equation (4.20) becomes

$$-\sum_{i=1}^{\nu} F_i^D (1 - \Delta'_{ip}) \left(\frac{m_i}{T}\right)^{1/2} = C_1 \left(\frac{1}{kT} \nabla p + n' \sum_{i=1}^{\nu} \sum_{j=1}^{\nu} x_i x_j \alpha'_{ij} \nabla \ln T\right)$$

$$\tag{4.21}$$

where

$$C_1 \equiv \frac{3(\pi k/2)^{1/2} Q_p}{8\pi r_p^2 n_p \Omega_{i,p}^{(1,1)*}} \qquad (4.22)$$

For isothermal conditions, equation (4.21) reduces to

$$\sum_{i=1}^{\nu} F_i^D (1 - \Delta_{ip}') m_i^{1/2} = -C_2 \nabla p \qquad (4.23)$$

where

$$C_2 \equiv C_1 / kT^{1/2} \qquad (4.24)$$

Equation (4.21) and (4.23) can be regarded as momentum balances for the collisions of the molecules with the particles and they give the relationship between the diffusive fluxes of the various components of the gas.

Note that the origin of $(m_i/T)^{1/2}$ in equation (4.21) or of $m_i^{1/2}$ in equation (4.23) is due to the molecule–particle collisions, i.e., they arise from the second term on the left-hand side of equation (4.12). Further, no matter how insignificant the role the particles play (i.e., no matter how small the ratio $[D_{ij}]_2/[D_{ii}^K]_2$), equations (4.19) and (4.21) are always valid. *Thus, unless there are no particles, the relationship between the diffusive fluxes always depends on $m_i^{1/2}$ and is independent of the prevailing diffusion regime (molecular or Knudsen).*

For the important case of an isobaric, isothermal system, equation (4.23) becomes

$$\sum_{i=1}^{\nu} F_i^D (1 - \Delta_{ip}') m_i^{1/2} = 0 \qquad (4.25)$$

Equation (4.25) should be compared with equation (2.71).

If we assume that in a binary gas mixture $\Delta_{Ap}' = \Delta_{Bp}'$, which is a good assumption for many gas pairs,[25] equation (4.25) reduces to

$$-\left(F_B^D / F_A^D\right) = (m_A/m_B)^{1/2} \qquad (4.26)$$

which is Graham's law of diffusion for porous media. This law (see Chapter 6) was reported for the first time in 1833,[33] regarded as incorrect a short time later, subsequently forgotten until it was rediscovered by Hoogschagen[48, 49] with the designation of "square root law," and has been experimentally verified by several authors.[26, 36, 95, 105]

Table 4.1. Relationships between the Diffusive Fluxes in a Multicomponent
Gas Mixture

| Relationship | Condition | Equation number |
|---|---|---|
| $\sum_{i=1}^{\nu} \mathbf{J}_{iM} = \sum_{i=1}^{\nu} \mathbf{J}_{iM}^{m} = \sum_{i=1}^{\nu} \mathbf{j}_i = 0$ | By definition | (1.7) |
| $\sum_{i=1}^{\nu} M_i \mathbf{J}_i = \sum_{i=1}^{\nu} m_i \mathbf{J}_i^{m} = 0$ | By definition | (1.12) |
| $\mathbf{F}_i^{N} = \mathbf{N}_i^{N} = 0$ | Inviscid flux or system without walls | (3.133) |
| $\sum_{i=1}^{\nu} \mathbf{F}_i^{D}(1 - \Delta'_{ip}) m_i^{1/2}$ $= \sum_{i=1}^{\nu} \mathbf{N}_i^{D}(1 - \Delta'_{ip}) M_i^{1/2} = 0$ | Isobaric, isothermal system with walls | (4.25) |

Equation (4.25) gives another important result: for the mass flux we can have $\mathbf{G}^{D} \neq 0$ even though $\nabla p = 0$, i.e., there can be a nonviscous, nonseparative net mass flux due to composition gradients.

Two points are to be noted: first, the procedure followed here to obtain the relationship between the diffusive fluxes cannot be applied to a system without walls; second, Graham's law of diffusion is the expression of a momentum balance for the diffusing species.

We have completed our determination, begun in Chapter 1, of the relationships between the different kinds of diffusive fluxes. These relationships are summarized in Table 4.1.

### 4.1.5.  Viscous Flux

The constitutive equation most commonly used for the viscous flux in a porous medium is the Darcy equation [equation (1.68)]. Brinkman has proposed another expression [equation (1.68a)], of which Darcy's is a particular case.

Slattery[101] has demonstrated theoretically the validity of Brinkman's equation for the creeping flow of an incompressible Newtonian fluid, and this equation has received statistical justification.[21, 68, 87] The Brinkman equation corrects for something lacking in the Darcy equation since it shows the existence of an extra shear stress in the zones next to the walls of the porous medium; this shear stress has been verified experimentally.[9] However, for low-porosity porous media, the contribution of this shear stress is negligible and outside a "layer" next to the boundaries it becomes insignificant[80]; as the thickness of this layer is actually quite small (it is of the order of the square root of the permeability[87]), Brinkman's equation effectively reduces to Darcy's equation within the largest portion of the porous medium outside the above-mentioned thin boundary zones.

For a porous medium such as that postulated in the DGM, there are many theoretical and experimental studies of the flow of an incompressible fluid. Good reviews can be found in Scheidegger[93] and Happel and Brenner.[37] Models for calculating the permeability can be found in related literature.[12, 18, 19, 61] Recently, Neale and Nader[80] predicted the relationship between the permeability in the Darcy equation and the porosity of a medium such as that of the DGM by computing the influence of shear stress; in other words, they introduce a correction to the Darcy permeability equation. Azzam and Dullien[4] also calculated the permeability of porous media from the Navier–Stokes equation by modeling the flow by a set of cubic networks consisting of tubes with periodic step changes in diameter. They used the experimental bivariate pore-volume distribution function, and the permeabilities found agree theoretically and experimentally within $\pm 20\%$.

Mason et al.[73] also calculated the permeability by applying the DGM. They assumed that Stokes' law is still valid for one reference sphere placed in a swarm of spheres. Neale and Nader's study[80] shows that the force $\mathbf{f}_p$ that must be exerted on one sphere placed in a set of spheres suspended in a fluid, when the fluid is moving at velocity $\boldsymbol{\vartheta}^V$, is

$$\mathbf{f}_p = -6\pi r_p \mu \boldsymbol{\vartheta}^V f(\varepsilon) \tag{4.27}$$

where $f(\varepsilon)$ is a complex function of the porosity $\varepsilon$ such that $f(\varepsilon) \geqslant 1$ and $f(\varepsilon) \to 1$ when $\varepsilon \to 1$, thus verifying Stokes' law.[†] The creeping flow around a sphere is then

$$\mathbf{F}^V \equiv n\boldsymbol{\vartheta}^V = -n\mathbf{f}_p/6\pi r_p \mu f(\varepsilon) \tag{4.28}$$

By applying equation (4.28) to the whole set of particles and eliminating $\mathbf{f}_p$ through equation (4.2), we obtain

$$\mathbf{F}^V = -(B_k p/kT\mu)\nabla p \tag{4.29}$$

where, for the DGM as corrected by Neale and Nader,

$$B_k \equiv [n_p 6\pi r_p f(\varepsilon)]^{-1} \tag{4.30}$$

## 4.1.6. Viscous Flux and Total Diffusive Flux

A given pressure gradient simultaneously gives rise to the viscous flux and the total diffusive flux (both nonseparative) of the gas mixture; in fact,

---

[†]The objective of Neale and Nader's study was to calculate the function $f(\varepsilon)$. If the shear stress is neglected, the expression for $f(\varepsilon)$ is different but we still have $f(\varepsilon) \geqslant 1$.

this is an alternative way of saying that $\mathbf{F}_i = \mathbf{F}_i^V + \mathbf{F}_i^D$. Consequently, we can obtain a relationship between the viscous and total diffusive fluxes from equations (4.29), (4.20), and (4.14):

$$\mathbf{F}^V \mu / B_k p = \sum_{i=1}^{\nu} \mathbf{F}_i^D / \left[ D_{ii}^K \right]_{2,\,\mathrm{eff}} = -\frac{\nabla p}{kT} \qquad (4.31)$$

where isothermal conditions have been assumed. Equation (4.31) indicates that the relationship between the viscous flux and the total diffusive fluxes is independent of the molecule–molecule collisions, a result not obvious at first sight. This can be understood, however, if we consider that such a relationship must be governed by the momentum transferred by the different species to the walls.

For a binary, isothermal gas mixture, assuming that $\Delta'_{Ap} = \Delta'_{Bp}$, equation (4.31) gives

$$\left( F_A^D + F_B^D m_{BA}^{1/2} \right) / F^V = \left[ D_{AA}^K \right]_{2,\,\mathrm{eff}} \mu / B_k p \qquad (4.32)$$

where $m_{BA} \equiv m_B / m_A$.

## 4.1.7. The Temperature Dependence of the Effective Knudsen Diffusivity

From equations (4.18) and (4.14) it follows that

$$\left[ D_{ii}^K \right]_{2,\,\mathrm{eff}} \propto \begin{cases} x_i^0 \\ p^0 \\ T^{1/2} / \Omega_{i,p}^{(1,\,1)*}(T) \end{cases} \qquad (4.33)$$

In particular, combining equations (4.14) and (4.18) we obtain

$$\left[ D_{ii}^K \right]_{2,\,\mathrm{eff}} = C_3 T^{1/2} / \Omega_{i,p}^{(1,\,1)*}(T) \qquad (4.34)$$

where

$$C_3 \equiv \frac{3(\pi k / 2 m_i)^{1/2}}{8\pi r_p^2 n_p (1 - \Delta'_{ip})} \qquad (4.35)$$

Note that we assume for the moment that $\Delta'_{ip}$ is independent of temperature; we leave the verification of this assumption to the Appendix (where we will see that $\Delta'_{Ap} = $ constant).

The variation of $\Omega_{i,p}^{(1,1)*}$ with temperature can be determined by observing that[48]

$$\Omega_{i,p}^{(1,1)*} = C_4\Omega_{i,p}^{(2,2)*} \tag{4.36}$$

where $\Omega_{i,p}^{(2,2)*}$ is the collision integral for viscosity and is a function of temperature[39]:

$$\Omega_{i,p}^{(2,2)*} = \frac{1 + S_d T^{(a-b)/b}}{C_5 T^{2/b}} \tag{4.37}$$

$C_4$ and $C_5$ are proportionality constants, $S_d$ is a generalized Sutherland constant, and $a, b$ are the parameters of a generalized potential function of the Lennard-Jones type,

$$\Xi = a_1(\Xi°/d)^b - a_2(\Xi°/d)^a \tag{4.38}$$

where $\Xi°$ is a potential energy length parameter, $d$ is the distance between two molecules, and $a_1 = a_2$, $a = 6$, $b = 12$ for the Lennard-Jones 6–12 model, and $a_1 = (\pi/45)(a_{1,LJ})^{ab}(\Xi°)^3$ (the subscript LJ denotes Lennard-Jones), $a_2 = 15a_1$, $a = 3$, $b = 9$ for Hill's model.[46] The constant $C_5$ depends on $a_1$, $a_2$, $a$, $b$, and $\Xi°$.

Using equations (4.36) and (4.37), we can write equation (4.34) as

$$[D_{ii}^K]_{2,\text{eff}} = C_6 \frac{T^{(4+b)/2b}}{1 + S_d T^{(a-b)/b}} \tag{4.39}$$

where $C_6 \equiv C_3 C_5/C_4$.

Equation (4.39) has been experimentally verified[88] for the diffusion of helium in porous glass[51, 52]; for $a = 6$, $b = 12$ as well as for $a = 3$, $b = 9$, equation (4.39) agrees with the results within the experimental error.†

---

†To interpret these results, Hwang and Kammermeyer[51, 52] proposed that surface diffusion was occurring along with Knudsen diffusion; an expression with an exponential in the temperature reproduced the data as well as did equation (4.39). However, surface diffusion of helium is almost impossible (see Section 1.9.2 and Reference 104). This was pointed out by Satterfield[89] and Brown and Haynes[16] although Hwang and Kammermeyer offered a rebuttal.[53] We cannot, then, use the experimental temperature dependence of the Knudsen flux to verify the existence of surface diffusion.

### 4.1.8. The Transport Coefficients and the Effect of a Porous Medium

Walls affect transport in a gas in several ways:

(i) They give rise to the Knudsen diffusivity, which, using equations (4.14), (4.18), (4.6), (4.8), and (4.22), we can write as

$$[D_{ii}^K]_{2,\,\text{eff}} = Q_p^\circ (kT/m_i)^{1/2} \tag{4.40}$$

where $Q_p^\circ \equiv C_1/k^{1/2}(1 - \Delta'_{ip})$ and $C_1$ is given by equation (4.22).

(ii) The molecular diffusivity in the presence of walls is, using equations (4.13), (4.6), and (4.7),

$$[D_{ij}]_{2,\,\text{eff}} = Q_m [D_{ij}]_1 / (1 - \Delta'_{ij}) \tag{4.41}$$

(iii) They give rise to viscosity; the permeability $B_k$ plays the role that $Q_m$ plays for the molecular diffusivity.[†]

Note that the first-degree approximations to the effective diffusivities are given by equations (4.40) and (4.41) with $\Delta'_{ij} = 0$ and $Q_m$ and $Q_p^\circ$ constant.

### 4.1.9. Fluxes in Multicomponent Isothermal Mixtures

For isothermal systems, the fluxes of the $\nu$ species of the mixture are, from equations (4.11) and (4.29),

$$[\mathbf{F}] = -[x]B_k(p/\mu)\nabla n - [f_{\text{eff}}]^{-1}[\nabla n] \tag{4.42}$$

where $[\mathbf{F}]$, $[x]$, and $[\nabla n]$ are column matrices of $\nu$ elements $F_i$, $x_i$, and $\nabla n_i$, respectively, and $[f_{\text{eff}}]$ is the $\nu \times \nu$ matrix formed by the coefficients of equation (4.12):

$$f_{ij,\text{eff}} \equiv (\delta_{ij} - 1)\frac{x_i}{[D_{ij}]_{2,\,\text{eff}}} + \delta_{ij}\left[ [D_{ii}^K]_{2,\,\text{eff}}^{-1} + \sum_{\substack{h=1 \\ h \neq i}}^{\nu} \frac{x_h}{[D_{ih}]_{2,\,\text{eff}}} \right] \tag{4.43}$$

[†] The reader should observe that although $B_k$ and the $Q$'s play the same role in their constitutive equations of transport, $B_k$ is called "permeability" while the $Q$'s are called "obstruction" factors. The reason for this is merely historical: the first parameter to be proposed was the permeability of a porous bed since it was seen as a measure of the amount of fluid the bed let through per unit time; on the contrary, diffusion within a porous solid was compared with that in a system devoid of walls, and this gave rise to the concept of obstruction.

Consequently, it would be more consistent to (1) refer to the $Q$'s as permeabilities for molecular and Knudsen diffusion, and continue to call $B_k$ the permeability for laminar flow, or, instead (2) refer to $B_k$ as the obstruction factor for laminar flow, and continue to call the $Q$'s obstruction factors.

where $\delta_{ij}$ is the Kronecker delta,

$$\delta_{ij} = \begin{cases} 1 & \text{for } i = j \\ 0 & \text{for } i \neq j \end{cases} \tag{4.44}$$

## 4.2. Particular Cases of the DGM

### 4.2.1. Pure Isothermal Gas

A pure gas A is represented in the DGM by a binary mixture of components A and p. For the isothermal case, $x_A = 1$, $\nabla x_A = 0$, $x_B = 0$ and equation (4.12) becomes $F_A^D = - [D_{AA}^K]_{2,\text{eff}} \nabla n_A$. When there are pressure gradients, the viscous flux given by equation (4.29) is superimposed on the diffusive flux and the total flux of component A is given by the sum of the diffusive and viscous contributions:

$$\mathbf{F}_A = \mathbf{F}_A^D + \mathbf{F}_A^V = - \left( [D_{AA}^K]_{2,\text{eff}} + \frac{B_k k T n}{\mu} \right) \nabla n \tag{4.45}$$

### 4.2.2. Binary Isothermal Gas Mixture

#### 4.2.2.1. Constitutive Equation of Diffusion

The binary gas mixture is represented in the DGM by a ternary mixture. Equation (4.12) with $i = A$, $j = B$, p becomes

$$\frac{F_B^D x_A - F_A^D x_B}{[D_{AB}]_{2,\text{eff}}} - \frac{F_A^D}{[D_{AA}^K]_{2,\text{eff}}} = \nabla n_A \tag{4.46}$$

The constitutive equation for the diffusion of B is the same as equation (4.46) but with the subscripts interchanged. These two constitutive equations for the diffusion of A and B are coupled through equation (4.21), which for an isothermal binary gas becomes

$$\mathbf{F}_B^D = - [D_{BB}^K]_{2,\text{eff}} \left( \nabla n + \frac{F_A^D}{[D_{AA}^K]_{2,\text{eff}}} \right) \tag{4.47}$$

Introducing equation (4.47) into equation (4.46), using $x_B = 1 - x_A$ and

$\nabla n_A = n \nabla x_A + x_A \nabla n$, and then rearranging, we readily find

$$\mathbf{F}_A^D = -\frac{1}{[D_{AB}]_{2,\,\text{eff}} + [D_{AB}^K]_{2,\,\text{eff}}} \left( [D_{AB}]_{2,\,\text{eff}} [D_{AA}^K]_{2,\,\text{eff}} n \nabla x_A \right.$$

$$\left. + [D_{AA}^K]_{2,\,\text{eff}} ([D_{BB}^K]_{2,\,\text{eff}} + [D_{AB}]_{2,\,\text{eff}}) x_A \nabla n \right) \quad (4.48)$$

where

$$[D_{AB}^K]_{2,\,\text{eff}} \equiv [D_{AA}^K]_{2,\,\text{eff}} x_B + [D_{BB}^K]_{2,\,\text{eff}} x_A \quad (4.49)$$

Using equation (4.40) and assuming $Q_p^\circ$ is the same for both species,[†] we can also write equation (4.48) as

$$\mathbf{F}_A^D = -\frac{n \nabla x_A + \{1 + ([D_{BB}^K]_{2,\,\text{eff}}/[D_{AB}]_{2,\,\text{eff}})\} x_A \nabla n}{1/[D_{AA}^K]_{2,\,\text{eff}} + [1 - (1 - m_{AB}^{1/2}) x_A]/[D_{AB}]_{2,\,\text{eff}}} \quad (4.50)$$

For $\nabla n = 0$, equation (4.50) and its analog for component B lead directly to Graham's law, so the term inside the braces in equation (4.50) represents the deviation from Graham's diffusion law as a result of pressure gradients. Indeed, we can write equation (4.46) as

$$\mathbf{F}_A^D = -\left\{ \frac{1}{[D_{AA}^K]_{2,\,\text{eff}}} + \frac{1 - (1 + F_{BA}^D) x_A}{[D_{AB}]_{2,\,\text{eff}}} \right\}^{-1} \nabla n_A \quad (4.51)$$

where $F_{BA}^D \equiv F_B^D/F_A^D$; this equation is useful only when the gas mixture is isobaric, for then $F_{BA}^D = -m_{AB}^{1/2}$.

### 4.2.2.2.  The Constitutive Equations for the Total Diffusive and Viscous Fluxes

When there are pressure gradients, the viscous flux is superimposed on the diffusive flux and the total flux of component A is $\mathbf{F}_A^D + \mathbf{F}_A^V$, with $\mathbf{F}_A^V$ given by equation (4.29)[‡]:

$$\mathbf{F}_A = -\frac{[D_{AA}^K]_{2,\,\text{eff}} [D_{AB}]_{2,\,\text{eff}} n \nabla x_A}{[D_{AB}]_{2,\,\text{eff}} + [D_{AB}^K]_{2,\,\text{eff}}}$$

$$-\left\{ \frac{[D_{AA}^K]_{2,\,\text{eff}} ([D_{BB}^K]_{2,\,\text{eff}} + [D_{AB}]_{2,\,\text{eff}})}{[D_{AB}]_{2,\,\text{eff}} + [D_{AB}^K]_{2,\,\text{eff}}} + \frac{B_k kTn}{\mu} \right\} x_A \nabla n \quad (4.52)$$

---

[†] That this assumption is usually valid is implied by the experimental checks of the equations derived from it (see Section 4.2.2.2); note, however, that this assumption is not *always* valid (see Section 4.2.2.5).

[‡] Equation (4.52) can also be written in a form similar to that of equation (4.46) if generalized

In equation (4.52) the first term is the diffusive flux due to molar-fraction gradients, the second is the diffusive flux due to pressure gradients, and the third is the viscous flux (also called Darcy or forced flux). The corresponding equation for $F_B$ can be obtained from equation (4.52) by interchanging the subscripts.

The applicability of the individual terms in equation (4.52) and of its complete form has been experimentally verified. The viscous flux term was the first to be verified.[1, 19, 57, 84] † The diffusive flux term due to molar-fraction gradients has been verified in isobaric diffusion experiments as follows: by Scott and Dullien[96] for the interdiffusion of argon and oxygen in microporous porcelain and for nitrogen–hydrogen in diatomaceous earth and in porcelain of unglazed kaolin; by Henry, Cunningham, and Geankoplis[43] for hydrogen–nitrogen in three different samples of micro-macroporous alumina over a pressure range of 1300 : 1, from the Knudsen to the transition regime; by Gunn and King[36] for helium–nitrogen and helium–argon in fritted glass.

Gunn and King[36] also verified the second and third terms together for a pure gas (nitrogen and helium) and for a binary mixture (nitrogen-helium, neglecting composition gradients) in fritted glass.

Finally, the complete equation was verified by Mason, Malinauskas, and Evans[73] for the interdiffusion of helium and argon under pressure

---

coefficients $E_{AB}, E_{AA}^K$ for the diffusive and viscous fluxes are introduced,

$$\frac{F_B x_A - F_A x_B}{E_{AB}} - \frac{F_A}{E_{AA}^K} = \nabla n_A \tag{4.52a}$$

We find from (4.52a) and its analog for component B,

$$E_{AB} \equiv \frac{[D_{AB}]_{2,\text{eff}}\left\{1 + \dfrac{B_k nkT}{\mu}\left(\dfrac{x_A}{[D_{AA}^K]_{2,\text{eff}}} + \dfrac{x_B}{[D_{BB}^K]_{2,\text{eff}}}\right)\right\}}{1 + \dfrac{B_k nkT}{\mu}\left(\dfrac{x_A}{[D_{AA}^K]_{2,\text{eff}}} + \dfrac{x_B}{[D_{BB}^K]_{2,\text{eff}}} + \dfrac{[D_{AB}]_{2,\text{eff}}}{[D_{AA}^K]_{2,\text{eff}}[D_{BB}^K]_{2,\text{eff}}}\right)} \tag{4.52b}$$

$$E_{AA}^K \equiv [D_{AA}^K]_{2,\text{eff}}\left\{1 + \frac{B_k nkT}{\mu}\left(\frac{x_A}{[D_{AA}^K]_{2,\text{eff}}} + \frac{x_B}{[D_{BB}^K]_{2,\text{eff}}}\right)\right\} \tag{4.52c}$$

Equation (4.52a) also follows from equation (4.46) and its analog for component B if in the Knudsen term we replace $F_i^D$ by $F_i - F^V x_i$, with $F^V$ as given by equation (4.29), and use $\nabla n = \nabla n_A + \nabla n_B$; $\nabla n_B$ can be eliminated from these two equations, and rearrangement leads to equation (4.52a).

The reader should observe that, for a pure gas ($x_B = 0$), equation (4.52c) is simply the sum of two coefficients: one for Knudsen collisions and the other for viscous collisions. We discussed the problem of the additivity of these two mechanisms at the beginning of this chapter.

†Recall that the Darcy equation is consistent with the Brinkman equation if shear stress effects are included in the Darcy permeability (see Section 4.1.5).

gradients in a low-permeability graphite,[27] and also by Gunn and King[36] for the interdiffusion of oxygen and nitrogen in a porous graphite under pressure gradients in the transition regime and for molar fluxes of oxygen that varied over four orders of magnitude.[44]

We will now consider three particular cases of equation (4.52).

(a) *Molecular Regime.* In this case, (i) the Knudsen diffusion terms [the second term on the left-hand side of equation (4.46) and its analog for component B] are negligible, i.e., $[D_{AB}]_{2,\,eff} \ll [D_{ii}^K]_{2,\,eff}$ for $i = A, B$; and (ii) the viscous flux is not in general negligible, so that from equation (4.31) we have $B_k p / \mu \gg [D_{ii}^K]_{2,\,eff}$ for $i = A, B$. Equation (4.52) then reduces to

$$\mathbf{F}_A = -\frac{[D_{AB}]_{2,\,eff}\,\nabla n_A}{1 - (1 - m_{AB}^{1/2})x_A} - \frac{B_k kTn}{\mu}x_A\nabla n \qquad (4.53)$$

where the terms $m_{AB}^{1/2}$ arises from the ratio $[D_{BB}^K]_{2,\,eff}/[D_{AA}^K]_{2,\,eff}$.

Equation (4.46) becomes for this case

$$\frac{\mathbf{F}_B x_A - \mathbf{F}_A x_B}{[D_{AB}]_{2,\,eff}} = \nabla n_A \qquad (4.54)$$

since $\mathbf{v}_A^D - \mathbf{v}_B^D = \bar{\mathbf{v}}_A - \bar{\mathbf{v}}_B$ [cf. equation (1.21)]. Equation (4.54) can be re-arranged as

$$\mathbf{F}_A = -[D_{AB}]_{2,\,eff}\,\nabla n_A + (\mathbf{F}_A + \mathbf{F}_B)x_A \qquad (4.55)$$

*For the isobaric case, equation (4.55) is the form given in textbooks on transport phenomena for a system without walls.*

(b) *Knudsen Regime.* In this case, (i) the molecular diffusion terms [the first term on the left-hand side of equation (4.46) and its analog for B] are negligible compared to the Knudsen diffusion term, i.e., $[D_{AB}]_{2,\,eff} \gg [D_{ii}^K]_{2,\,eff}$ for $i = A, B$; and (ii) the viscous flux is in general negligible, i.e., $B_k p / \mu \ll [D_{ii}^K]_{2,\,eff}$ for $i = A, B$. Equation (4.52) reduces to

$$\mathbf{F}_A = -[D_{AA}^K]_{2,\,eff}\,\nabla n_A \qquad (4.56)$$

(c) $m_A = m_B$. In this case, $[D_{AA}^K]_{2,\,eff} = [D_{BB}^K]_{2,\,eff}$ so, equation (4.52) can be written as

$$\mathbf{F}_A = -[D_{AB}^R]_{2,\,eff}\,n\nabla x_A - \left([D_{AA}^K]_{2,\,eff} + \frac{B_k kTn}{\mu}\right)x_A\nabla n \qquad (4.57)$$

where $(1/[D_{AB}^R]_{2,\,eff}) \equiv (1/[D_{AA}^K]_{2,\,eff}) + (1/[D_{AB}]_{2,\,eff})$. The addition of the reciprocals of molecular and Knudsen diffusivities to give a resultant diffusivity is usually ascribed to Bosanquet (see page 770 of Reference 84).

### 4.2.2.3.  Relationships between the Diffusive Fluxes

From the definition of the diffusive flux of a given component on a molecular basis [equation (1.18) on a molecular basis], it follows that[†]

$$F_A^D = J_{AM}^m \left[ 1 - (1 + F_{BA}^D) x_A \right]^{-1} \tag{4.58}$$

Similarly, from equation (1.10) it follows that

$$F_A = J_{AM}^m \left[ 1 - (1 + F_{BA}) x_A \right]^{-1} \tag{4.59}$$

On the other hand [cf. equation (1.17)],

$$F_A^N = F_A^D (1 + F_{BA}^D) x_A \tag{4.60}$$

Whenever $F_{BA}^D$ can be calculated from equation (4.26), the calculation of the fluxes relative to $F_A^D$, i.e., $J_{AB}^m / F_A^D$ and $F_A^N / F_A^D$, will be immediate. The fluxes of B relative to $F_A^D$ also follow immediately since $J_{AM}^m = -J_{BM}^m$ and $F_B^N = F_A^N x_B / x_A$. We shall see important applications of these simple relationships for some cases of practical interest in Chapter 5.

### 4.2.2.4.  Total Diffusive Flux

By adding equation (4.48) and its analog for B and rearranging, we obtain

$$F^D = \left( [D_{AB}]_{2,\text{eff}} + [D_{AB}^K]_{2,\text{eff}} \right)^{-1}$$

$$\times \left\{ [D_{AB}]_{2,\text{eff}} ([D_{BB}^K]_{2,\text{eff}} - [D_{AA}^K]_{2,\text{eff}}) n \nabla x_A \right.$$

$$- [[D_{AB}]_{2,\text{eff}} ([D_{AA}^K]_{2,\text{eff}} x_A + [D_{BB}^K]_{2,\text{eff}} x_B)$$

$$\left. + [D_{AA}^K]_{2,\text{eff}} [D_{BB}^K]_{2,\text{eff}}] \nabla n \right\} \tag{4.61}$$

---

[†]If we combine equations (4.51) and (4.58), we obtain

$$J_{AM}^m = -[D_{AB}]_{2,\text{eff}} \nabla n_A - ([D_{AB}]_{2,\text{eff}} / [D_{AA}^K]_{2,\text{eff}}) F_A^D \tag{4.58a}$$

If wall effects are negligible, the second term on the right-hand side of equation (4.58a) is negligible compared to the first one, and $J_{AM}^m$ can be calculated using only one constitutive equation. If the second term is not negligible, then the calculation of $J_{AM}^m$ from equation (4.58a) requires that $F_A^D$ be known [through equation (4.48)]. Note that equation (1.7) can be obtained from equation (4.58) and its analog for B.

We consider three particular cases of this equation:
(a) When $m_A = m_B$, equation (4.61) becomes

$$\mathbf{F}^D = - \left\{ \frac{([D_{AB}]_{2,\,\text{eff}} + [D_{AA}^K]_{2,\,\text{eff}}) \nabla n}{1 + [D_{AB}]_{2,\,\text{eff}}/[D_{AA}^K]_{2,\,\text{eff}}} \right\} \tag{4.62}$$

Note that when the system is isobaric $\mathbf{F}^D = 0$.
(b) In the Knudsen regime, equation (4.61) reduces to

$$\mathbf{F}^D = - [D_{AA}^K]_{2,\,\text{eff}} \nabla n_A - [D_{BB}^K]_{2,\,\text{eff}} \nabla n_B \tag{4.63}$$

(c) In the molecular regime, equation (4.61) becomes

$$\mathbf{F}^D = - \frac{[D_{AB}]_{2,\,\text{eff}}(1 - m_{AB}^{1/2})n \nabla x_A + [D_{BB}^K]_{2,\,\text{eff}} \nabla n}{1 - (1 - m_{AB}^{1/2})x_A} \tag{4.64}$$

where the term $[D_{BB}^K]_{2,\,\text{eff}} \nabla n$, which is negligible compared to the viscous contribution, gives the deviation from Graham's law of diffusion as a result of nonisobaric conditions [cf. the discussion following equation (4.50)].

When the pressure is uniform throughout the system, equation (4.64) becomes [using equation (4.13)]

$$\mathbf{F}^D = D_{sAB}[D_{AB}]_{1,\,\text{eff}} n \nabla x_A \tag{4.65}$$

where

$$D_{sAB} = (1 - \Delta'_{AB})^{-1} \frac{m_A^{1/2} - m_B^{1/2}}{m_A^{1/2} x_A + m_B^{1/2} x_B} \tag{4.66}$$

We thus see that under isobaric conditions there is a net molar flux of diffusive origin, i.e., it is driven by a concentration gradient (cf. our discussion in Section 4.1.4). This flux, given by equation (4.65), is what we have called the nonequimolar flux; it is usually called the *diffusive slip flux*[†] and the coefficient $D_{sAB}$ is usually called the *slip diffusion coefficient*.

According to equations (4.65) and (4.66), the nonequimolar flux is equal to zero if $m_A = m_B$, and, since $\Delta'_{AB} < 1$, it is in the direction of the concentration gradient of the more massive molecules [in Section 4.2.2.5 we will see, however, that that prediction does not always hold and that

---

[†] The slip contribution due to temperature gradients is called *thermal creep* (see Annis and Mason[2]; Annis and Mason also provide all the expressions for a multicomponent mixture).

some of the assumptions used in arriving at equation (4.65) are not always valid].

### 4.2.2.5. Total Diffusive Flux, DGM, and Other Models

While it is difficult to check equation (4.65) directly by experiment, it can be checked indirectly by observing that $D_{sAB}[D_{AB}]_{1,\,eff} \nabla x_A \equiv v_{aer} = \vartheta_M^N = \vartheta_M$[cf. equation (1.21)] is the velocity at which suspended aerosol particles would move in an isobaric isothermal binary gas mixture. Schmitt and Waldmann[94] carried out an experimental study of such an aerosol flow, calling the phenomenon *diffusiophoresis*. They observed that, if Graham's law of diffusion applied (i.e., $\nabla n = 0$) and $\Delta'_{AB} = 0$, the expression for $v_{aer}$ predicts a diffusiophoretic velocity of a sign opposite to that experimentally observed for a nitrogen–ethane gas mixture. Schmitt and Waldmann proposed an empirical expression for $D_{sAB}$ that involved molecular diameters and which represented the experimental results of several gas pairs very well. Their expression for $D_{sAB}$ predicted a sign inversion for the argon–carbon dioxide pair and this was experimentally verified in a further study[107] in a capillary. The sign inversion was also observed for the neon–ethylene pair in fritted glass and capillaries.[24, 63] The behavior of the argon–carbon dioxide pair is particularly interesting since the sign is that of Graham's law at low pressures, but it inverts in the transition region and becomes normal at high pressures.

This effect should appear in the DGM through the correction term $\Delta'_{AB}$ as suggested by Mason and Marrero[74] and Zhdanov,[112] and the fact that $Q^\circ_{pA}$ and $Q^\circ_{pB}$ are not exactly equal as was assumed in deriving equation (4.64). However, it is almost impossible to provide a complete analytical correction within the framework of this theory. We face a similar situation with the results of Brock,[15] who employed an improved "ansatz" for the Chapman–Enskog approximation far from the wall. In general, we expect the correction factor to Graham's law of diffusion [equation (4.66) with $\Delta'_{AB} = 0$] to be a function of the molecular masses, intermolecular forces, accommodation coefficients, and concentrations.

Other rigorous solutions of the Boltzmann equation have been proposed for calculating $D_{sAB}$; among them are relaxation models for solving the linearized Boltzmann equation (BGK model, Section 3.8.1) by the half-range method,[63, 80] by a variational technique,[97] and by the half-range method applied to rigid spheres.[13]

For gas mixtures near isobaric conditions, the calculation of $D_{sAB}$ must take into account the interaction between the molecules and any boundary by some general law of force. This can be achieved by variational techniques applied to the Boltzmann equation[67] and a modification of Maxwell's method.[64, 66] The results of these methods agree satisfactorily with the experimental observations. Lang and Loyalka[64] clearly

Table 4.2. Comparison of Calculated and Experimental Slip Diffusion Coefficients[a] at $T = 293°K$ and for $x_A = x_B = 0.5$

| Gas A | Gas B | $D_{sAB,exp}$ | $D_{sAB,calc}$ [b] | | | | | | | |
|---|---|---|---|---|---|---|---|---|---|---|
| | | | G | Z | SW | B | LL | LL $(m_A \simeq m_B)$ | L | S |
| $N_2$ | $H_2$ | 0.9 | 1.15 | 1.18 | 0.65 | 1.08 | 1.21 | 0.95 | 0.61 | 0.69 |
| | $C_2H_2$ | 0.13 | 0.037 | 0.035 | 0.11 | 0.11 | 0.13 | 0.13 | 0.14 | 0.13 |
| | $C_2H_4$ | 0.073 | -0.0007 | -0.0038 | 0.073 | 0.07 | 0.102 | 0.100 | 0.12 | 0.11 |
| | $C_2H_6$ | 0.085 | -0.035 | -0.040 | 0.062 | 0.06 | 0.088 | 0.086 | 0.13 | 0.11 |
| | $O_2$ | -0.10 | -0.067 | -0.070 | -0.10 | -0.10 | -0.093 | -0.094 | -0.062 | -0.063 |
| | Ar | -0.22 | -0.18 | -0.19 | -0.21 | -0.22 | -0.22 | -0.22 | -0.13 | -0.14 |
| | $CO_2$ | -0.20 | -0.22 | -0.24 | -0.17 | -0.18 | -0.19 | -0.19 | -0.06 | -0.08 |
| | $C_3H_8$ | -0.13 | -0.23 | -0.24 | -0.05 | -0.07 | -0.09 | -0.09 | 0.08 | 0.04 |
| $CO_2$ | $C_3H_8$ | 0.11 | -0.0010 | -0.0015 | 0.12 | 0.12 | 0.11 | 0.11 | 0.13 | 0.12 |

[a] Adapted from Lang and Loyalka,[64] by permission of the publisher.
[b] G = Graham's law [equation (4.66) with $\Delta'_{AB} = 0$]; Z = Zhdanov[112]; SW = Schmitt and Waldmann[94]; B = Breton[12]; LL = Lang and Loyalka[64]; L = Lang and Eger[63] and Lang[62]; S = Shendalman.[97]

showed the influence of the collision cross sections, as well as of the accommodation coefficients, on $D_{sAB}$.

In Table 4.2 (extracted from Reference 64) the values for $D_{sAB}$ calculated using various models are compared with the experimental values obtained by Schmitt and Waldmann for the diffusiophoresis of different gas pairs.[94] The table shows that Graham's law of diffusion does not adequately predict the experimental results when the molecular masses of the diffusing species are nearly the same. When the heavier mass has the smaller collision cross section, the collision cross section effect has the same sign as the mass effect, as can be seen from the formula of Schmitt and Waldmann[94] and from Lang and Loyalka.[64] Thus, the values of $D_{sAB}$ predicted by Graham's law of diffusion for the nitrogen–acetylene and the nitrogen–oxygen pairs have the correct sign, though the quantitative agreement is poor.

Neither Zhdanov's theoretical result nor Brock's formula lead to an improvement of Graham's law. All these equations are derived using Maxwell's momentum balance, which does not correctly account for the influence of the collision cross section. The eighth and ninth columns in Table 4.2 present the results obtained by the modified Maxwellian method[64]; except for the nitrogen–propylene and nitrogen–hydrogen pairs, the theoretical results are in excellent agreement with experiment.


## 4.3. Constitutive Equations of Total Diffusive and Viscous Fluxes for Porous Media

### 4.3.1. Application of the DGM to a Capillary

A very important contribution of the DGM is the rigorously derived expression (4.12) for the constitutive equations of diffusion. This equation is compatible with those obtained by other authors through less rigorous developments for multicomponent[99] and binary gas mixtures[96, 103] and for a pure gas in a capillary.[84]

While equation (4.12) is formally valid in a capillary since it includes the influence of the molecule–wall collisions, the Knudsen diffusivity, which measures the contribution of the molecule–wall collisions, would have to involve new parameters (such as the capillary radius) that are outside the framework of the DGM. We will introduce such parameters *ad hoc* and use a semiempirical approach based on the DGM.

Before going ahead with the application of equation (4.12) to a capillary or a porous medium, we will consider the range of capillary radii over which the DGM is valid.

### 4.3.1.1.   The Molecule–Wall Interaction and the Capillary Radius

We have assumed that the molecule–wall collisions are simple (elastic), i.e., that there is no change in the internal (configurational, vibrational, rotational, electronic) energy of the molecule, and further we have not taken into account the fact that the shape (configuration) of a molecule can have an effect on the molecule–wall interaction potential. There may, however, be some change in the configurational internal energy of a molecule during a collision, since under ordinary laboratory conditions of diffusion this energy is of the order of the kinetic energy of a collision, and further, experiment indicates that the shape of a molecule does have a definite effect on the interaction potential. While present in all systems with walls, these configurational effects become significant only when the dimensions of the system through which the gas is diffusing are of the order of the size of the molecules.

Bird et al.[10] state that for two molecules separated by about three times the collision diameter, the attractive force between them is somewhat less than 1% of its maximum value, and "configurational effects" can occur at this distance. This means that there will be a kind of layer close to the capillary walls, with a thickness of the order of the collision diameter, inside which there can be a molecule–wall interaction that is not a simple collision. The heavier molecules, since they are slower, will remain inside the layer for a longer time than the lighter ones, and so these molecules will have a greater concentration in the layer and will collide more frequently with the walls. The flux of these molecules will therefore be decreased. These configuration effects represent a diffusion mechanism, called *configurational diffusion* by Weisz,[108] not foreseen in the DGM. Configurational diffusion is a very new aspect in the field of diffusion.

On the basis of this discussion, we can say that equation (4.12) can be applied to a capillary insofar as we can neglect the configurational aspects of the molecule–wall interaction. In general, these effects can be neglected if the capillary is not too narrow.[†] We will return to this problem in Section 5.4.1.

### 4.3.1.2.   Equations

We now apply equation (4.12) to a capillary. Since we are not dealing here with a porous medium (cf. our discussion in Section 4.1.1.), we can write $Q_m = Q_p = 1$; $[D_{ij}]_2$ varies with pressure, since $[D_{ij}]_1$ does [equations (3.88) and (3.89); the variation of $\Delta'_{ij}$ with pressure is given by equation

---

[†]For example, a molecule of nitrogen has a collision diameter of 3.68 Å, so that the layer would have a thickness of approximately 10 Å. This means that in a capillary with a 50-Å diameter, the volume of the layer will be 66% of the total.

(3.128)]; finally, $[D_{ii}^{K}]_2$ will depend on the capillary radius $R$ [equation (2.65)] and on pressure through the variation of $\Delta'_{ip}$ with $p$, which we will consider in the next section.

If the capillary axis is oriented along the direction $w$, equation (4.12) becomes for isothermal conditions

$$\sum_{\substack{j=1 \\ j \neq i}}^{\nu} \frac{F_{wj}^{D} x_i - F_{wi}^{D} x_j}{[D_{ij}]_2(p)} - \frac{F_{wi}^{D}}{[D_{ii}^{K}]_2(R,p)} = \frac{dn_i}{dw} \tag{4.67}$$

where we have explicitly indicated the dependence of the diffusivities on $R$ and $p$.

The expression for the fluxes of all the species, including the viscous contribution [which is given by equation (1.67) written on a molecular basis], can be written in a form similar to equation (4.42):

$$[F_w] = -[x] \frac{pR^2}{8\mu} \frac{dn}{dw} - [f(R)]^{-1} \left[ \frac{dn}{dw} \right] \tag{4.68}$$

where an element of the matrix $[f(R)]$ is given by

$$f_{ij}(R) \equiv (\delta_{ij} - 1) \frac{x_i}{[D_{ij}]_2} + \delta_{ij} \left[ [D_{ii}^{K}]_2^{-1} + \sum_{\substack{h=1 \\ h \neq i}}^{\nu} \frac{x_h}{[D_{ih}]_2} \right] \tag{4.69}$$

and $\delta_{ij}$ is given by equation (4.44).

### 4.3.1.3. The Second-Approximation Knudsen Diffusivity: Its Physical Meaning

Equation (4.67) involves the second-approximation Knudsen diffusivity, which, as we saw above, is a function of the capillary radius and also of the pressure, because it depends on the correction factor $\Delta'_{ip}$, which is a function of pressure [equation (3.128)]. We will determine the explicit relationship between $[D_{ii}^{K}]_2$ and $p$ for a capillary using the DGM.

We first proceed semiempirically, making use of Knudsen's empirical relationship for the flow of a pure gas through capillaries.[58] [equation (1.71)]. We now write this equation as

$$F_z = -\frac{K}{kT} \frac{dp}{dz} \tag{4.70}$$

where

$$F_z = F_z^{D} + F_z^{V} \tag{4.71}$$

In terms of the DGM, Knudsen's experiments correspond to a binary system so that $i = A$ = pure gas, $j = $ p, $Q = 1$; then equation (4.11) becomes for isothermal conditions and flux along the $z$ axis

$$F_z^D = -\frac{[D_{AA}^K]_2}{kT}\frac{dp}{dz} \tag{4.72}$$

where we have taken into account equations (4.6), (4.8), and (4.14) and used $Q_p = 1$; equation (4.72) can also be obtained from equation (4.45) when applied to a pure gas in a capillary.

Using equations (4.70), (4.72), and (4.29), we find that

$$K = [D_{AA}^K]_2 + \frac{B_k p}{\mu} \tag{4.73}$$

Comparing equations (1.71) and (4.73), we see that

$$B_k/\mu = a^K \tag{4.74}$$

$$[D_{AA}^K]_2 = \frac{[D_{AA}^K]_1}{1 - \Delta'_{Ap}} = b^K\frac{1 + c_1^K p}{1 + c_2^K p} \tag{4.75}$$

Since $[D_{AA}^K]_1$ is independent of the pressure [equation (2.65)], we see from equation (4.75) that $\Delta'_{Ap}$ must vary with $p$ in order that the theoretical results of the DGM be consistent with Knudsen's experimental results. We will now show this.

For the particular case in hand, the expression for $\Delta'_{Ap}$ is simply[72] (see the appendix at the end of this chapter)

$$\Delta'_{Ap} = \frac{Cn_p}{5n'}\alpha'_{Ap} \tag{4.76}$$

where $C$ is a coefficient on the order of one and can be left adjustable; $\alpha'_{Ap}$ is a linear function of the molar fraction,[72]

$$\frac{n'}{\alpha'_{Ap}} = \frac{n_p}{\alpha_{L_r}} + \frac{n}{\alpha_{Q_a}} \tag{4.77}$$

where $\alpha_{L_r}$ and $\alpha_{Q_a}$ are the limiting values of $\alpha'_{Ap}$ for a Lorentzian and a quasi-Lorentzian gas, respectively (see the appendix to this chapter), and $\alpha_{L_r} = 0.5$. Substituting equation (4.77) into equation (4.76), we obtain

$$\Delta'_{Ap} = \frac{C_7 C}{10(C_7/2\alpha_{L_r} + p)} \tag{4.78}$$

where

$$C_7 \equiv 2\alpha_{Q_a} n_p kT \tag{4.79}$$

Using equation (4.78) we can write

$$\frac{1}{1 - \Delta'_{Ap}} = \frac{c_2^K}{c_1^K} \frac{1 + c_1^K p}{1 + c_2^K p} \tag{4.80}$$

where

$$c_1^K = C_7^{-1}; \qquad c_2^K = (C_7 - C_7 C / 10)^{-1} \tag{4.81}$$

By introducing equation (4.80) into equation (4.75), we obtain

$$\left[ D_{AA}^K \right]_1 = b^K c_1^K / c_2^K \tag{4.82}$$

And finally, by introducing equation (4.82) into equation (4.81), we obtain

$$b^K = \frac{\left[ D_{AA}^K \right]_1}{1 - C / 10} \tag{4.83}$$

We have therefore expressed the four empirical coefficients of equation (1.71) in terms of the parameters of the DGM.

We can now assert rigorously (within the DGM) that the variable coefficient $\Delta'_{Ap}$ is responsible for the variation of $[D_{AA}^K]_2$ with pressure, and, based on our discussions following equation (1.71), that the variation of $[D_{AA}^K]_2$ with pressure explains the minimum in the Knudsen data when the observed flux is plotted as a function of pressure (Figure 1.14). *The DGM therefore explains the slip flux in terms of a Knudsen diffusion.*

We shall complete this development by considering the limiting forms of equation (4.75). In the Knudsen regime, $p \to 0$, and equation (4.75) reduces to

$$\left[ D_{AA}^K \right]_2 \to b^K \tag{4.84}$$

where, for Knudsen, $b^K = D_{AA}^K = \frac{2}{3} \bar{v}_A R$ [this is equation (2.65) for $\zeta = 1$]. In the slip regime, $c_1^K p$ and $c_2^K p \gg 1$ and equation (4.75) reduces to

$$\left[ D_{AA}^K \right]_2 \to b^K c_1^K / c_2^K = \left[ D_{AA}^K \right]_1 \tag{4.85}$$

We thus see that, for the DGM, the slip flux coefficient is equal to the first-approximation Knudsen diffusivity. From equation (4.83) with $C = 1$, we see that the DGM predicts a value of $c_1^K/c_2^K = 0.9$, which agrees very well with experimental results (Table 1.6).

These results apply to a long, straight capillary since we have neglected end effects and we took $Q = 1$. The behavior of short capillaries is intermediate between that of long ones and orifices. For a given length-to-radius ($L/R$) ratio, there will be a point at which the hindering effect due to an increase in the length will be balanced by the enhancing effect due to an increase in the radius; this "balance point" occurs for a value of $L/R$ between 8 and 15.[7,45] The shape of the flux-vs.-pressure curve will also depend on the geometry of the walls,[7] a fact known many years ago.[32]

A porous medium is a more complicated case. If we represent a porous medium by an array of short tubes of various radii interconnected at random, the flux-vs.-pressure curve, as we will see below (in Section 4.3.2.1), will rise linearly, which is the case for most real porous media. For some porous media, the curve is flattened,[3–6, 8] and in some special cases a minimum will appear.[35]

In terms of the DGM, a linear flux-vs.-pressure curve means that the influence of pressure on the coefficient $\Delta'_{Ap}$ has decreased considerably. This will be the case if, in equation (4.75), both $c_1^K$ and $c_2^K$ are much greater than $p^{-1}$ (if $c_1^K \simeq c_2^K$ we would also have $\Delta'_{Ap} =$ constant, but from equation (4.81) we see that this requires $C/10 \ll 1$, a condition that cannot be satisfied since $C \simeq 1$).

From equations (4.79) and (4.81) and using $\alpha_{L_r} = 0.5$ we observe that

$$c_1^K \gg p^{-1} \text{ implies } \frac{\alpha_{L_r}}{\alpha_{Q_a}} \gg \frac{n_p}{n} \qquad (4.86)$$

and

$$c_2^K \gg p^{-1} \text{ implies } \frac{\alpha_{L_r}}{\alpha_{Q_a}}\left(1 - \frac{C}{10}\right) \gg \frac{n_p}{n} \qquad (4.87)$$

Equation (4.87) does not essentially modify the condition involved in equation (4.86).

Nevertheless, taking into account that $\alpha_{Q_a} \gg \alpha_{L_r}$,[71, 106] equations (4.86) and (4.87) can only be fulfilled if $n_p \ll n$, which is a realistic condition, even in the Knudsen regime. The opposite condition, $n_p \gg n$, is a mathematical expression of the Knudsen limit as $p \to 0$.

## 4.3.2.  Application to a Porous Medium

### 4.3.2.1.  General Equations

Our task now is to apply equation (4.68), valid for a capillary of uniform radius, to a porous medium, which is considered as a network of capillaries interconnected at random and in general of varying radii.

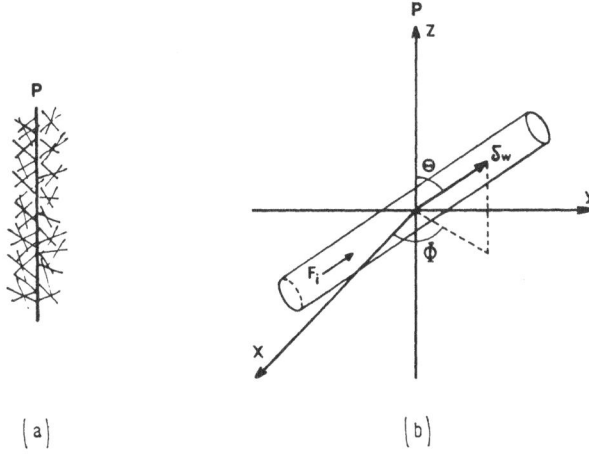

Figure 4.1. (a) Side view of a planar cross section through a porous medium. (b) The planar cross section $P$ is assumed to lie in the $xz$ plane. The angles $\Theta, \Phi$ specify the orientation of a capillary.

The gradients along a pore axis can be related to the gradients along the macroscopic coordinate of diffusion through

$$\frac{dn}{dw} = \delta_w \cdot \nabla n \tag{4.88}$$

$$\frac{dn_i}{dw} = \delta_w \cdot \nabla n_i \tag{4.89}$$

where $\delta_w$ is a unit vector along the pore segment under consideration.

Consider a planar cross section $P$ through the porous medium (Figure 4.1a). Many short capillaries of varying lengths, orientation, and radii cross the planar section. To determine the total flux crossing $P$, consider a capillary with orientation $\mathbf{W}$ (specified, e.g., by the angles $\Theta, \Phi$ shown in Figure 4.1b) and let $\delta_w$ be a unit vector along the capillary. Let the total flux of species $i$ through this capillary be $F_i \delta_w$. We will assume that $F_i$ depends only on $R$ and $\mathbf{W}$; $\delta_w$ is, of course, a function of $\mathbf{W}$. To obtain the total flux of $i$ across $P$ due to all the capillaries, we have to add the elemental contributions of all of them, taking account of the distribution of radii and orientation appropriate for a particular medium, and taking account of the fact that only a certain fraction—the void fraction—of $P$ is open. If, then, $g(R, \mathbf{W}) \, dR \, d\mathbf{W}$ is the number of capillaries per unit area of $P$ with radius $R$ in element $dR$ and with orientation $\mathbf{W}$ in $d\mathbf{W}$ (e.g., $d\mathbf{W} = d\Theta \, d\Phi \sin \Theta$), then the net flux of $i$ across $P$ is

$$\mathbf{F}_i = \int_R \int_\mathbf{W} \delta_w(\mathbf{W}) F_i(R, \mathbf{W}) g(R, \mathbf{W}) \, dR \, d\mathbf{W} \tag{4.90}$$

The function $g(R, \mathbf{W})$ is the void fraction per unit interval of pore radius $R$ and orientation $\mathbf{W}$.

We assume $F_i(R, \mathbf{W})$ satisfies equation (4.68) and obtain from (4.90)

$$[\mathbf{F}] = -\int_R\int_{\mathbf{W}}[\,x\,]\frac{pR^2}{8\mu}\delta_w(\delta_w \cdot \nabla n)g(R, \mathbf{W})\,dR\,d\mathbf{W}$$

$$-\int_R\int_{\mathbf{W}}[\,f(R)\,]^{-1}\delta_w(\delta_w \cdot \nabla[\,n\,])g(R, \mathbf{W})\,dR\,d\mathbf{W} \qquad (4.91)$$

Equation (4.91) can be written in terms of measurable parameters if we assume that $n$ and $[n]$ are independent of $\mathbf{W}$ and $R$ and that the pore orientations are independent of their radii; in such a case we obtain

$$[\mathbf{F}] = -\mu^{-1}[\,x\,]\mathbf{B}_k\,p\,\nabla n - \int_{R=0}^{\infty}[\,f(R)\,]^{-1}\kappa(R)\cdot\nabla[\,n\,]\,d\varepsilon(R) \qquad (4.92)$$

where

$$\mathbf{B}_k \equiv \frac{1}{8}\int_R\int_{\mathbf{W}}R^2\delta_w(\mathbf{W})\delta_w(\mathbf{W})g(R, \mathbf{W})\,dR\,d\mathbf{W}$$

$$\kappa(R) \equiv \int_{\mathbf{W}}\delta_w(\mathbf{W})\delta_w(\mathbf{W})g(R, \mathbf{W})\,d\mathbf{W}\Big/\int_{\mathbf{W}}g(R, \mathbf{W})\,d\mathbf{W}\,dR$$

$$\varepsilon(R) \equiv \int_0^R\int_{\mathbf{W}}g(R, \mathbf{W})\,d\mathbf{W}\,dR$$

The function $\varepsilon(R)$ is the observed void fraction for the pores with radii less than $R$; the permeability $\mathbf{B}_k$ and the tortuosity $\kappa$ are symmetric second-order tensors. If the porous medium is isotropic, the permeability and tortuosity are also isotropic and become $B_k$ and $\kappa$, respectively. The quantities $\mathbf{B}_k$, $\kappa$, and $R$ may vary with position since, as has been experimentally verified, the diffusion parameters may vary with position.[92] A porous medium thus may be inhomogeneous as well as anisotropic.

In practice, equation (4.92) is usually written for flow along only one coordinate (e.g., along $z$ in a diffusion cell or along $r$ in a spherical pellet); equation (4.92) then simplifies to

$$[F_z] = -\mu^{-1}[\,x\,]B_{kzz}p\frac{dn}{dz} - \int_{R=0}^{\infty}[\,f(R)\,]^{-1}\left[\frac{dn}{dz}\right]\kappa_{zz}(R)\,d\varepsilon(R)$$

$$(4.93)$$

Equation (4.93) cannot be directly used in practice, primarily because of the term $\kappa_{zz}(R)$.

Based on the characteristics of actual porous media, and following Feng and Stewart,[29] we can propose three particular cases of equation (4.93):

a. *For an isotropic porous medium with uniform pore radius*, equation (4.93) reduces to

$$[F_z] = -[x]\frac{B_k}{\mu}p\frac{dn}{dz} - \epsilon\kappa[f(R_1)]^{-1}\left[\frac{dn}{dz}\right] \qquad (4.94)$$

where the terms $B_k$, $\epsilon\kappa$, and $R_1$ are left as adjustable parameters; the two terms on the right-hand side of equation (4.94) are of the same form as those corresponding to the DGM [equation (4.42)].

b. *If $\kappa(R)$ is independent of $R$ and the porous medium is isotropic,* equation (4.93) simplifies to

$$[F_z] = -[x]\frac{B_k}{\mu}p\frac{dn}{dz} - \kappa\int_{R=0}^{\infty}[f(R)]^{-1}\left[\frac{dn}{dz}\right]d\epsilon(R) \qquad (4.95)$$

where $B_k$ and $\kappa$ are treated as adjustable parameters and $\epsilon(R)$ is obtained from porosimetric data. The second term on the right-hand side of equation (4.95) includes, as a particular case, a parallel-pore model[90] and a randomly oriented, noninterconnected-pores model[†] that yields an isotropic porous medium for which $\kappa = \frac{1}{3}$,[54] a value within a factor of 2 of the value for many catalyst pellets.[16, 50, 54]

Johnson and Stewart[54] have shown that equation (4.95) is also obtained from equation (4.91) provided we can write $g(R, \mathbf{W}) = g_1(R)g_2(\mathbf{W})$; this is the case for a bivariate distribution of discrete variables as well as for a bivariate normal distribution.[78]

c. *If the distribution of pore radii is not uniform,* as is often the case (in general, there are two very frequently occurring radii, corresponding to the macro- and the micropores), we can replace the integral of equation (4.93) by a summation, so that

$$[F_z] = -[x]\frac{B_k}{\mu}p\frac{dn}{dz} - \sum_{k=1}^{m}W_k[f(R_k)]^{-1}\left[\frac{dn}{dz}\right] \qquad (4.96)$$

where we have assumed that a pore radius can have only one of $m$ different values, and $W_k$ and $R_k$ are lumped contributions to the distribution of void fractions corrected for the tortuosity. The tortuous porosity $W_k$ and the pore radii $R_k$ are left as adjustable parameters.

---

[†]For brief descriptions of these models, see Table 4.3 below.

**Table 4.3.** Models for Predicting

| Model | Characteristics of porous medium |
| --- | --- |

**a. MODELS BASED ON TOTAL POROSITY (UNIMODAL[b])**

a.1. Wheeler[110]                                    Two-dimensional pile of parallelepipeds

a.2. Weisz and Schwartz[109]                  Set of interconnected spherical cells

**b. MODELS BASED ON A DISTRIBUTION OF PORE SIZES**

b.1. Capillary models

b.1.1. Parallel-pores model[41]              Three bundles of capillaries oriented along the three spatial coordinates. Each bundle of capillaries has a distribution of radii. Each capillary has a uniform radius. The bundle parallel to the macroscopic direction of diffusion contains 1/3 of the total capillaries; the remaining 2/3 do not contribute to diffusion

b.1.2. Randomly oriented pores model[54]                    Randomly oriented pores

b.1.3. Series-pores model[41]                  The pores are connected in series and the radius of each pore varies with its total length according to a distribution function. Only 1/3 of the pores is available for diffusion

b.1.4. Convergent–divergent-pores model (bimodal[b])[30]           Two parallel ducts formed by capillary segments; in one duct the radius decreases toward a symmetry axis while in the other it increases (convergent–divergent array)

[a]Haynes,[40] Haynes and Brown,[42] Satterfield and Cadle,[91] Brown et al.,[17] and Horák and Schneider[50] analyzed and compared these models, concluding that models b.1.1 or b.1.2 with an adjustable tortuosity are the most reliable. Feng and Stewart[29] observed that equation (4.96) with $m = 2$ is the best.

In these models it is customary to assume that the volumetric porosity is equal to the superficial

Transport Properties of Porous Media[a]

| Predicted parameters; equations; methods of calculation | Comments |
|---|---|
| a.1. | $\kappa = 1/2;\ R = 2V_p C_r(1-\varepsilon)/S_{BET}$ <br> Equation (4.94) | First model for diffusion. For a three-dimensional pile, $\kappa = 1/3$[109] |
| a.2. | $\kappa = 3^{-1/2}\varepsilon;\ R = 2V_p/S_{BET}$ <br> Equation (4.94) | Originally developed for the Knudsen regime; extended by Haynes[41] and verified for gel oxides |
| b.1.1. | $\kappa = 1/3$ <br> Equation (4.95) | Verified for various commercial catalysts with $\kappa$ as an adjustable parameter,[90, 91] aluminum oxides,[50] activated carbons alumina gel, and molecular sieves.[55] |
| b.1.2. | $\kappa = 1/3$ <br> Equation (4.95) | Verified for alumina pellets[54] |
| b.1.3. | Simultaneous solution of <br><br> $-F_{zA} = \left[\dfrac{nR^2}{3L}\displaystyle\int_0^{V_p} R^{-2}\,dV\right]$ <br><br> $\times\left[\dfrac{1}{D^K_{AA,\,eff}} + \dfrac{1-(1-m_{AB}^{1/2})x_A}{D_{AB,\,eff}}\right]$ <br><br> $\times \dfrac{dx_A}{dz}$ <br><br> $z = L\displaystyle\int_0^{V_p(R)} R^{-2}\,dV\Big/\int_0^{V_p} R^{-2}\,dV$ | Pores of variable cross section were also analyzed by Petersen[83] and Michaels[77] |
| b.1.4. | The volume of pores of a given radius is assigned to capillary segments of that radius | Verified for pellets of nickel oxide supported on kieselguhr and for molybdenum oxide and the alumina cited below for model b.2.3[31] |

*————— continued overleaf*

one. Relationships between these two porosities for different pore structures can be found in Foster *et al.*[31] The models can be used to predict the obstruction factor $Q$ referred to in Section 4.1.1. In all of these models, diffusion is assumed to take place in one direction.
[b]The terms *unimodal* and *bimodal* here mean the distribution curves for the pore radii have one and two peaks, respectively.

Table 4.3

| Model | Characteristics of porous medium |
|-------|----------------------------------|
| b.2. Random pore models | A bundle of parallel capillaries is cut into elements of uniform length; the elements are rejoined with random displacements from their initial position but they are kept aligned along the direction of macroscopic diffusion |
| b.2.1. Childs and Collis-George[22] | The pores are of different radii; while the pore axes are along the same coordinate in all the elements, the pore mouths on two adjoining parts do not necessarily coincide |
| b.2.2. Wyllie and Gardner [111] | As in b.2.1, but in the rejoined plane the void section decreases in proportion to $\varepsilon$; this permits the definition of an equivalent capillary radius |
| b.2.3. Wakao and Smith[105] | The porous medium is assumed to consist of an agglomeration of microporous particles with interparticle voids—the macropores. To model this medium, the bundle described in b.2.1 above is assumed to have capillaries of two different radii, corresponding to the micro- and macropores. At the rejoined plane there are three diffusion mechanisms: macro–macro (indicated by the subscript aa in the equations given on page 165); macro–micro (subscript ai); micro–micro (subscript ii). The probability of encounter of two void fractions $\varepsilon$ is $\varepsilon^2$. The increment of length for diffusion is taken as $\Delta z = \Delta z_1 + \Delta z_2 \simeq dz$, where 1 and 2 indicate the elements on each side of the rejoined plane. |
| b.2.4. Wakao and Smith modified b.2.4.1. Haynes[41] | Modifies the probability $\varepsilon^2$ of model b.2.3 to a new value P and proposes $\kappa = \frac{1}{3}$ |

*—continued*

| Predicted parameters; equations; methods of calculation | Comments |
| --- | --- |

b.2.1.

Equation like (4.95) with $\kappa = 1$ but with a bivariate pore-size distribution function for both faces at the rejoined plane and with a factor $\varepsilon^2$

The first model for predicting permeability that improved on the Kozeny–Carman equation for porous media with a pore-size distribution; Marshall[69] arrived at the same results independently.

b.2.2.

Equation like (4.95) with $\kappa = 1$, with a pore-size distribution function, a factor $\varepsilon^2$, and two adjustable parameters

b.2.3.

$F_{zA}^D = \varepsilon_a^2 F_{zAaa}^D + \varepsilon_i^2 F_{zAii}^D + 2\varepsilon_a(1 - \varepsilon_a) F_{zAai}^D$; $F_{zAaa}^D$ is given by equation (4.67) for a binary mixture with $R = \bar{R}_a$; $F_{zAii}^D$ is like $F_{zAaa}^D$ but with $R = \bar{R}_i$; $F_{zAai}^D = -2n(D_a^{-1} + D_i^{-1})\dfrac{dx_A}{dz}$

$D_i = \varepsilon_i^2/(1 - \varepsilon_a)^2$
$\times \left[ \dfrac{1}{D_{AAi}^K} + \dfrac{1 - (1 - m_{AB}^{1/2})x_A}{D_{AB}} \right]^{-1}$

$D_a = \left[ \dfrac{1}{D_{AAa}^K} + \dfrac{1 - (1 - m_{AB}^{1/2})x_A}{D_{AB}} \right]^{-1}$

$D_{AAi}^K = (2/3)\bar{R}_i\bar{v}_A$
$D_{AAa}^K = (2/3)\bar{R}_a\bar{v}_A$

For unimodal porous media this model reduces to b.2.1; it has been verified in alumina pellets[105]

b.2.4.1.

$F_{zA}^D = \kappa(P_{aa}F_{zAaa}^D + P_{ii}F_{zAii}^D + P_{ai}F_{zAai}^D)$
$P_{aa} \geqslant \varepsilon_a(2 - \varepsilon_a)(1 - e^{-4\varepsilon_a})$
$P_{ii} = (1 - \varepsilon_a)^2$
$P_{ai} < \varepsilon_a(2 - \varepsilon_a)e^{-4\varepsilon_a}$

*continued overleaf*

Table 4.3

| Model | Characteristics of porous medium |
|---|---|
| b.2.4.2. Shimizu *et al.*[98] | The macropores are dispersed uniformly in a microporous medium. The model is similar to b.2.3 but the elements with micro- and macropores are joined together by elements consisting only of micropores. Across the join, there are four diffusion mechanisms: macro–macro (indicated by subscript aia); macro–micro (aii); micro–macro (iia); micro–micro (iii). The increment of length for diffusion is taken as $\Delta z = \Delta z_{a1} + \Delta z_i + \Delta z_{a2} \simeq dz$ where $\Delta z_i = 2(\Delta z_{a1} + \Delta z_{a2})(1 - \varepsilon_a)^{1/3}\varepsilon_a^{-1/3}$ |
| b.2.5. Kim and Smith[56] | The porous structure is represented in two forms: (a) by a three-dimensional network of "cells," each with 2 pores along $x$, 3 along $y$, and 3 along $z$; for application to sintered materials; (b) by agglomerated spheres; for application to materials formed by chemical reaction |
| b.3. Network models | Network formed by capillaries; each element of the network is a pore |
| b.3.1. Two-dimensional[28] | Two-dimensional network |
| b.3.2. Three-dimensional[41] | Three-dimensional network of capillaries randomly oriented, of arbitrary but constant radius and length, and in an element with a volume the same as that of the porous medium |
| b.3.3. Haring and Greenkorn[38] | A network with a distribution of pore lengths and radii |

—continued

| Predicted parameters; equations; methods of calculation | Comments |
|---|---|
| b.2.4.2.   $F_{zA}^D = \varepsilon_a^2 F_{zAaia}^D + \varepsilon_i^2 F_{zAiii}^D$ <br> $+ \varepsilon_a(1 - \varepsilon_a)(F_{zAiia}^D + F_{zAaii}^D)$ <br> $F_{zAaia} = \left[\dfrac{\Delta z_{a1} + \Delta z_{a2}}{D_a} + \dfrac{\Delta z_i}{D_i}\right]\Delta z\,\dfrac{dn_A}{dz}$ <br> $F_{zAiii}^D$ is identical to $F_{zAii}^D$ of b.2.3 <br> $F_{zAiia}^D = \left(\dfrac{\Delta z_{a1} + \Delta z_i}{D_i} + \dfrac{\Delta z_{a2}}{D_a}\right)^{-1} \Delta z\,\dfrac{dn_A}{dz}$ <br> $F_{zAaii}^D = \left(\dfrac{\Delta z_{a1}}{D_a} + \dfrac{\Delta z_{a2} + \Delta z_i}{D_i}\right)^{-1} \Delta z\,\dfrac{dn_A}{dz}$ | For $\Delta z_i = 0$ and $\Delta z_{a1} = \Delta z_{a2} = \Delta z/2$, it reduces to b.2.3. This model gives values of $F_{zA}^D$ below the observed ones for calcined conglomerates of carbon mixed with clay[98] |
| b.2.5. | The variation of tortuosity during the process is calculated on the basis of (a) sintering and (b) chemical reaction. In (a), the model determines the successive elimination of pores in the network and gives as a result the tortuosity as a function of the porosity. In (b), the tortuosity variation is computed as a function of the particle radius | Verified in pellets of nickel oxide that were sintered or reduced with hydrogen[56] |
| b.3.1. | Once the network has been defined, the flux is computed through every pore individually by applying Kirchhoff's circuit law. The size of the network, the degree of interconnection of pores, and the ratio length/(pore radius) are left adjustable | Developed in the field of petroleum engineering to predict pressure gradients and permeabilities in rocks and sands |
| b.3.2. | The parameters are predicted statistically. The number of interconnections per pore is $V_p\varepsilon/2\pi\bar{R}^3$; the mean distance between interconnections is $\bar{R}/\varepsilon$ | Haynes[41] showed that the model cannot be applied to a macroporous network since it predicts a pore length less than $0.01 \times$ the particle diameter |
| b.3.3. | The size distribution is given as a function of two parameters that are related to the first and second moments of the distribution | Pakula and Greenkorn[82] calculated the permeability of a bed of spheres by applying this model |

In equations (4.94)–(4.96) the quantities $B_k$, $\varepsilon\kappa$, $\kappa$, $R$, and $W_k$ are *transport parameters of the porous medium*, and, unless the pores are very narrow, are independent of the gaseous components.[81]

Note that these equations are consistent with the suppositions expressed by equations (4.6)–(4.8).

From equations (4.6)–(4.8), (4.68) written with coefficients written in the first approximation, (4.94), (2.65), (2.17), (4.40), and (4.41), we find $Q_m = \varepsilon\kappa$, and $Q_p = \frac{2}{3} R(8/\pi)^{1/2}\varepsilon\kappa/\zeta = \varepsilon\kappa_K$. We thus see that, if we define a Knudsen tortuosity, it is different from the molecular tortuosity.

### 4.3.2.2. Prediction of the Transport Parameters of a Porous Medium

We noted above that some parameters of equations (4.94)–(4.96) are left adjustable. However, it is possible to construct models for the geometry of the porous structure from which these parameters can be predicted without any information about the transport phenomena in the porous medium and with only the aid, in some cases, of the porosity and pore-size data. These models are related to both terms on the right-hand sides of equations (4.94)–(4.96). The models related to the first term have arisen mainly in studies in the field of petroleum engineering and soil mechanics, while those related to the second term arose in the field of chemical engineering. Table 4.3 gives the classification of these models according to Haynes[41] and gives brief descriptions of each.

# Appendix.    The Correction Factor $\Delta'$ for the System of Pure Gas and Particles in the DGM

*Lorentzian Gas.* A binary gas mixture of species A and B is called a Lorentzian gas if

$$m_A \gg m_B \text{ and } x_A \gg x_B \quad \text{or} \quad \sigma_{AA} \gg \sigma_{BB} \text{ and } \sigma_{AA} \gg \sigma_{AB} \qquad (4A.1)$$

where $\sigma$ is the collision diameter. Under the conditions (4A.1), collisions between the lighter molecules have a negligible effect, so only the A–B collisions are important. This gas, which was first studied by Lorentz,[20] has the advantage that it is one of the few cases in which the second-approximation transport coefficients can be calculated exactly.

The Knudsen regime for a pure gas B in the DGM represents the Lorentzian gas B–p (i.e., A = p).

*Quasi-Lorentzian Gas.* A binary gas is quasi-Lorentzian if

$$m_A \gg m_B \quad \text{and} \quad x_A \ll x_B \qquad (4A.2)$$

so that the composition limit here is opposite that for a Lorentzian gas. In the DGM, this case corresponds to the viscous flux of a pure gas.

*The Coefficient* $\alpha'_{Ap}$. For the case of a pure gas A in the DGM, the coefficient $\alpha'_{ij}$ becomes $\alpha'_{Ap}$. This coefficient varies appreciably with composition.[34] There are no exact expressions for $\alpha'_{Ap}$, only different degrees of approximation. The first approximation depends in a complex way on composition.[72] It consists of a linear numerator of molar fractions and a quadratic denominator of molar fractions. However, Laranjeira[65] has shown that, if the inverse of the expression is taken, a two-term equation for $1/\alpha'_{Ap}$ can be obtained, with one of the terms linear; the second term is more complex, almost quadratic, but it is small compared to the first term unless the molecular masses of both components are nearly the same.

It can indeed be demonstrated that in the DGM the first approximation to $1/\alpha'_{Ap}$ is exactly linear in the limit $m_A/m_p \to 0$. Although such an expression is not exact in the Lorentzian limit, Mason, Evans, and Watson[72] have assumed that the linear expression is exact in that limit, and, by appropriate choice of an expression for $1/\alpha'_{Ap}$, force it to have the exact value in the Lorentzian limit and to achieve a very approximate value in the quasi-Lorentzian limit. The expression assumed by Mason *et al.*[89] was

$$\frac{1}{\alpha'_{Ap}} = \frac{x_p}{\alpha_{L_r}} + \frac{x_A}{\alpha_{Q_a}} \tag{4A.3}$$

where $\alpha_{L_r}$ is the value of $\alpha'_{Ap}$ for the Lorentzian limit and $\alpha_{Q_a}$ is the value for the quasi-Lorentzian limit; Mason *et al.*[72] showed that

$$\alpha_{L_r} = \tfrac{1}{2} \tag{4A.4}$$

and that, for elastic collisions between molecules and particles,[71, 106]

$$\alpha_{Q_a} = \frac{\lambda^{tr}_{KAA}}{5kn_p[D_{Ap}]_{1,\,eff}} \tag{4A.5}$$

where $\lambda^{tr}_{KAA}$ is the translational component of the thermal conductivity of the gas; if molecule–molecule collisions are also elastic,

$$\lambda^{tr}_{KAA} = \frac{15}{4}\frac{\mu_{AA}k}{m_A} \tag{4A.6}$$

For inelastic collisions, the value of $\lambda^{tr}_{KAA}$ decreases with the relaxation time of the collision.[70, 75]

The value of $\alpha_{Q_a}$ for a dusty gas is very high, but finite.

We should point out that equations (4A.4) and (4A.5) are not particular cases of the general expression for $\alpha_{ij}$ as given by equation (3.129). But the expression for $\alpha_{L_r}$ is exact, while that for $\alpha_{Q_a}$ involves an error of probably less than 2%.[47, 72]

*The Coefficient $\Delta'_{Ap}$.* The theoretical expression for $\Delta'_{Ap}$ consists of a numerator that is quadratic in molar fractions and the same denominator as that for $\alpha'_{Ap}$. In the limit $m_A/m_p \to 0$, the expression for $\Delta'_{Ap}$ reduces to[71]

$$\Delta'_{Ap} = \frac{Cn_p}{5n'} \alpha'_{Ap} \qquad (4A.7)$$

where the coefficient $6C^*_{AB} - 5$ of the theoretical expression was made equal to unity by assuming elastic collisions between molecules and particles (but not necessarily specular collisions). This expression is not exact, with only the approximation in the quasi-Lorentzian limit[72] being satisfactory. To compensate for this lack of accuracy, Mason et al.[72] used a trick similar to that which led to equation (4A.3), leaving $C$ as an adjustable parameter, so that equation (4A.7) provides the correct value in the Lorentzian limit. Thus, for the Chapman–Cowling first approximation,[20] $\Delta'_{Ap} = 1/12$ and $\alpha_{L_r} = 5/13$ in the Lorentzian limit; thus, $C = 13/12 = 1.08$. In higher approximations, $\Delta'_{Ap} = (32/9\pi) - 1$ and $\alpha_{L_r} = \frac{1}{2}$; thus, $C = 1.32$. Higher values of $C$ can be found experimentally.

# 5

# Analysis of Applications

## 5.1 Introduction

In the preceding chapters we analyzed the diffusive and viscous fluxes in capillaries and porous media from both a physical and a mathematical viewpoint. The goal of this chapter is to apply the equations for such fluxes to the systems most commonly used.

The analysis of a particular problem requires the statement of the equations of change of mass for the $\nu$ components of the system and the solution of these equations with suitable boundary and initial conditions; this is sufficient to give a complete description if the system is isothermal. To include nonisothermal systems requires in addition the statement of the equation of change of thermal energy; however, *we will restrict our discussion to the isothermal case*.

The equations of change of mass on a molar basis are of the general form (see Section 1.5.3.2)

$$\nabla \cdot \mathbf{N}_i + \epsilon \frac{\partial c_i}{\partial t} + R_{vi} = 0, \qquad i = 1, \ldots, \nu \tag{5.1}$$

where $\epsilon$ is the void fraction[†] and we have assumed a chemical reaction of the form

$$\sum_{i=1}^{\nu_r} a_i \mathbf{A}_i = 0 \tag{5.2}$$

with $a_i$ the stoichiometric coefficient for component $\mathbf{A}_i$ on a molar basis and $\nu_r$ the number of reaction components; $R_{vi}$ is the reaction rate of species $i$ per unit volume of the porous medium—it is expressed in moles

---

[†]$\epsilon$ enters this expression because (see Section 4.3.2.1) the flux $\mathbf{N}_i$ is expressed in terms of the unit area of the porous medium (rather than per unit void area) while $c_i$ is the number of moles per unit void volume.

and is given by

$$R_{vi} = a_i \rho_p R_{ve}^s S_{BET}^\circ \tag{5.3}$$

where $\rho_p$ is the mass density of the porous medium, $S_{BET}^\circ$ is the specific surface area of the porous medium (area per unit mass), and $R_{ve}^s$ is the reaction rate per unit surface area of the porous medium expressed in equivalents,

$$R_{ve}^s = - \frac{1}{S_{BET}} \frac{1}{a_i} \left( \frac{dN_{Mi}}{dt} \right) \tag{5.4}$$

with $S_{BET}$ the surface area of the porous medium and $N_{Mi}$ the number of moles of $i$.

The molar flux $\mathbf{N}_i$ of each component is the sum of the viscous, volumetric diffusive, and surface diffusive contributions. This total flux is given by equation (4.42) provided we add to it the surface flux contribution, which is given by (see Section 1.9.5)

$$\mathbf{N}_i^s = - D_i^s \boldsymbol{\nabla} c_i^s \tag{5.5}$$

The total flux along the $z$ coordinate can thus be written as follows [see equation (4.42)][†]:

$$[N_z] = - [c] \frac{B_k}{\mu} \frac{dp}{dz} - [f_{eff}]^{-1} \left[ \frac{dc}{dz} \right] - [D^s] \left[ \frac{dc^s}{dz} \right] \tag{5.6}$$

where $[D^s]$ is a $\nu \times \nu$ diagonal matrix of the surface diffusivities $D_i^s$.

The $\nu$ equations (5.1) contain, through the constitutive equations (5.6), $\nu$ dependent variables: the pressure $p$ and $\nu - 1$ molar fractions $x_i$; the solution of these equations thus completely describes an isothermal system.

It is important to point out that the constitutive equations [equations (5.6)] that we will use in equation (5.1) already contain the influence of walls. We also note that the constitutive equation for $\mathbf{N}_i^V$ may be regarded as arising from the equation of change of momentum when wall effects are included and when the velocity is averaged over the flow cross section; this constitutive equation can therefore be considered as an expression of a differential macroscopic balance. The constitutive equation for $\mathbf{N}_i^D$ may also be regarded as a differential macroscopic balance, but since for all practical purposes $\mathbf{N}_i^D$ does not depend on the coordinates transverse to the direction of flow (flat velocity profile; see Section 4.1), the microscopic and differential macroscopic balances are the same.

The $\nu$ equations (5.1) are usually solved to give the flux of a given species at a given point of the system. However, these equations can also be used to determine those parameters of the equation that cannot be

---

[†]In the molecular diffusion regime, the total flux may be calculated from the Stefan–Maxwell equations [equations (3.86) written in terms of total fluxes]. See the appendix to this chapter.

predicted by theoretical arguments. Further, an analysis of the results obtained under different working conditions permits verification of the applicability of the equations.

In the following, we will first describe some basic characteristics of binary systems. We will then analyze different types of systems of diffusion and viscous flux, emphasizing the case of binary mixtures, but with some consideration of ternary systems. Our study of binary systems will lead to a clear conceptual picture of the phenomena involved. Finally, we will discuss the experimental and predicted values of the transport parameters of a porous medium.

## 5.2. Isothermal Binary Systems

For binary systems, there are only two equations in (5.1), one for gas A and the other for gas B. The two equations are coupled through their dependent variables (e.g., pressure, composition), which affect the transport coefficients that appear in the equations. The molecular diffusivity depends on pressure and, to a smaller degree, on composition; the Knudsen diffusivity shows a similar dependence for a capillary, but may be taken as constant in a porous medium; finally, the viscosity depends on pressure and composition. Because of the resulting complexity of the equations, they must in general be solved by some numerical method.

### 5.2.1. The Relationship $x_A(p)$

The solution of the coupled equations of change can be simplified if we know *a priori* a relationship $x_A(p)$ that uncouples them. We would then only have to solve the equations of change for one of the species.

The relationship $x_A(p)$ between the molar fraction and pressure can be determined, at least implicitly, from the constitutive equations for the total fluxes of both components along one direction (e.g., along the $z$ axis). If we neglect surface diffusion and write equation (4.52) and its counterpart for B on a molar basis, form the ratio $N_{zA}/N_{zB}$, and rearrange, we obtain

$$\frac{dx_A}{dp} = \left[ \frac{1}{[D_{AB}]_2 p \{ [D_{BB}^K]_2(-N_{zA}/N_{zB}) - [D_{AA}^K]_2 \}} \right]$$

$$\times \left\{ [x_A + (N_{zA}/N_{zB})(1 - x_A)][B_k p / \mu ([D_{AB}]_2 + [D_{AB}^K]_2) \right.$$

$$+ [D_{AA}^K]_2 [D_{BB}^K]_2]$$

$$\left. + [D_{AB}]_2 [x_A [D_{AA}^K]_2 + (1 - x_A)(N_{zA}/N_{zB})[D_{BB}^K]_2] \right\} \quad (5.7)$$

Since equation .(5.7) is valid for any type of system (porous medium, capillary, etc.), the diffusivities involved may be the simple or the effective diffusivities. In the following, we will generalize by omitting the subscript eff.

The relationship $x_A(p)$ implicit in equation (5.7) can be readily obtained if we make some simplifying working hypotheses:

1. The first-approximation diffusivities can be used (in the following we will omit the subscript indicating the order of the approximation).
2. The viscosity $\mu$ is constant.
3. $-N_{zA}/N_{zB} = $ constant.

The first hypothesis leads to $D_{AB}p = $ const and $D_{ii}^K = $ const (see Sections 3.6.2 and 4.1.8). We are thus restricting ourselves to a porous medium or a capillary in the absence of slip flux. The second hypothesis introduces an error that depends on the nature of the system. The third hypothesis can be fulfilled by appropriate selection of the experimental conditions.

We can write equation (5.7) in dimensionless form as

$$\frac{dx_A^*}{dp^*} = [D^{K^*} + N^*]^{-1} \left\{ \left[ x_A^*(1-N^*) + \frac{N^*}{x_{A0}} \right] \right.$$

$$\times \left[ D^* - B_k^* p^* D^{K^*} \left( 1 + \frac{1}{D^* p^*} - x_{A0} \frac{1+D^{K^*}}{D^{K^*}} x_A^* \right) \right]$$

$$\left. + p^{*-1} \left[ \frac{N^*}{x_{A0}} - x_A^*(D^{K^*} + N^*) \right] \right\} \tag{5.8}$$

where

$$x_A^* \equiv x_A/x_{A0}; \quad p^* \equiv p/p_0; \quad D^* \equiv D_{AA}^K p_0/D_{AB}^\circ;$$
$$B_k^* \equiv B_k p_0^2/\mu D_{AB}^\circ; \quad D^{K^*} \equiv -D_{AA}^K/D_{BB}^K = -(M_{BA})^{1/2};$$
$$N_*^* \equiv (-N_{zA}/N_{zB}); \quad D_{AB}^\circ \equiv D_{AB}p = \text{constant}$$

and $x_{A0}$, $p_0$ are arbitrary constant values of the molar fraction and pressure; we will when appropriate take them as the values maintained at a boundary of the system. Following hypothesis 1, the diffusivities in equation (5.8) are taken to be the first-approximation diffusivities and we have dropped the subscripts on them. According to hypothesis 3, $N^*$ is a constant.

Equation (5.8) is a Riccati equation and has analytical solutions for $N^* = 0$ and $N^* = 1$, and, as we shall see, these solutions represent real cases.

If in the derivation of equation (5.7) we set $dp/dz = 0$, we obtain the condition $N^* = -D^{K^*} = M_{BA}^{1/2}$, i.e., the ratio of the diffusive fluxes must obey Graham's law of diffusion. The condition $N^* = -D^{K^*}$ is both necessary and, as can be seen by inserting it in equation (5.8), sufficient in order for the pressure in a binary mixture to be uniform.

### 5.2.2. Some Systems for Which the Ratio of Fluxes is Constant

There are several binary mixtures that fulfill the condition that the ratio of the fluxes of the components be a constant; such mixtures include, as we shall see, those arising from the drying of solids, from catalyzed as well as noncatalyzed gas–solid reactions, and in the cooling of nuclear reactors. With binary mixtures of this type, the problem usually is to find the flux of one of the components, with the transport coefficients known. Conversely, it is customary to employ such systems in the laboratory, with the fluxes known, to determine the transport coefficients.

## 5.3. Typical Isothermal Systems for Diffusion and Flow

### 5.3.1. Closed System

Essentially, a closed binary system is made of two chambers that initially contain different gases and that, at a given moment, may be connected. We may represent such a system by two bulbs connected by a capillary or a porous medium that has a valve, or some similar device, to keep the gases separate and then to allow them to come in contact (cf. System 2 in Table 1.1). In general, the volume of the bulbs is much greater than that of the capillary or porous medium. In the limiting case, the joining volume is zero and both chambers are merely separated by a valve.

Such a two-chamber system is a typical transient system that evolves toward the equilibrium state.

#### 5.3.1.1. Binary Mixture

Let us assume that the bulbs originally contain the pure gases A and B at the same pressure, with $M_A < M_B$ (see Figure 5.1). In Section 1.3.2.3 we described the phenomena that develop when both gases are placed in contact. We now have a similar situation. The molecules of A, since they are lighter than those of B, fly faster and give rise to an increase of pressure within the chamber of B. This pressure gradient generates a nonsegregative nondiffusive flow, usually a viscous flow, toward the chamber of A. The diffusive and viscous fluxes are thus coupled and any

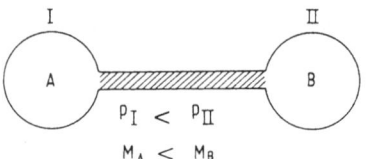

Figure 5.1. A closed system.

tendency toward an increase in the pressure is counterbalanced by the viscous flux. Consequently, once the pressure gradient has developed, a quasi-steady-state must be reached in which the composition in the bulbs varies slowly. Under such conditions, the net molecular flux crossing a given section of the capillary or porous medium must be zero, since otherwise the pressure within one of the chambers would still be increasing, implying that the quasi-steady-state had not yet been reached.

Although this system is not very frequently employed in the pure Knudsen regime, it is of interest to consider such an application. In the Knudsen regime, the flux of each species is given by [see equation (4.56)]

$$N_{zi} = -D_{ii}^{K} \frac{dc_i}{dz} \tag{5.9}$$

and is independent of the presence of the other fluxes (at least in the first approximation of the constitutive equations of diffusion). It is therefore impossible to couple the diffusive and nondiffusive fluxes, a coupling that makes it possible for the net molecular flux to be zero in non-Knudsen regimes.

We return now to the most general case of diffusion. Kramers and Kistemaker[71] are usually considered the pioneers in measuring the pressure gradient generated during the quasi-steady-state. The existence of such a gradient was also verified by McCarty and Mason,[80] who used a horizontal differential manometer that was connected between both chambers, and used a small oil drop as a hydraulic seal. The phenomenon has subsequently been studied by many authors,[69, 70, 78, 113, 122, 125] has been verified by use of the BGK model,[97, 116] and, by using a method in which pressure differences can be detected with an accuracy of $10^{-4}$ torr,[114] has even been detected for gas pairs with similar molecular masses such as argon–carbon dioxide, helium–deuterium, and xenon–silicon hexa-fluoride.[115]

We can think of the development of the pressure gradient in terms of a moving diffusion interface. This interface may be made visible, e.g., in the interdiffusion of ammonia and hydrogen chloride the moving front consists of a cloud of ammonium chloride, [81] and in some cases the moving front can be followed by the movement of aerosol drops. [100]†

According to the above discussion, for a closed system in the quasi-steady-state the net molecular flux is zero, or

$$N^* = 1 \tag{5.10}$$

with

$$N_{zA} > 0, \qquad N_{zA}^D > 0, \qquad N_{zA}^V < 0 \tag{5.11}$$

$$N_{zB} < 0, \qquad N_{zB}^D < 0, \qquad N_{zB}^V < 0 \tag{5.12}$$

If we set $N^* = 1$ (i.e., $N_{BA} = -1$) in equation (4.59), we obtain

$$N_{zA} = J_{zAM} \tag{5.13}$$

When the diffusion regime is molecular, which is usually the case as a result of the working conditions in the capillary (or porous medium) joining the two chambers, the diffusive flux obeys Fick's law,

$$N_{zA} = -D_{AB}\frac{dc_A}{dz} \qquad \text{(molecular diffusion)} \tag{5.14}$$

Equation (5.14) asserts that the *total* flux can be expressed by Fick's law, an assumption that has been used since the last century to measure molecular diffusivities; obviously, when the law was originally formulated, the fact that the validity of equation (5.14) was the result of a coupling of the diffusive and viscous fluxes was unknown (see Chapter 6 for a more detailed analysis of this).‡

---

†Incidentally, the observation of such moving fronts in gases recalls those of Kirkendall and Smikelsgras [67, 107, 108] performed some years earlier with solids. In their experiments, a brass nucleus was kept in contact with a covering layer of copper, and molybdenum wires were placed as tracers where the brass and copper were in contact. After some time, it was found that the amount of zinc that had diffused outward from the nucleus was greater than the amount of copper that had migrated in the opposite direction; futhermore, the interface had moved.

‡A thorough review of the methods of measurement of diffusivities has been made by Mason and Marrero. [77, 79]

Stewart et al. [110] made a step-by-step integration of equation (5.1) with equation (5.14) for a closed system formed by two cylindrical chambers. The analysis takes into account variations in molecular diffusivities, in pressure (which, initially, is assumed uniform), and the absorption and desorption of each component at the seals of the cell (i.e., at the sliding surfaces that separate the chambers, which are generally sealed by a lubricant or by a gasket). The solutions obtained with a perturbation method give diffusion coefficients that compared excellently with the experimental results obtained continuously by means of an interferometric method. [39]

Let us consider the nonequimolar flux in the general case for this system. Using

$$N_{zA}^V = N_{zA} - N_{zA}^D \tag{5.15}$$

and using equations (4.58) and (5.13), we obtain

$$N_{zA}^V = -N_{zA}^D(1 + N_{zBA}^D)x_A \tag{5.16}$$

or, using equation (4.60),

$$N_{zA}^V = -N_{zA}^N \tag{5.17}$$

We thus see that the mutual cancellation of the viscous and nonequimolar fluxes is the natural condition of the system, and this cancellation makes the total flux equal to the diffusive flux.

Let us now consider the calculation of pressure variations that build up in this system. As noted earlier, equation (5.8) has an analytical solution for $N^* = 1$. With the boundary conditions $x_A^* = 1$, $p^* = 1$, the solution is given by

$$x_A^* = \left[ W_3 p^* \exp\left(-W_2 p^{*2}\right)\right]^{-1}\left\{\left(W_1 + D^{K^*}p^*\right)\exp\left(-W_2 p^{*2}\right)\right.$$
$$+ \left(W_3 - D^{K^*} - W_1\right)\exp\left(-W_2\right) + W_4\left[\text{erf } W_2^{1/2} - \text{erf } W_2^{1/2}p^*\right]\right\} \tag{5.18}$$

where

$$W_1 \equiv \frac{D^{K^*}}{D^*} - \frac{D^*}{B_k^*}; \qquad W_2 \equiv B_k^*/2;$$

$$W_3 \equiv x_{A0}(1 + D^{K^*}); \qquad W_4 \equiv \left(\frac{\pi}{2B_k^*}\right)^{1/2}(D^{K^*} - 1) \tag{5.19}$$

We will now consider equations (5.7) and (5.8) for $N^* = 1$ in the Knudsen, viscous, and transition regimes [the results obtained could be derived from equations (5.18) and (5.19), but we find it easier to work directly from equations (5.7) and (5.8)].

For the Knudsen regime (i.e., $p$ small), we substitute $D_{AB}^\circ = D_{AB}p$ (using hypothesis 1 in Section 5.2.1) into equation (5.7) in order to make the full $p$-dependence of the equation explicit, and we use $N^* = 1$ and keep only the lowest-order terms in $p$ (i.e., terms of order $p^{-1}$); we find

$$\frac{dx_A}{dp} = -\frac{1}{p}\frac{x_A(M_{BA}^{1/2} - 1) + 1}{M_{BA}^{1/2} - 1} \tag{5.20}$$

This equation may also be derived from equation (4.56) and the corresponding equation for B, both written on a molar basis. The limit $p \to 0$ used in the derivation given above is a practical condition for the operation of any given system in the Knudsen regime and, therefore, for the fulfillment of the conditions stated in deriving equation (4.56).

Equation (5.20) can be readily integrated, yielding the "small $p$" version of equation (5.18):

$$\frac{p}{p_0} = \frac{M_{BA}^{1/2}}{x_A\left(M_{BA}^{1/2} - 1\right) + 1} \tag{5.21}$$

with $p_0$ the pressure at the low-pressure end of the capillary or porous plug, and with $x_{A0} = 1$. The pressure drop $\Delta p$ across the capillary or porous plug is then

$$\Delta p = p_0 \frac{\left(M_{BA}^{1/2} - 1\right)(1 - x_{AL})}{x_{AL}\left(M_{BA}^{1/2} - 1\right) + 1} \qquad \text{(Knudsen regime)} \tag{5.22}$$

where $x_{AL}$ is the molar fraction of A at the high-pressure end of the capillary or porous plug. From equation (5.21) the average pressure $\bar{p}$ is given by

$$\bar{p} = \frac{p_0 + p_L}{2} = \frac{p_0}{2} \frac{x_{AL}\left(M_{BA}^{1/2} - 1\right) + M_{BA}^{1/2} + 1}{x_{AL}\left(M_{BA}^{1/2} - 1\right) + 1}$$

and substituting this into equation (5.22) yields

$$\Delta p = 2\bar{p} \frac{\left(M_{BA}^{1/2} - 1\right)(1 - x_{AL})}{M_{BA}^{1/2} + 1 + x_{AL}\left(M_{BA}^{1/2} - 1\right)} \qquad \text{(Knudsen regime)} \tag{5.23}$$

Thus for small pressures, i.e., in the Knudsen regime, the pressure drop is proportional to $p_0$ or to $\bar{p}$ for fixed $x_{AL}$ [this statement is valid, of course, only for values of $x_{AL}$ that correspond to a quasi-steady-state condition of the system, since we assumed such a condition in arriving at equation (5.20)].

For the viscous regime (i.e., large $p$), we again substitute $D_{AB}^\circ = D_{AB}p$ and $N^* = 1$ into equation (5.7); we now keep only the highest-order terms in $p$ (i.e., terms of order $p$). We find

$$\frac{dx_A}{dp} = -p \frac{B_k}{\mu D_{AB}^\circ} \frac{M_{BA}^{1/2} - x_A\left(M_{BA}^{1/2} - 1\right)}{M_{BA}^{1/2} - 1} \tag{5.24}$$

Like equation (5.20), this equation may also be derived from equations developed in Chapter 4, in this case equation (4.53) and the corresponding equation for B, both written on a molar basis. The limit $p \to \infty$ is a practical condition for the operation of a system in the molecular diffusion regime with viscous flow and, therefore, for the fulfillment of the conditions given in deriving equation (4.53).

Equation (5.24) can be readily integrated. Using the same boundary values as for the Knudsen regime, we find

$$(p_L^2 - p_0^2)/2 = (\mu D_{AB}^\circ/B_k) \ln\left[ M_{BA}^{1/2} - x_{AL}(M_{BA}^{1/2} - 1)\right] \quad (5.25)$$

We can write $(p_L^2 - p_0^2)/2 = (p_L - p_0)(p_L + p_0)/2 = \bar{p}\Delta p$, and equation (5.25) becomes

$$\Delta p = (\mu D_{AB}^\circ/\bar{p}B_k)\ln\left[ M_{BA}^{1/2} - x_{AL}(M_{BA}^{1/2} - 1)\right] \quad \text{(viscous regime)}$$
$$(5.26)$$

and

$$\bar{p} = \frac{p_L + p_0}{2} = \frac{1}{2}\left\{ p_0 + \left( p_0^2 + \frac{2\mu D_{AB}^\circ}{B_k} \ln\left[ M_{BA}^{1/2} - x_{AL}(M_{BA}^{1/2} - 1)\right]\right)^{1/2}\right\}$$

$$\text{(viscous regime)}$$

In the viscous regime, then, $\Delta p$ is inversely proportional to the pressure $\bar{p}$.

Since $\Delta p \propto \bar{p}$ for small $\bar{p}$ and $\Delta p \propto 1/\bar{p}$ for large $\bar{p}$, $\Delta p$ must have a maximum in the transition regime for some value of $\bar{p}$ (as had already been observed by Waldmann and Schmitt [125]). This phenomenon, which in closed systems is known as the *Kramers–Kistemaker effect*, is a property of all binary systems in which the ratio of the fluxes of both components is a constant.

Let us now take a look at equation (5.8) written as

$$\frac{dx_A^*}{dp^*} = a_1(x_A^*, x_{A0}, D^{K^*}, N^*)$$

$$\times \left[ D^* + \frac{B_k^*(-D^{K^*})}{D^*} + B_k^* p^* x_{A0} x_A^* + B_k^*(-D^{K^*})p^*(1 - x_{A0}x_A^*)\right]$$

$$+ a_2(x_A^*, p^*, x_{A0}, D^{K^*}, N^*) \quad (5.27)$$

where

$$a_1 \equiv (D^{K^*} + N^*)^{-1}\left[ x_A^*(1 - N^*) + N^*/x_{A0}\right]$$

$$a_2 \equiv (D^{K^*} + N^*)^{-1}p^{*-1}\left[ N^*/x_{A0} - x_A^*(D^{K^*} + N^*)\right]$$

According to the mean-value theorem, there exists an $x_A^*$, call it $\overline{x_A^*}$ such that $\Delta p^* / \Delta x_A^* = (dp^*/dx_A^*)_{\overline{x_A^*}}$, and using equation (5.27) we have

$$\Delta p^* = \Delta x_A^* \left\{ \overline{a_1} \left[ \overline{D^*} + \frac{B_k^*(-D^{K^*})}{\overline{D^*}} + B_k^* \overline{p^*} x_{A0} \overline{x_A^*} \right. \right.$$

$$\left. \left. + B_k^*(-D^{K^*}) \overline{p^*} (1 - x_{A0} \overline{x_A^*}) \right] + \overline{a_2} \right\}^{-1} \tag{5.28}$$

where $\overline{p^*} = p^*(\overline{x_A^*})$ and

$$\overline{a_1} \equiv a_1(\overline{x_A^*}, x_{A0}, D^{K^*}, N^*)$$

$$\overline{a_2} \equiv a_2(\overline{x_A^*}, \overline{p^*}, x_{A0}, D^{K^*}, N^*)$$

If we assume that $\overline{x_A^*}$ and $\overline{p^*}$ are not much affected by varying the parameters in equation (5.28), then $\Delta p^*$ decreases as $B_k^*$ increases (i.e., as the permeability of the porous medium increases); from the form of equation (5.28) when considered as a function of $D^*$, we see that it has an extremum for $D^* = [B_k^*(-D^{K^*})]^{1/2}$. Since $\Delta p^* \to -\Delta x_A^* D^* / \overline{a_1} B_k^* D^{K^*} \to 0$ as $D^* \to 0$ and $\Delta p^* \to \Delta x_A^* / D^* \overline{a_1} \to 0$ for $D^* \to \infty$, the extremum must be a maximum.

The curves $C$ of Figures 5.2 and 5.3 illustrate the actual [i.e., according to equation (5.18)] variations of the dimensionless pressure drop $\Delta p^*$ between the borders of the system as a function of the parameters $B_k^*$ and

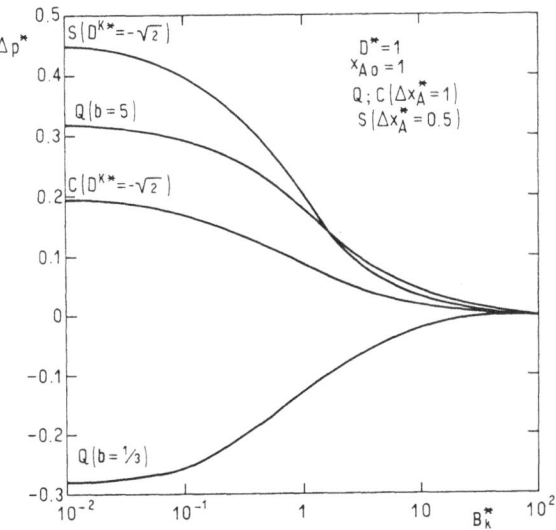

Figure 5.2. Variation of the dimensionless pressure drop $\Delta p^*$ between the borders of a system as a function of the dimensionless permeability $B_k^*$. Curve $C$ is for a closed system, curve $S$ for a semiopen system without reaction and the curves $Q$ are for semiopen systems with chemical reaction. Reprinted from Di Napoli, Williams, and Cunningham,[28] by permission.

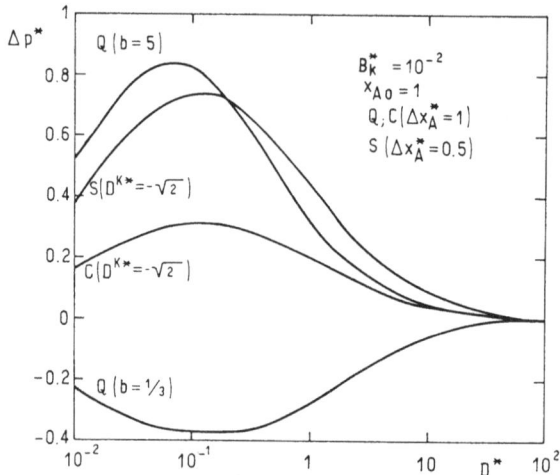

Figure 5.3. Variation of the dimensionless pressure drop $\Delta p^*$ between the borders of a system as a function of $D^*$ or, assuming all parameters except $D_{AA}^K$ are kept fixed, of the diffusion regime. $C$, $S$, and $Q$ are as in Figure 5.2. Reprinted from Di Napoli, Williams, and Cunningham,[28] by permission.

$D^*$, respectively. Figure 5.2 shows, as was to be expected from the mean-value approximation, that the system approaches the isobaric condition as the permeability of the porous medium increases. Figure 5.3 shows that there is a maximum when $\Delta p^*$ is plotted as a function of $D^*$, i.e., as a function of the diffusion regime in the system.[†] In the molecular diffusion regime, i.e., for large $D^*$, the pressure can be considered uniform, namely, equation (5.14) may be considered valid with $dc_A \simeq c\, dx_A$.

Mason et al. [78] calculated the relationship $\Delta p(p)$ by a step-by-step numerical solution of equation (5.7). The result, expressed as $\Delta p$ vs. $\bar{p}$, was compared with the experimental results of Evans et al. [31] for the interdiffusion of helium–argon in a low-permeability graphite, using a diffusion cell (see Section 5.3.4.1 below) at steady state and for various $\overline{\Delta p}$. Since under such conditions we usually have $N^* \neq 1$ (see Section 5.3.4) the experimental results were interpolated to $N^* = 1$. The calculation was based on independent measurements of the parameters. Thus, effective Knudsen diffusivities and permeabilities were obtained for the flow of pure argon with the aid of equation (4.45) and the effective molecular diffusivity

[†]Since the other parameters in equation (5.8) are kept constant, we may assume that only the factor $D_{AA}^K$ in $D^*$ is varied (i.e., the components A and B are varied in such a way that $M_{BA}$ and $D_{AB}^o$ are kept constant), and it will be recalled that $D_{AA}^K \gg D_{AB}$ corresponds to the molecular regime, while $D_{AA}^K \ll D_{AB}$ corresponds to the Knudsen regime (cf. Section 4.2.2.2).

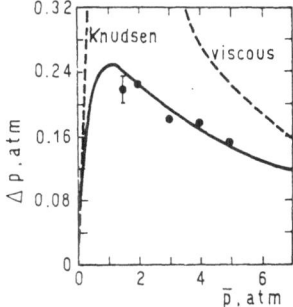

Figure 5.4. Variation of the pressure drop $\Delta p$ between the borders of a closed system as a function of the mean pressure $\bar{p}$: comparison between theory and experiment.[31] Reprinted from Mason, Malinauskas, and Evans,[78] by permission.

of argon in helium was obtained from measurements at uniform pressure. [30] Account was taken of the fact that the mean viscosity of the mixture was 1% higher than the viscosity of pure argon. The experimental and calculated results are compared in Figure 5.4.

For the same experiment, Mason and Marrero [79] calculated the flux of one component as a function of the pressure $\bar{p}$ by integrating the equation of change. Figure 5.5. shows a comparison of the calculated and experimental results. From the figure we see that the flux is directly proportional to pressure in the Knudsen regime and is independent of it in the molecular regime.

The excellent agreement between the calculated and experimental results shown by Figures 5.4 and 5.5 may be considered a verification of the equations of the dusty gas model.

We note that the two-bulb system can be made isobaric if as shown in Figure 5.6 (cf. System 2b in Table 1.1) we connect the bulbs by a second tube containing a frictionless piston which, by its movement, instantaneously dissipates any pressure gradient that builds up. If $M_A < M_B$, then when the bulbs are connected the piston will move in the direction indicated in Figure 5.6.

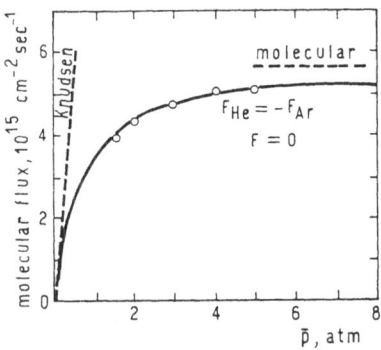

Figure 5.5. Molecular flux of one component in a closed system as a function of the pressure $\bar{p}$: comparison between theory and experiment. Reprinted from Mason and Marrero,[79] by permission.

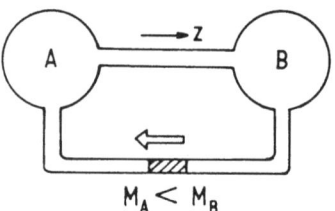

Figure 5.6. Isobaric closed system.

### 5.3.1.2. Multicomponent Mixtures

By reasoning similar to that for binary systems, we find that, for a multicomponent mixture that has reached the quasi-steady-state,

$$\sum_{i=1}^{v} N_{zi} = 0 \tag{5.29}$$

The validity of equation (5.29) makes it possible, as in the case of the binary system, to use total fluxes rather than diffusive fluxes in the corresponding constitutive equations. The closed system thus represents the system on which the classical development of diffusion is based.

Toor [118] analyzed a closed system of three components in the molecular diffusion regime assuming uniform pressure and assuming that the bulb volumes were sufficiently large so that a steady state was maintained in the capillary (of length $L$) that separates the bulbs. With these assumptions, the integration of equation (5.1) with equations (3.86) (see the appendix to this chapter) leads to

$$D_{AC}^{-1}(D_{AB}^{-1} - D_{BC}^{-1})N_{zA} + D_{BC}^{-1}(D_{AB}^{-1} - D_{AC}^{-1})N_{zB}$$
$$= \frac{p}{R_g TL}\left[(D_{AB}^{-1} - D_{BC}^{-1})(x_{A0} - x_{AL}) + (D_{AB}^{-1} - D_{AC}^{-1})(x_{B0} - x_{BL})\right] \tag{5.30}$$

$$(D_{AB}^{-1} - D_{BC}^{-1})N_{zA} + (D_{AB}^{-1} - D_{AC}^{-1})N_{zB}$$
$$= \frac{p}{R_g TL}\ln\left(\frac{x_{AL}/N_{zA} - x_{BL}/N_{zB} - W}{x_{A0}/N_{zA} - x_{B0}/N_{zB} - W}\right) \tag{5.31}$$

where

$$W \equiv \frac{D_{AC}^{-1} - D_{BC}^{-1}}{(D_{AB}^{-1} - D_{BC}^{-1})N_{zA} + (D_{AB}^{-1} - D_{AC}^{-1})N_{zB}} \tag{5.32}$$

The values of $N_{zA}$, $N_{zB}$, and $N_{zC}$ may be obtained from equations (5.29)–(5.31). Under certain conditions of composition, the flux of one component may be zero even though there may be a concentration gradient of that species (diffusive barrier); there also may be flux without a concentration gradient of the species (osmotic diffusion) and even against the concentration gradient (reverse diffusion).

We note that in the case of multicomponent mixtures, a closed system may give rise to instabilities if diffusion is along the vertical, i.e., if gravity plays a part. Imagine, for example, a ternary mixture of A, B, C in a two-bulb system with the capillary axis along the vertical and with $M_A < M_B < M_C$. Assume that gas A is originally in the upper bulb while B is in the lower one. Even if the composition of C is initially uniform throughout the system, this species will be carried up by the generated pressure gradient. If $x_A \gg x_B$ and $x_A \gg x_C$, a mass density gradient, opposite to the molar concentration gradient can be generated; this mass density gradient will give rise to a natural convection downward that will restore the uniform mass density. But the evolution of diffusion, now slower, will regenerate the mass density gradient, now smaller, and the phenomenon repeats itself until the system reaches equilibrium.

### 5.3.2. Semiopen Systems without Chemical Reaction

#### 5.3.2.1. Binary Systems

A semiopen binary system is a system in which one component (B) diffuses through another (A) at rest. The so-called Stefan method [63] illustrated in Figure 5.7 (and in Table 1.1, System 3) involves such a system and is used to measure the diffusivity of the vapor B in the gas A. The bottom of a tube is filled with liquid B; a gas A, which initially fills the tube, flows across the top of the tube. The liquid B evaporates and diffuses through A. The flow rate of A is generally high enough to keep the concentration of B in A in the tube mouth negligible, but low enough to prevent turbulence in the tube.

Under these conditions, a composition gradient develops along the tube in the gas phase (i.e., above $z = 0$); this implies that there must be a

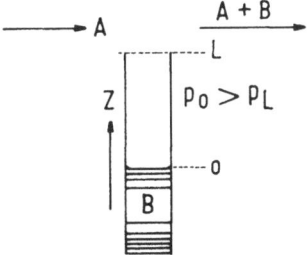

Figure 5.7. Stefan's system.

diffusive flux of A down toward the liquid and a diffusive flux of B up the tube. Nevertheless, if gas A is neither soluble in the liquid B (or if it saturates B at the initial time) nor reacts with B, then once the steady state has been reached, the total flux of A must be zero. For, in the steady state, $dN_{zA}/dz = 0$ at every point, and since $N_{zA} = 0$ at the liquid surface, we must have $N_{zA} = 0$ at every point. This means that the diffusive flux of A down the tube must be accompanied by a viscous flux of A of the same magnitude but opposite in direction so that the net flux of A is zero. A pressure gradient must therefore have developed along the tube.

Another type of semiopen systems is illustrated in Figure 5.8 (and in Table 1.1, System 5), where solid 1 decomposes according to the reaction

$$\text{solid } 1 = b B_{(g)} + \text{solid } 2 \qquad (5.33)$$

If solid 2 is porous and as it is generated is filled with the inert gas A, the gas B generated by the remainder of the reactant solid 1 will diffuse through the porous product layer of solid 2, i.e., through gas A, which is at rest in solid 2. A similar situation arises in the drying of porous materials, in which case B represents water vapor and A is usually air.

Systems in which the processes are the reverse of those described above are also examples of semiopen systems, for example, the consumption of a gas by a reactant solid (e.g., if, in Figure 5.8, solid 2 is surrounded by and filled with gases A and B, and B reacts with solid 1, then B will diffuse through A to solid 1) and the absorption of a gas by a liquid.

In all cases of semiopen systems, we have from equation (4.59), since $N_{zA} = 0$,

$$N_{zB} = \frac{J_{zBM}}{1 - x_B} \qquad (5.34)$$

Equation (5.34), like equation (5.13), is useful for the molecular diffusion regime, in which case $J_{zBM}$ can be expressed by Fick's law,

$$J_{zBM} = - D_{AB}(dc_B/dz) \qquad (5.35)$$

Equation (5.35) is usually valid in Stefan's method. By introducing

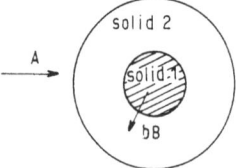

Figure 5.8. Decomposition of a solid with production of gas B.

equation (5.35) into equation (5.34), assuming $dc_B \simeq c\, dx_B$ (a hypothesis that we will consider later), using the equation of change (5.1) at steady state ($dc_B/dt = 0$, so $N_{zB} = $ constant), and integrating, we obtain

$$N_{zB} = \frac{D_{AB}c}{L} \ln \frac{p - p_{BL}}{p - p_{B0}} \tag{5.36}$$

where $p_{BL}$ is the partial pressure of B in the mouth of the tube ($p_{BL}$ approaches zero if the flow rate of gas A is sufficiently high), and $p_{B0}$ is the equilibrium vapor pressure at the gas–liquid interface [the validity of assuming equilibrium at the interface has been verified by Maa [75]]. Equation (5.36) is used to calculate the binary molecular diffusivity from experimental results obtained by Stefan's method. For a critical analysis of this method, see Lee and Wilke. [73]

Theoretical as well as experimental studies have shown the validity of assuming a one-directional flux in Stefan's system, [51, 89] with the exception of solvents of high volatility. To avoid having to correct for end effects [85] due to mixing in the tube mouth, a tube with a high length/radius ratio must be used.

When a semiopen system operates outside the molecular diffusion regime, which is usually the case in the decomposition or drying of solids, the analysis changes substantially: Fick's law, equation (5.35), is no longer valid. In such cases, the equation of change (5.1) has to be integrated using the constitutive equation (4.52), and it is then important to know the relationship $x_A^*(p^*)$. For systems in which $N^* = 0$, equation (5.8) can be solved analytically with the boundary condition $x_A^* = 1, p^* = 1$; we obtain

$$x_A^* = \frac{\exp(W_1 p^* - W_2^2 p^{*2})}{W_3 p^*[\operatorname{erf}(W_2 + W_4) - \operatorname{erf}(W_2 p^* + W_4)] + p^* \exp(W_1 - W_2^2)} \tag{5.37}$$

where

$$W_1 \equiv \frac{D^*}{D^{K^\bullet}} - \frac{B_k^*}{D^*}; \qquad W_2 \equiv \left(\frac{B_k^*}{2}\right)^{1/2};$$

$$W_4 \equiv 2^{-1/2}\left(\frac{B_k^{*1/2}}{D^*} - \frac{D^*}{B_k^{*1/2} D^{K^\bullet}}\right);$$

$$W_3 \equiv x_{A0}\frac{1 + D^{K^\bullet}}{D^{K^\bullet}}\left(\frac{B_k^* \pi}{2}\right)^{1/2} \exp\left[\frac{1}{2}\left(\frac{B_k^{*1/2}}{D^*} - \frac{D^*}{B_k^{*1/2} D^{K^\bullet}}\right)^2\right]$$

The curves $S$ of Figures 5.2 and 5.3 illustrate the variation of the dimensionless pressure drop $\Delta p^*$ between the borders of the system as a function of the parameters $B_k^*$ and $D^*$, respectively. We can observe the decrease of the pressure gradient as the permeability increases, the existence of a maximum in Figure 5.3 for $D^* \approx [B_k^*(-D^{K^*})]^{1/2}$, and the decrease of the pressure drop in the molecular diffusion regime. This permits the simplification $dc_B \simeq c\, dx_B$ in equation (5.35) even though in arriving at equation (5.34) we assumed that there was a pressure gradient that produced a viscous flux such that $N_{zA} = 0$.

Equation (5.37) can be used to facilitate the integration of equation (5.1) in the general case of semiopen systems (see Reference 42). The analysis is trivial for the Knudsen regime at steady state, for then $dc_A/dz = 0$, so that $N_{zA} = 0$, and the flux of B can be calculated from equation (5.9).

For the transition regime we may simplify the analysis as follows: Let us write the total flux of B as the sum of its viscous and diffusive contributions, with the dependence of each contribution on the concentration and pressure gradients explictly indicated:

$$N_{zB} = N_{zB}^D(\nabla x_B, \nabla p) + N_{zB}^V(\nabla p) \tag{5.38}$$

We also write

$$N_{zB}^V = x_B(N_z - N_z^D) = x_B(N_{zA} + N_{zB}) - x_B(N_{zA}^D + N_{zB}^D)$$

$$= x_B N_{zB}\left(1 + \frac{N_{zA}}{N_{zB}}\right) - x_B N_{zB}^D\left(1 + \frac{N_{zA}^D}{N_{zB}^D}\right) \tag{5.39}$$

By introducing equation (5.39) into (5.38) and rearranging, we obtain

$$N_{zB} = \frac{1 - (1 + N_{zA}^D/N_{zB}^D)x_B}{1 - (1 + N_{zA}/N_{zB})x_B}\, N_{zB}^D(\nabla x_B, \nabla p) \tag{5.40}$$

Equation (5.40), which is of general validity, becomes useful if the following three conditions hold: (1) $N_{zA}/N_{zB} = N^* = 0$, which is the case for semiopen systems; (2) $N_{zA}^D/N_{zB}^D$ can be expressed by Graham's law of diffusion, i.e., $N_{zAB}^D = D^{K^*}$; (3) the dependence of $N_{zB}^D$ on $\nabla p$ is much smaller than that on $\nabla x_B$. For these conditions, equation (5.40) becomes

$$N_{zB} = \frac{1 - (1 + D^{K^*})x_B}{1 - x_B}\, \frac{N_{zA}^D(\nabla x_A)}{D^{K^*}} \tag{5.41}$$

For the particular case of the molecular diffusion regime, the substitution of equation (4.58), written on a molar basis, into equation (5.41) leads to equation (5.34).

We now determine the error introduced when we apply equation (5.41) for the general semiopen system. We write equation (4.23) for a binary mixture, assume $\Delta'_{Ap} = \Delta'_{Bp}$ [as we did in arriving at equation (4.26)], and use equations (4.6), (4.8), (4.14), and (4.18); after some rearrangement we obtain

$$-\frac{N_{zB}^D}{N_{zA}^D} = \left(1 + \frac{D_{AA}^K}{N_{zA}^D}\frac{\nabla p}{R_g T}\right)\left(\frac{M_A}{M_B}\right)^{1/2} \tag{5.42}$$

The relative error $Er_G$ introduced by applying Graham's law of diffusion [hypothesis 2 used in arriving at equation (5.41)] is thus given by

$$Er_G = \frac{N_{zBA}^D - (N_{zBA}^D)_G}{(N_{zBA}^D)_G} = \frac{D_{AA}^K}{N_{zA}^D}\frac{\nabla p}{R_g T} \tag{5.43}$$

where $(N_{zBA}^D)_G = -(M_A/M_B)^{1/2}$ is the Graham's law ratio. By introducing equation (4.48) into equation (5.43) and writing the result in dimensionless form, we obtain

$$Er_G = \frac{D^* p^*\{[(1 + D^{K^*})/D^{K^*}]x_{A0}x_A^* - 1\} - 1}{x_{A0}p^*(dx_A^*/dp^*) + (1 - D^*p^*/D^{K^*})x_{A0}x_A^*} \tag{5.44}$$

The relative error in computing diffusive fluxes when we assume hypothesis 3, i.e., that the dependence of the diffusion on pressure is negligible, is obtained from equation (4.48),

$$Er_N = \frac{N_{zA}^D(\nabla x_A, \nabla p) - N_{zA}^D(\nabla x_A)}{N_{zA}^D(\nabla x_A)} = \left(1 + \frac{D_{BB}^K}{D_{AB}}\right)\frac{d\ln p}{d\ln x_A} \tag{5.45}$$

or, in dimensionless form,

$$Er_N = \left(1 - \frac{D^*p^*}{D^{K^*}}\right)\frac{d\ln p^*}{d\ln x_A^*} \tag{5.46}$$

The errors $Er_G$ or $Er_N$ can be evaluated with the aid of equation (5.37) [or equation (5.8)]. [28] It can be shown that the errors $Er_G$ and $Er_N$ are directly proportional to $D^*$ and inversely proportional to $B_k^*$, and depend exclusively on the ratio $B_k^*/D^*$. As a result, in any given case there will be a minimum value of $B_k^*/D^*$, call it $(B_k^*/D^*)_{min}$, such that for any value $B_k^*/D^* > (B_k^*/D^*)_{min}$ the error will be smaller than a predetermined value. The analytic expressions of these errors for the most unfavorable conditions are

$$Er_G = 1.30D^*/B_k^*; \qquad |Er_N| = (1 + B_k^*/D^*)^{-1} \qquad (5.47)$$

Note that, while $Er_G$ may have any value, the maximum for $|Er_N|$ is 1.

Equation (5.47) together with an estimation of $B_k^*/D^*$ can be used to determine the applicability of equation (5.41) for a given error.

The use of simplified equations such as equation (5.41) is not restricted to semiopen systems, but may be used in any system for which $N^* = \text{const.}$[28]

### 5.3.2.2. Multicomponent Systems.

Carty and Schrodt [17] studied the ternary diffusion of cocurrent methanol and acetone in stagnant air in a Stefan tube. By integrating equation (5.1) with the Stefan–Maxwell equations (see the appendix to this chapter), and letting $N_{zA} = 0$, we obtain

$$x_A = x_{A0} \exp\left[\left(\frac{z - z_0}{c}\right)\left(\frac{N_{zB}}{D_{AB}} + \frac{N_{zC}}{D_{AC}}\right)\right]$$

$$x_B = \frac{N_{zB}}{N_{zB} + N_{zC}} - \frac{x_{A0}N_{zB}D_{ABC}\exp\left[(z - z_0)/c\right](N_{zB}/D_{AB} + N_{zC}/D_{AC})}{N_{zB}D_{ABC} + N_{zC}D_{ACB}}$$

$$+ \left[x_{B0} + \frac{x_{A0}N_{zB}D_{ABC}}{N_{zB}D_{ABC} + N_{zC}D_{ACB}} - \frac{N_{zB}}{N_{zB} + N_{zC}}\right]$$

$$\times \exp\left(\frac{z - z_0}{c}\right)\left(\frac{N_{zB} + N_{zC}}{D_{BC}}\right) \qquad (5.48)$$

where $D_{ABC} \equiv (D_{AB} - D_{BC})/D_{AB}$, $D_{ACB} \equiv (D_{AC} - D_{BC})/D_{AC}$. These equations have been experimentally verified.

### 5.3.3. Semiopen Systems with Chemical Reaction

### 5.3.3.1. Binary Systems.

A typical example of a semiopen binary system with chemical reaction is the chemical reaction of A catalyzed by a porous solid,

$$A = bB, \qquad (5.49)$$

which, depending on the value of $b$, may represent a cracking $(b > 1)$, an isomerization $(b = 1)$, or a polymerization $(b < 1)$. In any case, the

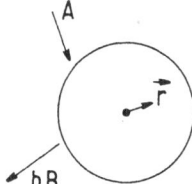

Figure 5.9. Chemical reaction catalyzed by a porous solid.

stoichiometric coefficient $b$ is related to the molecular masses of both reaction components,

$$b = (-D^{K^{\bullet}})^{-2} \qquad (5.50)$$

When the system reaches the steady state, the ratio of the fluxes of both species is given by (see Figure 5.9),

$$-N_{rA}/N_{rB} = b^{-1} \qquad (5.51)$$

To satisfy the stoichiometric requirement at steady state, the system develops a pressure gradient if $b \neq 1$ and the system is isothermal.[†]

If $b < 1$, the pressure decreases in the same direction as the diffusive flux of A; if $b > 1$ the opposite situation prevails. Here, in contrast to the situation in a closed system, the diffusive flux of the lighter component has the same direction as its viscous flux. This is perhaps why there have been some confused attempts to consider this system as isobaric but having a volume change, i.e., as a system in which there is no coupling of the viscous and diffusive fluxes (see Section 6.2.2).

For the Knudsen regime, the stoichiometric requirement of equation (5.51) is also satisfied at the steady state. If now $b = 1$, then according to equation (5.50), $D_{AA}^{K} = D_{BB}^{K}$, so the system is isobaric. When $b > 1$, $D_{AA}^{K} < D_{BB}^{K}$, and the pressure diminishes in the same direction as B diffuses, with $|\Delta c_{B}| > |\Delta c_{A}|$. The reverse occurs when $b < 1$.

For the general case of this system, equation (5.8) has no analytical solution. However, when $b = 1$, we see that $D^{K^{\bullet}} = -1$ and $N^{*} = 1$ and (see the last paragraph of Section 5.2.1) $\nabla p = 0$. Thus, in a catalyzed reaction such as equation (5.49), equimolarity is synonymous with uniform

[†]If the system is not isothermal, the temperature gradients generate pressure gradients, even for the case $b = 1$. However, as pointed out by Carberry, [16] external temperature gradients can be far more important than internal gradients. Thus, the catalyst pellet may be regarded to be at the uniform temperature imposed by external heat transport.

pressure. In a noncatalyzed gas–solid reaction of the type

$$A + \text{solid } 1 = bB + \text{solid } 2 \tag{5.52}$$

(which represents many reduction and oxidation reactions), there is also a binary gas mixture, but the value of $b$ does not determine the value of $D^{K^*}$ as it does for the reaction (5.49). In other words, even though the ratio of the fluxes in the steady state is still calculated from equation (5.51), the stoichiometric coefficient is not related in a simple way to the molecular masses of the gas species. The system described by equation (5.52) is thus generally nonisobaric, even for $b = 1$. For $b = 1$, stoichiometry requires equal and opposite fluxes but, since the molecular masses are different, a pressure gradient must develop to satisfy that requirement. In this case, as in a closed system, a pressure drop develops in the direction of the flow of the heavier molecules.

For the particular case $b = (M_A/M_B)^{1/2}$, we have $N^* = -D^{K^*}$ and the system will be isobaric even though $b \neq 1$. If $b > (M_A/M_B)^{1/2}$, the pressure will decrease toward the outer surface of the solid; the reverse will be true for $b < (M_A/M_B)^{1/2}$.

We will now analyze the catalyzed reaction (5.49) and then the noncatalyzed reaction (5.52).

5.3.3.1a.   *The Catalyzed Reaction.* For the system described by equation (5.49), equations (4.59) and (5.51) lead to

$$N_{rA} = \frac{J_{rAM}}{1 - (1 - b)x_A} \tag{5.53}$$

As for equations (5.13) and (5.34), in the molecular diffusion regime, when $J_{rAM}$ is expressed by Fick's law, there is a simple solution of equation (5.53). For this case, the integration of the equation of change (5.1) with the constitutive equation (5.53) has been carried out by Weekman and Gorring [126] and Lin and Lih.[74] In Section 5.3.3.2 we will analyze the extension of these results to a multicomponent mixture.

Unfortunately, the molecular diffusion regime is not usually encountered in a catalyst, so we must solve equation (5.1) with the aid of the constitutive equation (5.53) in the transition regime. The solution of the problem may be separated into two parts: the determination of the pressure drop between the center and the border of the catalyst, and the determination of the effectiveness of the catalyst.

The first part of the problem, the pressure drop, can be obtained by numerical solution of equation (5.8) applied to our case. The curves $Q$ of Figures 5.2 and 5.3 show the results. As in the previously analyzed cases, the system approaches isobaric behavior as the permeability of the catalyst increases or as the system approaches the molecular diffusion regime. This justifies using the simplification in the molecular diffusion regime of assuming uniform pressure and applying Fick's law together with $dc_A \simeq c \, dx_A$. The curves of Figure 5.3 also show an extremum (maximum if $b > 1$, minimum if $b < 1$), which occurs for $[D^*] \approx [(B_k^*/b^{1/2})^{1/2}]$.

For the second part of the problem we begin by defining the effectiveness factor $\eta$ for the catalyst pellet,

$$\eta \equiv \frac{R_{vA\,Obs}}{R_{vA}^{\circ}} \qquad (5.54)$$

where $R_{vA\,Obs}$ is the macroscopically observed reaction rate of A per unit volume of the pellet and the superscript $^{\circ}$ indicates a suitably selected reference state.

We will consider the effectiveness $\eta$ for a reaction of the type given by equation (5.49) with a kinetics of the form

$$R_{vA} = k_V p_A^m \qquad (5.55)$$

where $k_V$ is the reaction rate coefficient, $p_A$ is the partial pressure of A, and $m$ is the reaction order. The solution can be obtained by solving equation (5.1) for the steady state with equation (4.52) and using equation (5.8) with $N^* = b^{-1}$. If the result is expressed in terms of the molar flux of A at the outer surface of the catalyst, $N_{rA}|_s$, then the effectiveness factor can be calculated from

$$R_{vA\,Obs} = (-N_{rA}|_s)S/V \qquad (5.56)$$

$$R_{vA}^{\circ} = R_v|_s \qquad (5.57)$$

where $V$ is the volume of the catalyst in which the reaction is taking place, $S$ is the external area of the catalyst, and the subscript s indicates conditions on the external surface.

Apecetche et al.[4] calculated the effectiveness factor for this problem for a spherical geometry, a viscosity that varies with composition, and over the following ranges of the parameters involved: $D^*$, $10^{-2}$ to $10^2$; $B_k^*$, $10^{-2}$ to $10^2$; $b$, $10^{-1}$ to $10$; $x_{A0}$, 0.1 to 0.9; $m$, 1.0 and 0.5. For the sake of comparison, three effectiveness factors were defined:

(1) The nonisobaric effectiveness factor, denoted by $\eta_I$ and identical to the $\eta$ defined by equation (5.54).

(2) A "mathematically isobaric" effectiveness factor, $\eta_{II}$, based on the assumption that, in the constitutive equation for the total flux, $\nabla p = 0$ with $b \neq 1$.

This assumption is, of course, physically inconsistent and leads to two possibilities for generating simplified equations. If we write equation (4.52) for $N_{rA}$ and $N_{rB}$, using $dp/dr = 0$, and then form the ratio, we obtain

$$N_{rA}/N_{rB} = -D_{AA}^K/D_{BB}^K \tag{5.58}$$

But from equation (5.50) we have

$$D_{AA}^K/D_{BB}^K = 1/b^{1/2} \tag{5.59}$$

while equation (5.51) states that

$$N_{rA}/N_{rB} = -1/b \tag{5.51}$$

Thus, equation (5.58) has physical meaning only for $b = 1$, which is precisely the necessary condition for the system to remain isobaric. If, however, we go on and neglect the physical meaning of terms, we can substitute equation (5.59) into (4.52) written for $dp/dr = 0$ and obtain

$$N_{rA} = -\left\{\left[1 - (1 - b^{1/2})x_A\right]p + \frac{D_{AB}^\circ}{D_{AA}^K}\right\}^{-1} D_{AB}^\circ \frac{p}{R_g T} \frac{dx_A}{dr} \tag{5.60}$$

This equation, which is inconsistent with equation (5.51) for $b \neq 1$, is the one used by Apecetche *et al.* [4] to generate a "mathematically isobaric" $\eta_{II}$.

If, on the other hand, we use equations (5.58) and (5.51), following the same steps as before, we find

$$N_{rA} = -\left\{[1 - (1 - b)x_A]p + \frac{D_{AB}^\circ}{D_{AA}^K}\right\}^{-1} D_{AB}^\circ \frac{p}{R_g T} \frac{dx_A}{dr} \tag{5.61}$$

This equation, which is inconsistent with equation (5.59) for $b \neq 1$, has been used by Kehoe and Aris[65] to solve the same problem (see later in this section).

(3) *The isobaric effectiveness factor,* $\eta_{III}$, known in related literature as the no-volume-change effectiveness factor; $\eta_{III}$ is calculated by assuming that $\nabla p = 0$ and $b = 1$. These effectiveness factors were calculated as a function of a Thiele modulus, $h^*$, which arises from equation (5.1) and is given by

$$h^* \equiv \frac{R}{3}\left(\frac{m+1}{2} \frac{k_V R_g T p_{As}^{m-1}}{D_{AB}^R|_s}\right)^{1/2} \tag{5.62}$$

where

$$\frac{1}{D_{AB}^R|_s} \equiv \frac{p_s}{D_{AB}^\circ} + \frac{1}{D_{AA}^K} \tag{5.63}$$

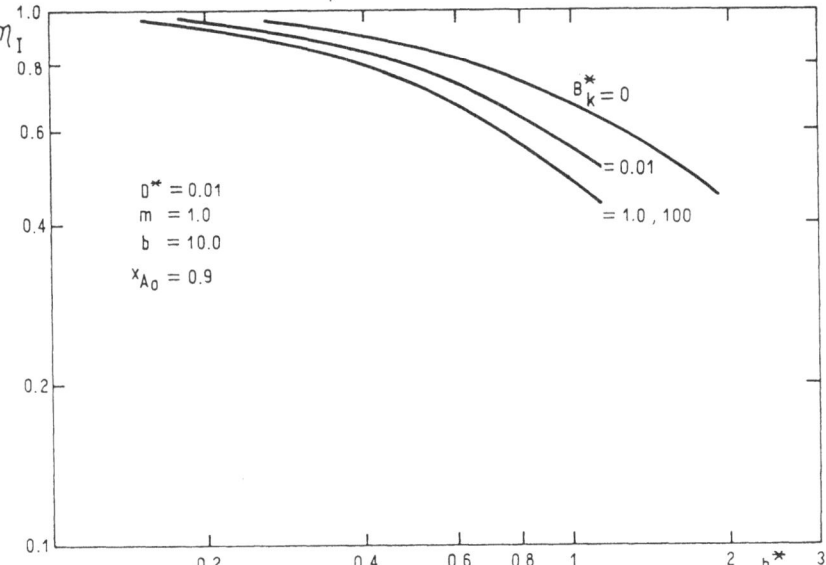

Figure 5.10. Effectiveness factors as a function of a Thiele modulus $h^*$ for a diffusion regime near the Knudsen regime. For $B_k^* = 0$ the three effectiveness factors are the same, $\eta_I = \eta_{II} = \eta_{III}$. Reprinted from Apecetche, González, Williams, and Cunningham,[4] by permission.

and $D_{AB}^{\circ}$ is given by $D_{AB}|_s p_s$. The Thiele modulus so-defined has the property that $\eta_{III}(h^*)$ is independent of the prevailing diffusion regime as specified by $D^*$.

The results of the calculation of the effectiveness factors, typical ones of which are shown in Figures 5.10 to 5.12, lead to the following conclusions:

(i) The nonisobaric effectiveness factor is significantly different from the isobaric effectiveness factor $\eta_{III}$ for any diffusion regime.

(ii) If $b < 1, \eta_{III} < \eta_I$; if $b > 1, \eta_{III} > \eta_I$; these correspond to the direction of the pressure gradient that develops in each case.

(iii) The difference between the effectiveness factors $\eta_I$ and $\eta_{III}$ is significant even for values of $b = 2$ and $b = 1/2$.

(iv) The effectiveness factor $\eta_{II}$ does not represent a simplification of the problem for any case; in the pure Knudsen regime, however, we have $\eta_I = \eta_{II} = \eta_{III}$.

(v) The dependence of $\eta_I$ on $B_k^*$ is negligible in the transition and molecular regimes (from Figure 5.11 we see that $\eta_I$ is independent of $B_k^*$ for $0.01 < B_k^* < 100$); in the Knudsen regime only values for $B_k^* \to 0$ have physical meaning.

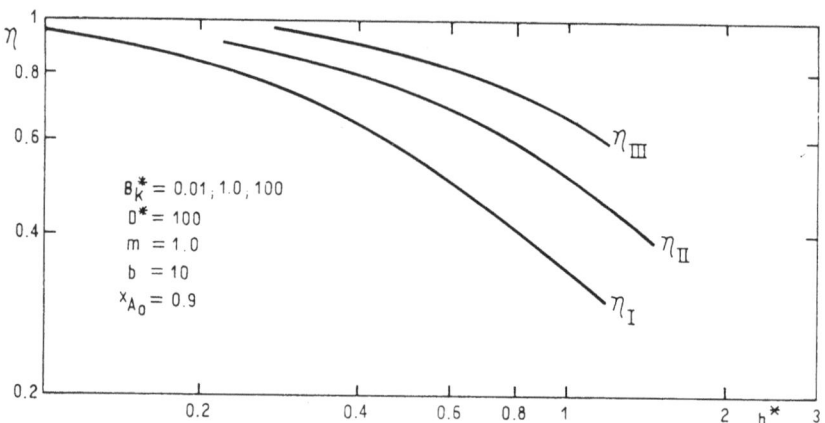

Figure 5.11. Effectiveness factors as a function of a Thiele modulus $h^*$ for a diffusion regime near the molecular regime. Reprinted from Apecetche, González, Williams, and Cunningham,[4] by permission.

We can understand the extremely slight dependence of $\eta_I$ on $B_k^*$ from Figure 5.13, which shows the effect of $B_k^*$ on the ratio of fluxes produced by pressure gradients to fluxes produced by molar fraction gradients [the latter fluxes are, of course, independent of $B_k^*$; see equation (4.48)]. We observe that, as $B_k^*$ increases, the viscous flux of A increases in the same proportion as the pressure diffusion flux decreases. In other words, the sum of the fluxes due to pressure gradients, $N_{rA}^V + N_{rA}^D(\nabla p)$, is independent of $B_k^*$.

Two other important conclusions, one physical and the other mathematical, may be drawn from Figure 5.13. First, in the transition diffusion regime and for the particular values for which Figure 5.13 was calculated, 37.5% (0.6/1.6) of the total flux of component A is pressure-gradient-induced and this percentage is independent of the value of this gradient

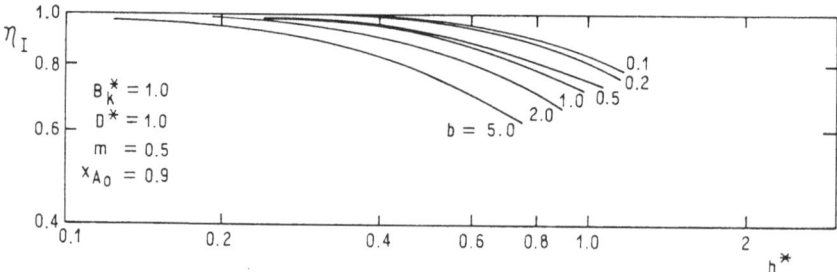

Figure 5.12. Influence of the stoichiometry of a chemical reaction on the nonisobaric effectiveness factor $\eta_I$. The numbers on the curves give the stoichiometric factor $b$. Reprinted from Apecetche, González, Williams, and Cunningham,[4] by permission.

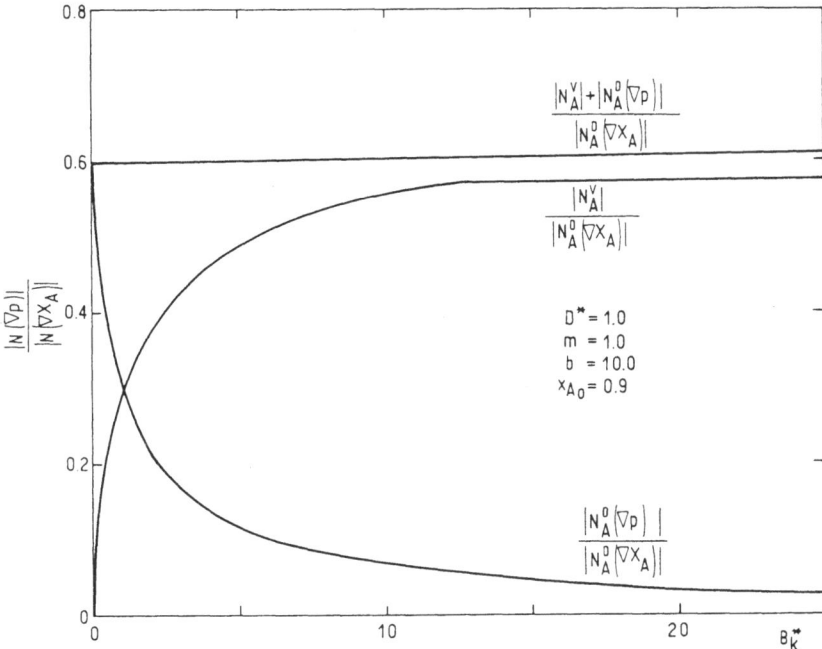

Figure 5.13. Influence of the permeability $B_k^*$ on flux ratios in a catalyst pellet. Reprinted from Apecetche, González, Williams, and Cunningham,[4] by permission.

(which depends on $B_k^*$). This independence has led to a commonly encountered erroneous statement in the literature that pressure gradients do not alter the effectiveness of the system; pressure gradients are in fact necessary so that fluxes can satisfy the stoichiometric requirement equation (5.51). The second conclusion is that, because the pressure-gradient-induced flux is independent of $B_k^*$ [recall $N_A^D(\nabla x_A)$ is also independent of $B_k^*$], the total flux $N_A = N_A^V + N_A^D(\nabla x_A) + N_A^D(\nabla p)$, and hence the effectiveness $\eta$, can be calculated assuming $B_k^* = 0$ (i.e., neglecting the viscous flux) or by any other convenient value.[†] This result does not mean that the viscous flux is in reality negligible, but only that the viscous flux is replaced mathematically by a virtual pressure diffusive flux.

Kehoe and Aris [65] have solved this problem for plane geometry and for $D^* = 10^{-2}, 1, 10^2$; $B_k^*(-D^{K^*})/D^* = 0, 10^3$; $b = 0.25, 4$: $x_{A0} = 1$; $m = 0, 1, 2$. Solutions were also obtained for plane and cylindrical geometries[2] for the following ranges of the parameters: $D^*, 10^{-8}$ to $10^8$;

[†]Recently, Hite and Jackson [56] have derived equations valid in the limit $B_k^* \to \infty$ and have shown that they correctly predict the effectiveness value.

$B_k^*(-D^{K^*})/D^{*2}$, $10^{-2}$ to $10^2$; $b$, 1 to 4; $x_{A0} = 1$; $m = 0$. These results were extended to an arbitrary reaction order $m$.[1] The results obtained by Kehoe and Aris [65] show that (1) $\eta_I \neq \eta_{III}$ for every diffusion regime; (2) the nonisobaric effectiveness factor is only slightly dependent on the group $B_k^*(-D^{K^*})/D^*$; (3) a mathematically isobaric effectiveness factor $\eta'_{II}$ may be defined by letting $\nabla p = 0$, $b \neq 1$, and using equation (5.61).

Results show that $\eta'_{II} \simeq \eta_I$ for any diffusion regime when both effectiveness factors are represented as a function of a Thiele modulus defined by

$$h^* \equiv L \left[ \frac{k_V R_g T p_{As}^{m-1}}{D_{AB|s}} \frac{b^{1/2}(1 + b^{1/2}D^*)}{D^*} \right]^{1/2} \tag{5.64}$$

It is to be noted that equation (5.61) reduces to equation (5.53) in the molecular diffusion regime, taking $dc_A \approx c\, dx_A$. The inconsistency in the value of the ratio of the Knudsen coefficients [equations (5.51) and (5.59)] is not relevant in the molecular diffusion regime, and it is clear that $\eta'_{II}$ must be equal to $\eta_I$ in this particular regime. However, both $\eta_I$ and $\eta'_{II}$ have similar values in every diffusion regime, and this provides a sound argument for the above simplification of the problem.[†]

It should be pointed out that Jackson [59] has shown that, for the molecular diffusion regime and high permeabilities, the equation to be solved is integrodifferential. Only for $b = 1$ or for highly symmetric pellets and for the steady state does the equation reduce to the conventional one in which the flux is given by equation (5.53). This fact is very important for the study of the stability, number, and symmetry of steady states in porous solids of high permeability.

5.3.3.1b.   *Noncatalyzed Gas–Solid Reaction.* An interesting application of a gas–solid reaction described by equation (5.52) is the conservation of fruits in controlled atmospheres.

By its metabolism, a fruit in effect absorbs oxygen (gas A) and liberates carbon dioxide (gas B). The stoichiometric coefficient $b$ depends on the species and variety of the fruit and on the temperature; $b$ ranges between 1 and 1.5. The process usually takes place in the presence of nitrogen (gas C), so the gaseous system is ternary.

This ternary system has the property that, if the atmosphere surrounding the fruit is kept at $p_A = 0.03$ atm and $p_B = 0.05$ atm, the fruit can be stored for a long time without decomposing. It is interesting to ask if there is a jacket with transport properties that permit maintenance of the inner

---

[†]See the discussion provided by Hite and Jackson. [56]

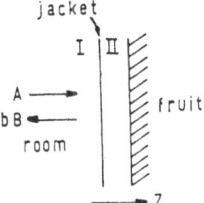

Figure 5.14. Schematic representation of a fruit-conservation device.

atmosphere at the desired composition while the outer composition corresponds to room atmosphere ($p_A = 0.21$ atm, $p_B \simeq 0$); the flux of A through the jacket would have to just compensate for that consumed reaction.

Assume that such a jacket has been found and that we surround the fruit with it, leaving an annular region (II in Figure 5.14) that initially has the same gas composition as the room; the fruit consumes the oxygen in II and produces carbon dioxide. If the consumption is faster than its supply through the jacket, $p_{AII}$ will decrease continuously while $p_{BII}$ will simultaneously increase until the required values are reached ($p_{AII} = 0.03$ atm, $p_{BII} = 0.05$ atm). This process may generate a pressure gradient with a simultaneous flux of the inert gas C (even at uniform pressure there may be osmotic diffusion of C). Although the general case may be analyzed numerically, for the sake of simplicity we will consider the Knudsen regime, where the flux of each component is independent of the presence of the other ones.

Thus, at the steady state we have

$$N_{zC} = 0 \tag{5.65}$$

and, from equations (5.51) and (4.56),

$$-\frac{N_{zB}}{N_{zA}} = b = \frac{D_{BB}^{K}}{D_{AA}^{K}} \frac{p_{BII} - p_{BI}}{p_{AI} - p_{AII}} \tag{5.66}$$

Since $D_{BB}^{K}/D_{AA}^{K} = -1/D^{K*} = (M_A/M_B)^{1/2} = 1.17$ and $1 \leqslant b \leqslant 1.5$, we have

$$1.17 \leqslant \frac{p_{BII} - p_{BI}}{p_{AI} - p_{AII}} \leqslant 1.76 \tag{5.67}$$

This result is independent of the type of porous medium as long as it ensures Knudsen diffusion for both components. Note that for $b = 1$, the system must be nonisobaric ($p_{II} > p_{I}$) in order to satisfy the stoichiometric requirement [see the discussion following equation (5.52)].

If in equation (5.67) we use the values of $p_{AII}$ and $p_{BII}$ required for the conservation of the fruit, we obtain

$$\frac{p_{BII} - p_{BI}}{p_{AI} - p_{AII}} = 0.278 \qquad (5.68)$$

In the Knudsen regime, then, the desired composition is incompatible with the stoichiometry of the process. We could have foreseen this by noting that the driving force (partial pressure gradient) on B (heavier component) is smaller than that on A (lighter component) but the flux of B was to be higher than that of A.

The solution of the problem, however, may lie in finding membranes for which other factors than the mass of the molecules affect their transport, i.e., membranes for which $D_{BB}^K / D_{AA}^K \neq (-1/D^{K^*})$. For example, we can assume that we have a jacket operating in the configurational diffusion regime (see Sections 4.3.1.1 and 5.4.1). If we assume that configurational diffusion is described by a Fick's-type law [or a Knudsen-type law such as equation (4.56)],

$$N_{zi} = -D_{ii}^C \nabla p_i$$

where $D_{ii}^C$ is the configurational diffusivity, then using equation (5.51) we find that the new requirement is given by

$$\frac{D_{BB}^C}{D_{AA}^C} = \frac{b}{0.278} = 3.6 \text{ to } 5.4 \qquad (5.69)$$

For a membrane satisfying equation (5.69) the flux of carbon dioxide relative to that of oxygen is considerably enhanced compared with an "ordinary" membrane.

Marcellin [76] has found a kind of polyethylene for which $D_{BB}^C / D_{AA}^C = 4.3$, a value that is independent of the temperature. The rate of consumption of oxygen by the fruit requires that the ratio of the area of the polyethylene film to its thickness be $2.4 \times 10^5$ cm. The thickness of the commercial form of this kind of polymeric material is $5 \times 10^{-3}$ cm, so an area of 1200 cm$^2$ is needed to fulfill the process requirements; fortunately, this is a practical value.

### 5.3.3.2.  Multicomponent Mixtures

Consider a catalyst pellet in which an isothermal multicomponent mixture reacts according to equation (5.2). The gas phase within the pellet will be isobaric if

$$\sum_{j=1}^{\nu} N_j^D M_j^{1/2} = 0 \tag{5.70}$$

which is equation (4.25) for $\Delta'_{ip} = 0$, using hypothesis 1 of Section 5.2.1. However, since the system is isobaric, $N_j^D = N_j$. For a steady state, the ratio of the fluxes for one-dimensional flow is given by

$$N_j/N_i = a_j/a_i \tag{5.71}$$

From equations (5.70) and (5.71) we find

$$\frac{N_i}{a_i} \sum_{j=1}^{\nu} a_j M_j^{1/2} = 0 \tag{5.72}$$

so that a more general condition for uniform pressure within the pellet is

$$\sum_{j=1}^{\nu} a_j M_j^{1/2} = 0 \tag{5.73}$$

This implies that a reaction in which $\sum_{j=1}^{\nu} a_j = 0$ (i.e., a reaction in which there is no change in the number of moles) will lead, in general, to pressure variations within the pellet. The viscous fluxes that develop must be of such magnitude that the stoichiometric condition, equation (5.71), is satisfied.

We now give an example of the calculation of the effectiveness factor for a multicomponent system in the molecular diffusion regime. This analysis, which may be considered as an extension of that of References 74 and 126 for binary mixtures, was given by Gros and Bugarel.[41]† The constitutive equation of diffusion for a multicomponent mixture is written in a form like that for a binary mixture,

$$\mathbf{N}_i = -D_{i,m} \nabla c_i + x_i \sum_{j=1}^{\nu} \mathbf{N}_j \tag{5.74}$$

where $D_{i,m}$ is a binary-type molecular diffusivity for a multicomponent

---

† Recently, Hesse[53] found a solution for a ternary mixture by making use of Stefan–Maxwell equations (see the appendix to this chapter).

mixture, and it is a function of composition.[†] Gros and Bugarel now assume, following a suggestion by Bird *et al.*,[(10)] that $D_{i,\,m}$ is constant; this considerably simplifies the problem.

For a radial flux in a catalyst pellet, equation (5.74) can be written as

$$N_{ri} = -D_{i,\,m}\frac{dc_i}{dr} + x_i N_{ri}\left[1 + \sum_{\substack{j=1 \\ j \neq i}}^{\nu} \frac{N_{rj}}{N_{ri}}\right] \qquad (5.75)$$

Using the reaction stoichiometry as given by equation (5.2), and assuming a steady state, we have

$$N_{rj}/N_{ri} = a_j/a_i \qquad (5.76)$$

From equation (5.75) and (5.76) we obtain

$$N_{ri} = -D_{i,\,m}\frac{dc_i}{dr}\bigg/\left[1 - \left[1 + \sum_{\substack{j=1 \\ j \neq i}}^{\nu} \frac{a_j}{a_i}\right]x_i\right] \qquad (5.77)$$

which is a generalization of equation (5.53) for multicomponent mixtures. When the reaction does not involve a change in the total number of moles, we have

$$1 + \sum_{\substack{j=1 \\ j \neq i}}^{\nu} (a_j/a_i) = 0$$

and the total flux is given by Fick's law.

We now substitute equation (5.77) into equation (5.1) written for spherical symmetry and the steady state, and then introduce the new

---

[†]The relationship between $D_{i,\,m}$ and $D_{ij}$ can be obtained from equation (4.12) written on a molecular basis for an isothermal molecular diffusion regime ($\nabla T = 0$, $D_{ij} \ll D_{ii}^K$) and from equation (5.74):

$$D_{i,\,m} = \left(N_i - x_i\sum_{j=1}^{\nu} N_j\right)\bigg/\sum_{\substack{j=1 \\ j \neq i}}^{\nu} \frac{N_i x_j - N_j x_i}{D_{ij}} \qquad (5.74a)$$

where we have used total fluxes rather than total diffusive fluxes; for the justification of this, see the appendix to this chapter. For a discussion of the assumption of constant multicomponent diffusion coefficients, see Reference 61.

variable $c_i^*$ and parameter $W_i$ by

$$c_i^* \equiv \frac{c_{is} - c_i}{c_{is}} \; ; \qquad W_i \equiv x_{is}\left[1 + \sum_{\substack{j=1 \\ j \neq i}}^{\nu} \frac{a_j}{a_i}\right] \qquad (5.78)$$

where $c_{is}$ is the value of $c_i$ at the surface of the spherical pellet, i.e., at $r = R$, and $x_{is} = c_{is}/c$. The following results:

$$r^{-2}\frac{d}{dr}\left(r^2\frac{dc_i^*}{dr}\right) - \frac{W_i}{1 - W_i(1 - c_i^*)}\left(\frac{dc_i^*}{dr}\right)^2$$

$$+ \frac{\rho_p\left[1 - W_i(1 - c_i^*)\right]}{D_{i,m}c_{is}}\hat{R}_i(c_1^*, \ldots, c_i^*, \ldots, c_\nu^*) = 0 \quad (5.79)$$

where $\rho_p$ is the mass density of the catalyst pellet and $\hat{R}_i = \rho_p R_{vi}$ is the molar reaction rate of species $i$ per unit mass of catalyst. The statement of the complete problem requires $\nu$ equations like equation (5.79) since they are coupled. These equations may be uncoupled by the following transformation:

$$X_i \equiv \frac{W_i}{1 - W_i(1 - c_i^*)}\frac{dc_i^*}{dr} \qquad (5.80)$$

Equation (5.79) then becomes

$$\frac{dX_i}{dr} + \frac{2}{r}X_i + \frac{\rho_p W_i}{D_{i,m}c_{is}}\hat{R}_i(c_1^*, \ldots, c_i^*, \ldots, c_\nu^*) = 0 \qquad (5.81)$$

Since, from equation (5.78),

$$W_i = (x_{is}/a_i)\sum_{j=1}^{\nu} a_j = (c_{is}/ca_i)\sum_{j=1}^{\nu} a_j$$

it follows from equation (5.81) that

$$\frac{D_{i,m}}{a_i}\left(\frac{dX_i}{dr} + \frac{2}{r}X_i\right) = \frac{-\left(\rho_p \sum_{j=1}^{\nu} a_j\right)\hat{R}_i(c_1^*, \ldots, c_\nu^*)}{ca_i} \qquad (5.82)$$

Since $\hat{R}_i/a_i$ is a constant for a reaction of the type given by equation (5.2), the right-hand side of equation (5.82) is a constant and independent of $i$.

Calling the constant $C$, we can write equation (5.82) as

$$\frac{D_{i,\mathrm{m}}}{a_i}\frac{d}{dr}r^2X_i = r^2C,$$

so that $(D_{i,\mathrm{m}}/a_i)r^2X_i = r^3C/3 + C'$. Using now equation (5.80), we can integrate again and find

$$\frac{D_{i,\mathrm{m}}}{a_i}\ln[1 - W_i(1 - c_i^*)]\Big|_r^R = \frac{Cr^2}{6} - \frac{C'}{r}\Big|_r^R$$

Using the boundary conditions

$$X_i = X_j = 0 \quad \text{at } r = 0$$
$$c_i^* = c_j^* = 0 \quad \text{at } r = R \tag{5.83}$$

we find $C' = 0$ and, since $C$ is independent of $i$,

$$\frac{D_{i,\mathrm{m}}}{a_j}\ln\frac{1 - W_i(1 - c_i^*)}{1 - W_i} = \frac{D_{j,\mathrm{m}}}{a_j}\ln\frac{1 - W_j(1 - c_j^*)}{1 - W_j} \tag{5.84}$$

Equation (5.84) permits the calculation of the concentrations $c_j^*$ for $j \neq i$ as a function of $c_i^*$, thus uncoupling the $v$ equations (5.79). The problem is therefore reduced to the solution of a system of two nonlinear first-order differential equations—equations (5.80) and (5.81)—with the boundary conditions (5.83).

The results obtained by numerical integration can be expressed in terms of the effectiveness factor. For the experimental data for the reaction

$$SO_2 + \tfrac{1}{2}O_2 = SO_3 \tag{5.85}$$

the influence of a change in the number of moles on the effectiveness factor is relatively small, so the maximum deviation of the effectiveness factor calculated with Fick's law is 10% less than the correct one.[41]

### 5.3.3.3. Reaction Rate and Diffusivity.

It is already known[25, 64, 87, 93, 104, 105, 109, 131] that, in fast chemical reactions, perturbation of the Maxwellian velocity distribution by the reaction can give rise to nonequilibrium corrections to the theoretical rate coefficient. Chemical reaction can affect other transport coefficients, and these effects are significant under the same conditions for which nonequilibrium corrections to the chemical rate coefficient are important.[47, 106]

Simons[106] studied an irreversible bimolecular gas-phase reaction of the type $A + B = C + C'$ in which each species possesses a single internal quantum state. The use of a modified BGK distribution function leads to

$$D_{AB} = (3\nu_{AB})^{-1}\overline{v_A^2} - \left(\overline{v_A^2 \Delta\nu_{AB}^+}/\nu_{AB}^2\right) \qquad (5.86)$$

where $\nu_{AB}$ is the nonreactive collision frequency, $\nu_{AB}^+$ is the reactive collision frequency, $\Delta\nu_{AB}^+ = \nu_{AB}^+(v) - \Delta\overline{\nu_{AB}^+}$, with the bar denoting an average over the Maxwellian velocity distribution function $f(v)$:

$$\overline{\nu_{AB}^+} = \int f(v)\nu_{AB}^+(v)\,dv$$

To illustrate the result given by equation (5.86), let us assume that in a short period of time the chemical reaction selectively consumes high-speed molecules of A. The velocity distribution of the molecules that remain after this period is no longer Maxwellian since the high-speed tail of the Maxwellian distribution has been reduced in magnitude. Immediately following this time period, collisions with molecules of type B will cause the velocity distribution to relax towards its equilibrium Maxwellian form. However, as this relaxation proceeds, both diffusion and further chemical reaction can occur. Thus, during the time $(\nu_{AB}^+)^{-1}$ it takes the velocity distribution to relax, chemical reaction and diffusion proceed under the influence of a velocity distribution that is deficient in high-speed molecules. This temporary deficiency causes both the diffusion and reaction rates to be lower than they would be if the velocity distribution were Maxwellian for all time. Of course, if the velocity relaxation is very much faster than the rate of depletion of high-speed molecules by the reaction, the nonequilibrium effects on the rate coefficient and the diffusivity will be negligible. However, for reactions with a bimolecular rate coefficient on the order of $10^8$ to $10^{10}$ sec$^{-1}$ g-mole$^{-1}$ liter, these corrections to the rate coefficient and the diffusivity can be important.

These nonequilibrium effects are insignificant for all but very fast reactions and have been neglected in the analyses of Sections 5.3.3.1 and 5.3.3.2.

### 5.3.4. Open Systems

In an open system, in contrast to a closed system, the component gases flow past the borders of the system, and by adjustment of the rate of flow the compositions and pressures can be changed arbitrarily. A well-known example of such a system is the diffusion cell (Table 1.1, System 4), which is used for the measurement of diffusivities in gaseous systems and porous media.

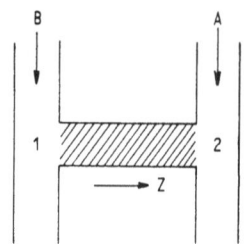

Figure 5.15. The diffusion cell.

### 5.3.4.1.  Binary Systems

In the diffusion cell, the gases flow past the borders of the system through which they diffuse (shaded zone in Figure 5.15); the gases may diffuse through a capillary or through a porous septum. The diffusion cell has an advantage over the closed system in that it permits work at the steady state and control of the pressure drop through the system by regulation of the flow rates of the gases. In particular, adjustments can be made to obtain the important case of uniform pressure.

The introduction of the diffusion cell is usually ascribed to Wicke [128] or to Wicke and Kallenbach, [130] though in fact it had been used by Buckingham at the beginning of this century [13] and, in essence, it is basically the same as a device employed by Graham in 1831 (System 1b in Table 1.1; see also Section 6.1.1.1). Mason and Marrero [79] give a review of the application of this cell.

Another example of an open system is encountered in helium-cooled nuclear reactors (Table 1.1, System 6). These reactors have a core of graphite and uranium, represented by II in Figure 5.16. The radioactive gases B are produced by fission of the uranium. The reactor must be cooled and, at the same time, the concentration of the contaminating gases B must be kept as low as possible. To accomplish this, the core is surrounded by a jacket made of a high-permeability graphite. An annular space is left to let the gases inside the jacket flow at 20 atm; the gas A (helium) outside the jacket is maintained at 21 atm and 600–800°C. As a consequence, the jacket supports a countercurrent diffusion under a pressure drop of 1 atm. The system is similar to the diffusion cell with the core II as the source of gases B. The study of this type of reactor seems to have been the prime motive for the development of the dusty gas model (see Section 6.2.4.1).

In an open system at steady state for one-dimensional flows and no reaction, equation (5.1) leads to

$$N^* = \text{const} \tag{5.87}$$

The value of this constant is unknown *a priori*, so equations such as (5.13), (5.34), or (5.53) are no longer useful for the special case of molecular diffusion.

Since the value of $N^*$ depends on the values of $D^{K^*}$ and of the pressure gradient, we have the following special cases:

(i)     isobaric, i.e., adjusting the pressure drop so that $\Delta p = 0$; then, $N^* = -D^{K^*}$ (fulfillment of Graham's law of diffusion; see Section 5.2.1);

(ii)    nonisobaric with $N^* = 1$, by establishing a $dp/dz > 0$ if $(-D^{K^*}) < 1$, or $dp/dz < 0$ if $(-D^{K^*}) > 1$;

(iii)   nonisobaric with $N^* = 0$, by establishing a $dp/dz < 0$ so that $N_{zA} = 0$.

From this we see that, by means of a suitable choice of the pressure gradient, open systems may simulate the behavior of closed and semiopen systems in the sense that the same conditions are used in applying equation (5.7). This possibility is illustrated in Figure 5.17, which shows experimental results obtained by Mason et al.[78] for the molecular fluxes $F_{He}$ and $F_{Ar}$ of helium and argon and for the total flux $F = F_{He} + F_{Ar}$. For $\Delta p = 0.23$ atm, the total flux $F$ is zero and this is the working condition for a closed system; a catalyzed cracking reaction with $b = 10$ is simulated at $\Delta p = -0.18$ atm, where $-F_{Ar}/F_{He} = 0.10$; the semiopen system without chemical reaction (Stefan's system) lies outside the figure since it corresponds to $F_{Ar} = 0$ with $\Delta p \approx -0.36$ atm. Finally, a gas–solid noncatalyzed reaction is more difficult, but not impossible, to simulate, since (a) we need to match the ratio of the molecular weights of the diffusing species and (b) the fluxes must be such that they satisfy the stoichiometric requirements (for example, the argon–helium system roughly simulates the reduction of an oxide with hydrogen since $M_{Ar}/M_{He} = 10$ and $M_{H_2O}/M_{H_2} = 9$; if we take these ratios as roughly equal and use the stoichiometric coefficient $b = 1$ for the oxide reduction, the behavior of the system lies on $F = 0$). Composition and pressure profiles (related to equations of change of a

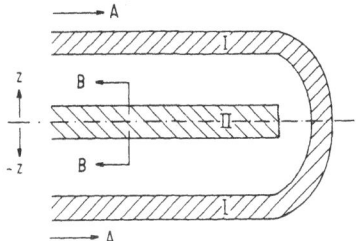

Figure 5.16. Schematic representation of a nuclear reactor cooled with helium (gas A). I is a graphite jacket, II is the core, which consists of graphite and uranium; B are the radioactive gases.

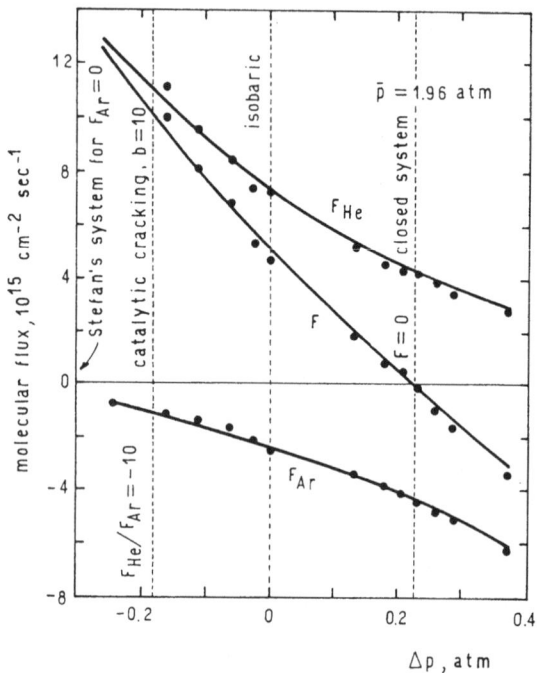

Figure 5.17. Simulating other systems with an open system. $F_{He}$ and $F_{Ar}$ are the fluxes of helium and argon and $F = F_{He} + F_{Ar}$; $\Delta p$ is the pressure difference between the two ends of the diffusion tube. Reprinted from Mason, Malinauska, and Evans,[78] by permission.

reference component) may also be simulated for those systems in which diffusion takes place without simultaneous reaction (closed system, Stefan's system, and ash layer of a shrinking-core model for a gas–solid noncatalyzed reaction).

We consider now some properties and applications of the diffusion cell.

*5.3.4.1a.　The Isobaric Case.* This is the most frequently encountered working condition in open systems, especially when used for the measurement of transport parameters.

It follows from equation (5.8) that in this case the ratio of the measured fluxes corresponds to that predicted by Graham's law of diffusion,

$$\frac{N_{zB}}{N_{zA}} = \frac{N_{zB}^{D}}{N_{zA}^{D}} = -\left(\frac{M_A}{M_B}\right)^{1/2} \qquad (5.88)$$

The fulfillment of this law in isobaric diffusion experiments has been widely verified.[†] When deviations are found that are higher than the

[†]See References 14, 24, 29, 43, 52, 58, 91, 102, 124, 129.

experimental error, they are usually ascribed, without any substantiating calculations, to the existence of surface diffusion[57, 98] or configurational diffusion[82] (see Section 5.4.1), but it is important to point out that in some cases the deviations cannot be easily justified[35, 36, 95] (see Section 5.3.4.1c).

The constitutive equation for diffusion in this case is obtained from equation (4.52) with $\nabla p = 0$. Introducing this equation into equation (5.1) for the steady state and for no reaction, and then integrating with the boundary conditions

$$x_A = \begin{cases} x_{A0} & \text{at } z = 0 \\ x_{AL} & \text{at } z = L \end{cases} \tag{5.89}$$

we obtain the following expression:

$$N_{zA} = \frac{cD_{AB}}{L(1 + D^{K^*})/D^{K^*}} \ln\frac{X_{AL}}{X_{A0}} \tag{5.90}$$

where

$$X_A \equiv \frac{D_{AB}}{D_{AA}^K} + 1 - \frac{1 + D^{K^*}}{D^{K^*}} x_A \tag{5.91}$$

Equation (5.90) was obtained within the framework of the dusty gas model by Evans et al.,[29] while Rothfeld[94] and Scott and Dullien[102] independently arrived at it by use of a momentum balance in a capillary.

Figure 5.18 shows theoretical curves for the flux of helium and argon as a function of pressure and shows the excellent agreement between theory and experiment.[79] Satisfactory agreement between the predictions of the equations and the experimental data was also reported in References 24, 52, 95, 102. Remick and Geankoplis[91] performed a rigorous experimental test of equation (5.90) in an isobaric open system of fine glass capillaries in parallel. For the pair helium–nitrogen, the results obtained confirmed the theoretical equation over a 675:1 pressure range (from 0.444 to 300.2 torr absolute) in the transition region between the Knudsen and molecular diffusion regimes.

In the molecular diffusion regime ($D_{ii}^K \gg D_{AB}$), equation (5.90) reduces to

$$N_{zA} = \frac{cD_{AB}}{L(1 + D^{K^*}/D^{K^*})} \ln\frac{1 - [(1 + D^{K^*})/D^{K^*}]x_{AL}}{1 - [(1 + D^{K^*})/D^{K^*}]x_{A0}} \tag{5.92}$$

Equation (5.92) predicts that, since $D_{AB}p$ is independent of pressure, the diffusive flux is independent of pressure.

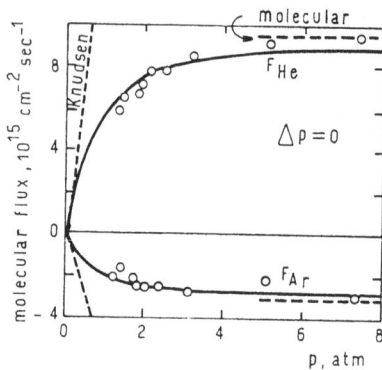

Figure 5.18. Variation of the fluxes of helium and argon with pressure in a diffusion cell. Reprinted from Mason and Marrero,[79] by permission.

In the Knudsen regime ($D_{AB} \gg D_{ii}^{K}$), we write

$$\ln \frac{X_{AL}}{X_{A0}} = \ln \frac{1 + x_{AL}^{\dagger}}{1 + x_{A0}^{\dagger}} \tag{5.93}$$

where

$$x_A^{\dagger} \equiv \frac{D_{AA}^{K}}{D_{AB}} \left( 1 - \frac{1 + D^{K \bullet}}{D^{K \bullet}} x_A \right) \tag{5.94}$$

with $x_A^{\dagger} \ll 1$. Consequently, if we expand the right-hand side of equation (5.93) in a series in the two variables $x_{AL}^{\dagger}$ and $x_{A0}^{\dagger}$ and keep only the first term, equation (5.90) becomes

$$N_{zA} = \frac{cD_{AA}^{K}}{L} (x_{A0} - x_{AL}) \tag{5.95}$$

Equation (5.90) and its particular cases, equations (5.92) and (5.95), agree with the experimental results of Evans *et al.*[30] as shown in Figure 5.18.

If we try to represent equation (5.90) by a Fick's-type law,

$$N_{zA} = \frac{cD_{AB}^{F}}{L} (x_{A0} - x_{AL}) \tag{5.96}$$

then, from a comparison of equations (5.90) and (5.96), we see that $D_{AB}^{F}$ is strongly dependent on composition,

$$D_{AB}^{F} = \frac{D_{AB}}{X_{A,ml}} \tag{5.97}$$

where the subscript ml indicates logarithmic mean:

$$X_{A,ml} \equiv \frac{X_{AL} - X_{A0}}{\ln(X_{AL}/X_{A0})}$$

in the Knudsen regime,

$$D_{AB}^F = D_{AA}^K \qquad (5.98)$$

Finally, it is interesting to analyze the influence of temperature on the flux at constant pressure. In the Knudsen regime, since $D_{AA}^K \propto T^{1/2}$ and $c \propto T^{-1}$, we have $N_{zA} \propto T^{-1/2}$. In the molecular diffusion regime, since $D_{AB} \propto T^c, c = 1.5-2$, we have $N_{zA} \propto T^{c'}, c' = 0.5-1$. Consequently, in the transition regime, the temperature influences the flux only very slightly. These trends have been verified experimentally.[36]

*5.3.4.1b.   Experimental Determination of the Transport Parameters of the Porous Medium*

*b.1 Determination of $Q_m$ and $Q_p^\circ$.* As we discussed in Chapter 4, the molecular and Knudsen diffusivities involve the obstruction factors $Q_m$ and $Q_p^\circ$ given by equations (4.6) and (4.7) and by (4.40), respectively.

The experimental determination of both of these factors with a diffusion cell can be performed in two ways. The simplest procedure would be to work in pure diffusion regimes at uniform pressure to measure each obstruction factor separately. However, it is unusual, in practice, for the limiting pure regimes to be reached. It is consequently necessary to work in the transition regime.

If we perform the experiment in the transition regime at uniform pressure, it is convenient to write equation (5.90) as follows[43] (since we will be dealing here with both effective and noneffective diffusion coefficients, we will again use the subscript eff to distinguish them):

$$\frac{D_{AB,eff}}{D_{AA,eff}^K} = \frac{(x_{A0}e^u - x_{AL})(1 + D^{K^\bullet})/D^{K^\bullet}}{e^u - 1} - 1 \equiv f(u) \qquad (5.99)$$

where

$$u \equiv \frac{LN_{zA}(1 + D^{K^\bullet})/D^{K^\bullet}}{cD_{AB,eff}} = \frac{LN_{zA}(1 + D^{K^\bullet})/D^{K^\bullet}}{cD_{AB}Q_m} \qquad (5.100)$$

so that $u$ is a function of $Q_m$.

We obtain another expression for $D_{AB,\,eff}/D^K_{AA,\,eff}$ from equations (4.6), (4.7), and (4.40):

$$\frac{D_{AB,\,eff}}{D^K_{AA,\,eff}} = \frac{D^\circ_{AB}Q_m}{(R_gT/M_A)^{1/2}Q^\circ_p p} = f(u) \tag{5.101}$$

where, it is to be noted, $D^\circ_{AB} = D_{AB}p$ is *not* an effective coefficient.

We assume that all quantities (including $D_{AB}$) in equations (5.99)–(5.101) are known or measured except for $Q_m$ and $Q^\circ_p$. If $Q_m$ were known *a priori* (for example, from measurements in the molecular diffusion regime), then $u$ could be found from (5.100), $f(u)$ from (5.99), and $Q^\circ_p$ from (5.101) written as

$$Q^\circ_p = \frac{CQ_m}{pf(u(Q_m))} \tag{5.102}$$

where $C = D^\circ_{AB}/(R_gT/M_A)^{1/2}$. If $Q_m$ is not known then we can carry out the experiment at various pressures (at constant temperature) and, noting that $u$ is a function of $p$ (through $N_{zA}$), write equation (5.102) as

$$pf(u(Q_m,p)) = CQ_m/Q^\circ_p = \text{constant}$$

A plot of $f(u)$ as a function of $p^{-1}$ must therefore be a straight line passing through the origin, and the solution to our problem means finding a value of $Q_m$ that when used in equation (5.99) will yield such a straight line. Any of several numerical procedures can be used to find such a $Q_m$. Once, of course, $Q_m$ is determined, $Q^\circ_p$ can be found from equation (5.102).

There is a particular case which must be considered. If we select the gas pair so that $M_A \simeq M_B$ (for example, if the pair is selected from carbon monoxide, nitrogen, and ethylene, or selected from nitrogen suboxide, carbon dioxide, and propane) the method of determination of $Q_m$ and $Q^\circ_p$ is greatly simplified. In fact the constitutive equation for the diffusive flux now follows from equation (4.57) with $\nabla p = 0$, implying $N^D_{zA} = N_{zA}$:

$$N_{zA} = \frac{cD^R_{AB}}{L}(x_{A0} - x_{AL}) \tag{5.103}$$

We observe that the classic relationship of Bosanquet[11] [equation (6.2)] is valid only in this case [see also Section 4.2.2.2, equation (4.57)].

Equation (5.103) may be rearranged, giving

$$u'^{-1} = Q_m^{-1} + (Q^\circ_p)^{-1}\frac{D^\circ_{AB}}{(R_gT/M_A)^{1/2}}p^{-1} \tag{5.104}$$

where

$$u' \equiv \frac{N_{zA} L R_g T}{(x_{A0} - x_{AL}) D_{AB}^{\circ}} \tag{5.105}$$

If we represent $u'^{-1}$ as a function of $p^{-1}$, we get a straight line and we can obtain $Q_m$ from the intercept and $Q_p^{\circ}$ from the slope. We must keep in mind the approximations used in arriving at equation (5.103): For the system $o$-hydrogen–$p$-hydrogen, equation (5.104) does not fit the experimental results for $p < 200$ torr[54]—the shape of the experimental curves suggests that second-approximation diffusivities must be used in that pressure range.

   b.2. *Determination of $Q_p^{\circ}$ and $B_k$.* The experimental measurement of $Q_p^{\circ}$ and $B_k$ may be performed in an open system with a pure gas flowing through the porous medium. In this particular case, equations (4.40) and (4.45) lead to

$$N_{zA} = -\left[\left(\frac{R_g T}{M_A}\right)^{1/2} Q_p^{\circ} + \frac{B_k p}{\mu}\right](R_g T)^{-1}\frac{dp}{dz} \tag{5.106}$$

We introduce equation (5.106) into equation (5.1) and (for the steady state and no reaction) integrate it with the boundary conditions

$$p = \begin{cases} p_0 & \text{at } z = 0 \\ p_L & \text{at } z = L \end{cases} \tag{5.107}$$

The integration leads to

$$\frac{(M_A R_g T)^{1/2}(-N_{zA})L}{\Delta p} = Q_p^{\circ} + \frac{B_k}{\mu}\bar{p}\left(\frac{M_A}{R_g T}\right)^{1/2} \tag{5.108}$$

where $\Delta p = p_L - p_0$ and $\bar{p} = \frac{1}{2}(p_0 + p_L)$. Equation (5.108) permits us to determine $Q_p^{\circ}$ and $B_k$ from experimental data. For this, it is customary to work with helium in order to avoid surface diffusion.

   In general, the value of $Q_p^{\circ}$ obtained using equation (5.108) is more accurate than that obtained using equation (5.99). Consequently, when the value of $Q_p^{\circ}$ is determined from equation (5.108), it can be introduced into equation (5.99) to obtain the value of $Q_m$

   5.3.4.1c. *The Isobaric Condition.* In the preceding analysis of the isobaric case (Section 5.3.4.1a) and of the determination of $Q_m$ and $Q_p^{\circ}$ (Section 5.3.4.1b) we have assumed that uniform pressure can be maintained in open systems. It is evident, however, that the requirement will be achieved in practice only to within a given experimental error. This error depends on the accuracy of the instrument used to measure pressure differences as well as on the controlling device used. Thus, if a differential manometer is used, the pressure difference can be controlled to within approximately $10^{-4}$ atm; if a pressure transducer is used, the control can be within a value of the order of $10^{-6}$ atm. In certain diffusion cells[15, 55] the pressure difference between the borders might have been significant in the experimental runs. It is therefore important to analyze the effect of a pressure difference on the fluxes.

First, we must point out that the pressure gradients in open systems are different in nature from those in closed and semiopen systems. The condition that a system is closed or semiopen implies that the steady-state (or quasi-steady-state) value of $N*$ has a natural characteristic value (i.e., $N* = 1$ for closed systems, $N* = 0$ for semiopen systems without chemical reaction), and the pressure gradients build up in order to maintain that value of $N*$ (for example, a very low pressure gradient does not necessarily mean that the system is isobaric since it is an inherent property of the system that in general the value of $N*$ will be different from that given by Graham's law of diffusion). The system is naturally isobaric only for the particular case of $N* = -D^{K*}$. On the other hand, in open systems the value of $N*$ is undetermined a priori—it depends on the nature of the capillary, the porous plug, the gases, etc., and on the pressure difference, which either develops spontaneously or is imposed externally, between the borders of the cell.

The experiments in an isobaric diffusion cell are interpreted by letting $\nabla p = 0$ in equations like (4.52), giving

$$N_{zA} = N_{zA}^{D}(\nabla x_A) \tag{5.109}$$

$$\frac{N_{zB}}{N_{zA}} = \left(\frac{N_{zB}^{D}}{N_{zA}^{D}}\right)_G \equiv (N_{zBA}^{D})_G \tag{5.110}$$

where the subscript G stands for Graham's value [cf. equation (5.88)]. The errors in the total flux and in the flux ratio as a result of a pressure gradient beyond our control can be expressed, respectively, as follows:

$$Er_N^T = \frac{N_{zA}(\nabla x_A, \nabla p) - N_{zA}^{D}(\nabla x_A)}{N_{zA}^{D}(\nabla x_A)} \tag{5.111}$$

$$Er_G^T = \frac{N_{zBA} - (N_{zBA}^{D})_G}{(N_{zBA}^{D})_G} = -\left(1 + \frac{D^{K*}}{N*}\right) \tag{5.112}$$

where the superscript T indicates total fluxes.

By introducing the constitutive equations (4.48) and (4.52) for the fluxes in equation (5.111) and writing the result in dimensionless form, we obtain

$$\text{Er}_N^T = \left[ 1 - \frac{D^* p^*}{D^{K^*}} + B_k^* \left( 1 + \frac{1}{D^* p^*} - x_{A0} \frac{1 + D^{K^*}}{D^{K^*}} x_A^* \right) p^{*2} \right] \frac{d \ln p^*}{d \ln x_A^*}$$

$$(5.113)$$

If we introduce equation (5.8) into equation (5.113) and calculate the maximum value of $\text{Er}_N^T$ (which, from numerical inspection is shown to correspond to $p^* = x_A^* = 1$ with $x_{A0} = 1$), we obtain

$$\left( \text{Er}_N^T \right)_{\max} = -\left( 1 + \frac{N^*}{D^{K^*}} \right) = -\frac{\text{Er}_G^T}{1 + \text{Er}_G^T} \qquad (5.114)$$

Consequently, once we know the error in assuming Graham's law of diffusion [which error is independent of position in the system, as can be seen from equation (5.112)], we can calculate the maximum relative error in the value of the flux of one of the components, i.e., calculate the error in the flux introduced by assuming isobaric behavior.

For a given system for which it is assumed all parameters $(D^{K^*}, B_k^*, D^*)$ are known, we can select a value of $\text{Er}_G^T$ and find, from equation (5.114), the corresponding value of $N^*$. The differential equation (5.8) can be solved using this value of $N^*$ and a selected value of $x_{AL}$; then the pressure difference $\Delta p^*$ corresponding to the selected $\text{Er}_G^T$ (or $N^*$) can be determined as a function of the parameters $B_k^*$, $D^*$, and $D^{K^*}$. It can be shown that with the calculation performed in this way for $\pm \text{Er}_G^T$, [28] plots of $\Delta p^*$ as a function of $[B_k^* + D^* + (B_k^*/D^*)]$, with $D^{K^*}$ as a parameter, include all points in a very narrow band. Figure 5.19 shows these results[28] for a deviation of $\pm 10\%$ from Graham's law of diffusion; we see that for this case $\Delta p^*$ decreases as the group $[B_k^* + D^* + (B_k^*/D^*)]$ increases. The curves shown in Figure 5.19, with $D^{K^*}$ as a parameter, may be represented by the following functions:

$$|\Delta p^*| \simeq \left\{ 10 \left( 1 + B_k^* + D^* + \frac{B_k^*}{D^*} \right) \right\}^{-1}, \qquad D^{K^*} = -1, \, -2^{1/2}$$

$$(5.115)$$

$$|\Delta p^*| \simeq \left\{ 10 \left[ 1 + 2.1 \left( B_k^* + D^* + \frac{B_k^*}{D^*} \right) \right] \right\}^{-1}, \qquad D^{K^*} = -5 \quad (5.116)$$

It is obvious, and these results confirm it, that diffusion experiments in

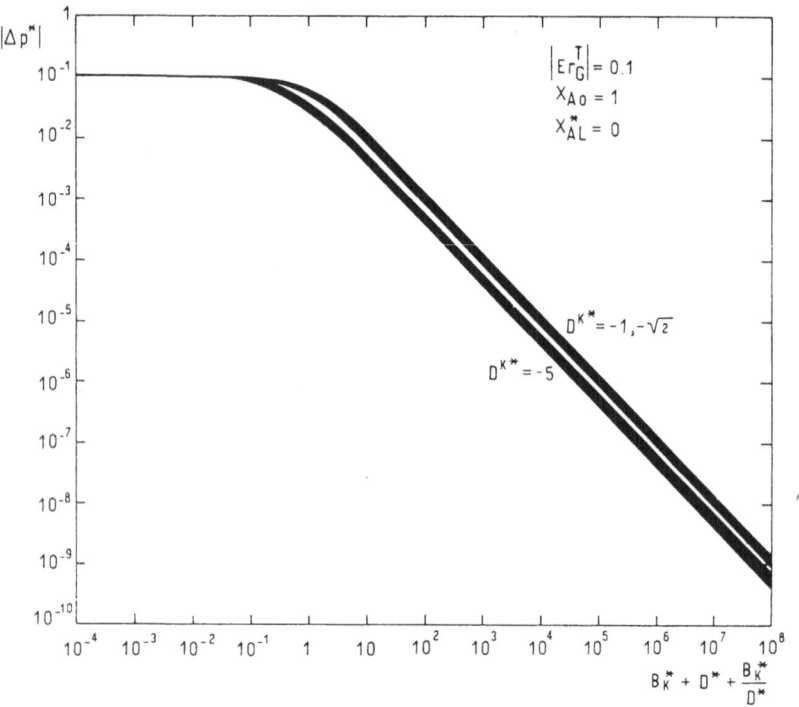

Figure 5.19. The pressure drop in a diffusion cell must be kept less than the values given by the curves in order to keep the error in Graham's law of diffusion less than 10%. Reprinted from Di Napoli, Williams, and Cunningham,[28] by permission.

porous media of high permeability, and particularly in capillaries, will have to be very carefully designed in order to avoid errors arising from a false assumption of isobaric behavior.

We now see that experimental deviations from Graham's law of diffusion may be simply the result of an undetectable pressure gradient. This fact could explain the high dispersion of some experimental results reported in the literature[57, 94] as well as unexpected deviations of curves from the theoretical predictions of the measured flux as a function of pressure.[35, 36] It is to be remembered, however, that in analyzing these deviations, surface diffusion must also be taken into account.

*5.3.4.1d. Experimental Errors in Flux Measurements.* Another important error to be analyzed is that intrinsic to the measurement of fluxes in a diffusion cell. First of all, as is well known, fluxes are not directly measured, but are calculated by using a material balance and other measured properties (flow rates and compositions) as the following example shows. Assume that in the diffusion cell shown in Figure 5.15, pure gas

A is fed through side 2 at a molar flow rate $Q_{MA2}$ and a mixture of gases A and B are fed through side 1 at molar flow rates $Q_{MA1}$ and $Q_{MB1}$. Let us call $x_{B1}$ and $x_{B2}$ the molar fractions of gas B in the streams leaving the diffusion cell. A molar balance of component A at side 2 of the cell can be written as follows:

$$Q_{MA2} = N_{zA}S + (1 - x_{B2})(Q_{MA2} - N_{zA}S + N_{zB}S) \qquad (5.117)$$

where $S$ is the cross section of the porous pellet or capillary and $N_{zA}, N_{zB}$ are the molar fluxes, both taken as positive values, across section $S$. Similarly, the molar balance of component B at side 1 of the cell is given by

$$Q_{MB1} = N_{zB}S + x_{B1}(Q_{MA1} + Q_{MB1} + N_{zA}S - N_{zB}S) \qquad (5.118)$$

From equations (5.117) and (5.118) we find

$$N_{zA} = S^{-1}$$

$$\times \left[ Q_{MA2} - \frac{(Q_{MA1} + Q_{MA2})x_{B1}(1 - x_{B2}) - Q_{MB1}(1 - x_{B1})(1 - x_{B2})}{x_{B1} - x_{B2}} \right]$$

$$(5.119)$$

$$N_{zB} = \frac{(Q_{MA1} + Q_{MA2})x_{B1}x_{B2} - Q_{MB1}x_{B2}(1 - x_{B1})}{S(x_{B1} - x_{B2})} \qquad (5.120)$$

For small experimental errors, the relative error in the diffusion flux may be calculated from

$$Er = N_{zA}^{-1} \sum_i \frac{\partial N_{zA}}{\partial s_i} \Delta s_i \qquad (5.121)$$

where the $s_i$ are the experimentally determined quantities. Horák and Schneider[58] have used equation (5.121) to calculate errors for the following working conditions: (i) $x_{B1} = 0.3$, $0.6$, $0.9$, $Q_{MA1} = Q_{MB1} = 0.5$ mmole/min, $Q_{MA2} = 5$ mmole/min; (ii) $x_{B1} = 0.1$, $Q_{MA1} = Q_{MA2} = 1$ mmole/min, $Q_{MB1} = 0.1$ mmole/min; it was also assumed that the experimental error in the measurement of compositions was 4% and in the measurement of flow rates 2%. The results obtained are summarized in Table 5.1. The relative error in $N_{zAB} \equiv N_{zA}/N_{zB}$ was calculated as a sum of the relative errors in $N_{zA}$ and $N_{zB}$.

Table 5.1. Errors in Total Diffusive Fluxes and the Individual Contributions[a]

| $x_{B1}$ | $\Delta x^b$ | Error in $N_{zA}$, % | Error contribution (%) to $N_{zA}$ due to error in | | | | | Error in $N_{zB}$, % | Error contribution (%) to $N_{zB}$ due to error in | | | | | Error in $N_{zAB}$, % |
|---|---|---|---|---|---|---|---|---|---|---|---|---|---|---|
| | | | $x_{B1}$ | $x_{B2}$ | $Q_{MA1}$ | $Q_{MA2}$ | $Q_{MB1}$ | | $x_{B1}$ | $x_{B2}$ | $Q_{MA1}$ | $Q_{MA2}$ | $Q_{MB1}$ | |
| 0.9 | 0.1 | 77.7 | 49.5 | 26.0 | 1.1 | 1.0 | 0.1 | 69 | 30.4 | 36.0 | 0.7 | 1.4 | 0.1 | 147 |
| 0.9 | 0.2 | 37.5 | 25.6 | 9.6 | 1.4 | 0.7 | 0.2 | 33 | 12.5 | 18.0 | 0.7 | 1.4 | 0.1 | 70 |
| 0.9 | 0.3 | 22.3 | 15.2 | 4.7 | 1.6 | 0.5 | 0.2 | 21 | 6.5 | 12.0 | 0.7 | 1.4 | 0.1 | 43 |
| 0.9 | 0.7 | 4.1 | 1.2 | 0.4 | 2.1 | 0.1 | 0.2 | 8 | 0.4 | 5.2 | 0.7 | 1.4 | 0.1 | 12 |
| 0.1 | 0.01 | 81.2 | 36.0 | 39.6 | 2.0 | 1.8 | 1.8 | 77.7 | 32 | 40.0 | 1.8 | 1.8 | 1.6 | 159 |
| 0.1 | 0.05 | 16.1 | 0.7 | 7.2 | 3.5 | 1.6 | 3.1 | 14 | 0.4 | 8.0 | 1.8 | 1.8 | 1.6 | 30 |
| 0.1 | 0.09 | 41.3 | 18.3 | 2.3 | 10.4 | 0.9 | 9.4 | 13 | 3.2 | 4.4 | 1.8 | 1.8 | 1.6 | 54 |
| 0.9 | 0.2 | 37.5 | 25.6 | 9.6 | 1.4 | 0.7 | 0.2 | 33 | 12.5 | 18.0 | 0.7 | 1.4 | 0.1 | 70 |
| 0.6 | 0.2 | 24.1 | 9.8 | 10.2 | 1.6 | 1.4 | 1.1 | 20 | 5.1 | 12.0 | 0.9 | 1.7 | 0.6 | 44 |
| 0.3 | 0.2 | 25.4 | 14.7 | 1.3 | 2.5 | 1.3 | 5.7 | 40 | 18.0 | 6.0 | 3.0 | 6.0 | 7.0 | 65 |

[a] Reprinted from Horák and Schneider,[58] by permission.
[b] $\Delta x = x_{B1} - x_{B2}$.

It can be seen that the error in the different fluxes can be considerable, although the errors in the experimentally determined quantities is never higher than 4%.

### 5.3.4.2. Multicomponent Systems

The case of a ternary isobaric mixture in an open system has been analyzed.[22, 23] In this case, the constitutive equations of diffusion are obtained from equations (4.12) and (4.29) and, using equation (4.19), can be written

$$-\frac{p}{R_g T}\frac{dx_A}{dz} = \frac{N_{zA}}{D_{AA}^K} + \frac{x_B N_{zA} - x_A N_{zB}}{D_{AB}}$$
$$+ \frac{(1 - x_A - x_B)N_{zA} + x_A\left[N_{zA}M_{AC}^{1/2} + N_{zB}M_{BC}^{1/2}\right]}{D_{AC}}$$

$$\text{(5.122)}$$

$$-\frac{p}{R_g T}\frac{dx_B}{dz} = \frac{N_{zB}}{D_{BB}^K} + \frac{x_A N_{zB} - x_B N_{zA}}{D_{AB}}$$
$$+ \frac{(1 - x_A - x_B)N_{zB} + x_B\left[N_{zA}M_{AC}^{1/2} + N_{zB}M_{BC}^{1/2}\right]}{D_{BC}}$$

$$\text{(5.123)}$$

and the following relationship holds for the fluxes:

$$N_{zA}M_A^{1/2} + N_{zB}M_B^{1/2} + N_{zC}M_C^{1/2} = 0 \qquad \text{(5.124)}$$

Equation (5.124) is Graham's law of diffusion extended to a ternary system.

The preceding equations have been solved analytically[23] and the solution illustrated with an argon–neon–helium system contained in a capillary of $10^3$ Å radius at 25°C and in the pressure range $10^{-2}$–$10^2$ atm.[90] Even though the flux plotted as a function of pressure has the same shape as in binary systems, calculations show the possibility of the existence of osmotic diffusion in the transition and molecular diffusion regimes (in the Knudsen regime the independence of the fluxes makes this phenomenon impossible). Composition profiles also show extrema (maximum or minimum). Although it is not explicitly shown, we may also find reverse diffusion and diffusive barriers outside the Knudsen regime.

The theoretical solutions have been experimentally verified for the above-mentioned system in the pressure range 0.450–303.2 torr (the use of low pressures is important in order to minimize the possible influence of surface diffusion.)[92] Studies have also been made for a ternary gas mixture in twelve porous pellets prepared from seven commercial catalysts.[84] We will discuss this further in Section 5.4.2.

A general analytic solution for the multicomponent isobaric case has recently been provided.[72]

## 5.4.  Diffusion and Porous Structure

We have seen that the representation of a porous medium by a dusty gas has enabled us to obtain the constitutive equations of diffusion by a simple but rigorous procedure. When these equations are written in the first approximation, two adjustable parameters, $Q_m$ and $Q_p^\circ$, appear, and these depend on the structure of the porous medium. If these parameters are to be experimentally determined, then the equations can be used for any porous medium, provided the viscous and surface fluxes are also taken into account (this can be done if the porous medium does not contain too-narrow pores; see Sections 4.3.1.1 and 5.4.1). With this approach, it is not necessary to develop a model for the porous structure of the solid.

The first-approximation equations of the dusty gas model have been verified over a wide range of working conditions and for many porous solids, and it seems that there are few systems which require the second-approximation equations (*o*-hydrogen–*p*-hydrogen is one example of a system requiring the second approximation; see Section 5.3.4.1b); even in the case of markedly bimodal (with micro- and macropores) porous media, reported differences between theory and experiment are small (though some improvements in the use of the model have been suggested[20]).

However, there are two points which must be emphasized. On the one hand, the first-approximation equations of the dusty gas model do not necessarily provide the best agreement with experimental data (see, e.g., Section 5.4.2), and, as noted above, with these equations we cannot predict even the order of magnitude of the diffusive fluxes without experiments to determine the two parameters $Q_m$ and $Q_p^\circ$ of the porous medium. Consequently, as we saw in Chapter 4, there have been attempts to develop models for the prediction of transport properties of porous media. In this section we will compare the predictions of the models described in Chapter 4 with experimental results. But we shall first deal briefly with configurational diffusion.

### 5.4.1.  Configurational Diffusion

Consider an isobaric flux through a capillary in a diffusion cell. For this case the diffusivities in equation (5.90) can be predicted *a priori* [see

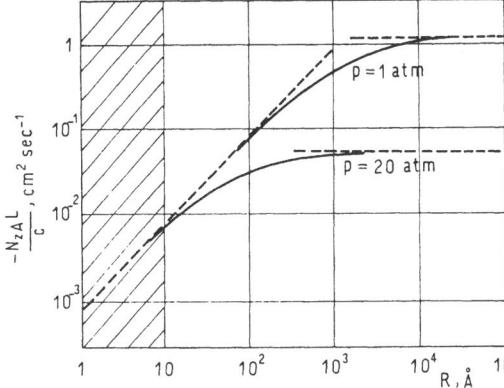

Figure 5.20. Isobaric flux in a capillary as a function of its radius.

equation (2.65) and Section 4.3.2.1]. If gas A is helium, gas B nitrogen, and $x_{AL} = 1$, $x_{A0} = 0$, we may represent $N_{zA}$ as a function of the capillary radius, as shown in Figure 5.20. Observe that, as indicated in Section 4.3.1.1, for very narrow pores, i.e., in approximately the region indicated by the hatched area in Figure 5.20, the Knudsen mechanism of diffusion is no longer valid. The diffusion regime in this region has been called configurational.[127] This type of diffusion was first observed in studies of the diffusion of large molecules in polymers (i.e., dyes in cellulose and man-made fibers).[37, 86]

In the configurational regime, diffusion depends on the size and configurational structure of the molecules (i.e., the dynamics of the rotations and vibrations of the different groups constituting the molecule) because the spaces through which they diffuse are of near-molecular dimensions. Thus, for example, the difference in configuration of the molecules of *cis*- and *trans*-butene produces a difference of two orders of magnitude in their diffusivities in Ca-A zeolite.[19] The results of the diffusivity of *n*-paraffins as a function of the number of carbon atoms are also surprising since when the diffusivity is plotted as a function of the number of carbons in the *n*-paraffin for the range $C_2H_6$–$C_{13}H_{28}$, two distinct maxima are observed (for $C_4H_{10}$ and $C_{11}H_{24}$).[18, 38] These facts illustrate the possibility of controlling the selectivity of complex reactions by the use of suitable catalysts (primarily zeolites).[127]

The configurational diffusion regime is bordered on one side by the Knudsen regime and on the other side by a region that cannot be clearly defined, but which is encountered in diffusion through membranes. We can expect that in membrane diffusion the chemical (e.g., the acid–base properties) as well as the configurational characteristics of the molecules will influence the diffusion phenomenon. Barrer and co-workers,[5, 6, 9] for example, have studied the efficiency of separation of binary mixtures by membranes of microporous carbon.

Some work by Omata and Brown[82, 83] and by Abed and Rinker[3] gives an indication of the location of the precise border between the Knudsen and configurational diffusion regimes. Omata and Brown[83] found that the dusty gas model equations were not applicable for pores with radii less than 50 Å; however, these authors did not consider the surface diffusion present in their study and did not consider configurational diffusion separately. Later, Abed and Rinker[3] studied the validity of the first approximation of the dusty gas model for $0.25 \times 0.25$-inch commercial catalysts. Two types of solids were studied: pellets that were highly microporous since pores with diameters below 50 Å made up 61% of the pellet volume, and pellets that were moderately macroporous with 65% of their pore volume consisting of pores with diameters between 800 and 1800 Å. Experiments on the isobaric interdiffusion of the gas pair helium–nitrogen show excellent agreement with theory: the maximum relative deviation of the predicted helium flux is 5% for the microporous solid and 1.8% for the macroporous one, so that for the first of the solids configurational diffusion is of minor importance for the gas pair helium–nitrogen.

### 5.4.2.  Agreement of the Constitutive Equations of Diffusion with Experimental Results

Equations (4.92) to (4.96) can be used to calculate total diffusive fluxes in porous media. We now consider the suitability of these equations on the basis of the agreement between experiment and theory.

First of all, equation (4.96) can be written in terms of two lumped pore radii, $R_1$ and $R_2$, which must be known from experiment in order to use the constitutive equation in subsequent applications. Feng and Stewart,[33] from calculations based on the results obtained by applying their equation to various solids, concluded that $R_1$ and $R_2$ may be calculated provided we know the pore-volume distribution of the porous solid; the values of both pore radii can be determined from

$$V_{R_1}/V_\infty = 0.676 \tag{5.125}$$

$$V_{R_2}/V_\infty = 0.977 \tag{5.126}$$

where $V_{R_1}$ is the volume of pores with radii up to $R_1$, $V_{R_2}$ is the volume of pores with radii up to $R_2$, and $V_\infty$ is the total pore volume. Equations (5.125) and (5.126) enable us to calculate transport parameters from porosimetric information.

A comparison and analysis of equations (4.92) to (4.96) was carried out by Feng and co-workers.[32, 33] The equations were tested with the following series of experiments:

1. Helium fluxes measured in binary isobaric systems in a great variety of porous solids and over a wide range of pressures (the data used are from References 14, 22, 34, 48).

2. Fluxes of helium and nitrogen in isobaric and nonisobaric experiments taking into account the surface diffusion of nitrogen (data from Reference 98).

3. Fluxes in the ternary system helium–nitrogen–methane in isobaric experiments with $\gamma$-alumina catalyst pellets over a wide range of pressures $(1 - 70$ atm) in the temperature range $300–390°K$ and over a wide range of compositions and composition gradients.[32] Surface diffusion of nitrogen and methane was taken into account.

Table 5.2 gives the percentage standard error (SE%) for each of the models described by equations (4.94)–(4.96). This error corresponds to $\ln N_{zi}$ computed with $n_k - p_k$ degrees of freedom, $n_k$ being the number of

Table 5.2.   Comparison of Various Theoretical Equations with Experimental Data for the Total Diffusive Flux in Several Porous Media[a,b]

| Reference | $n_k$ | $\kappa$ | SE% | | | |
|---|---|---|---|---|---|---|
| | | | eq. (4.94) | eq. (4.95) | eq. (4.96) | eqs. (4.96) and (5.125), (5.126) |
| 14 | 10 | 0.1472 | 4.11 | 5.37 | 1.51 | 1.99 |
| 14 | 10 | 0.3042 | 1.13 | 18.82 | 1.08 | 4.07 |
| 14 | 7 | 0.1346 | 3.23 | 5.09 | 0.84 | 1.16 |
| 14 | 7 | 0.2217 | 1.71 | 3.69 | 1.10 | 1.30 |
| 22 | 7 | 0.5331 | 6.65 | 18.80 | 6.65 | 6.65 |
| 34 | 27 | 0.1959 | 3.11 | 7.89 | 2.90 | 2.89 |
| 34 | 26 | 0.1895 | 2.51 | 4.21 | 2.50 | 2.55 |
| 48 | 12 | 0.4341 | 8.91 | 2.19 | 1.89 | 2.63 |
| 48 | 12 | 0.5326 | 2.33 | 4.46 | 0.75 | 4.18 |
| 48 | 7 | 0.5414 | 0.20 | 4.48 | 0.15 | 4.43 |
| 48 | 14 | 0.3147 | 6.44 | 3.37 | 0.79 | 4.37 |
| 48 | 6 | 0.3116 | 4.42 | 2.95 | 0.68 | 2.55 |
| 48 | 11 | 0.2511 | 3.92 | 4.04 | 0.77 | 2.51 |
| 48 | 12 | 0.0287 | 3.88 | 3.18 | 0.56 | 2.98 |
| 48 | 14 | 0.3957 | 8.62 | 2.64 | 1.35 | 1.66 |
| 48 | 12 | 0.4118 | 2.71 | 3.01 | 2.02 | 2.50 |
| 48 | 13 | 0.2544 | 7.26 | 1.05 | 2.09 | 4.87 |
| 48 | 12 | 0.2840 | 5.90 | 5.77 | 0.76 | 1.66 |
| 98 | 52 | 0.1449 | 8.70 | 5.08 | 3.48 | 5.25 |
| 98 | 20 | 0.3279 | 1.95 | 15.77 | 1.26 | 3.02 |
| 98 | 14 | 0.1389 | 3.18 | 2.81 | 1.60 | 1.56 |
| 98 | 14 | 0.2222 | 2.64 | 4.80 | 1.63 | 2.14 |
| 32 | 283 | 0.2029 | 5.3 | 5.1 | 2.1 | 2.7 |

[a] Prepared from results that appeared in, and that are reprinted with permission from, C. F. Feng, V. V. Kostrov, and W. E. Stewart, *Ind. Eng. Chem. Fund.*, **13**, 5 (1974)[32] and C. F. Feng and W. E. Stewart, *Ind. Eng. Chem. Fund.*, **12**, 143 (1973).[33] Copyright by the American Chemical Society.
[b] $n_k$ is the number of experimental results.

experimental results and $p_k$ the number of adjustable parameters in the equations.

It is seen from Table 5.2 that all the tested equations agree satisfactorily with experiment. If we arbitrarily give the same statistical weight to each of the values in the table, the average percent standard errors for each model are, arranged according to size,

| ave. SE%: | 1.67 | 3.03 | 4.30 | 5.85 |
|---|---|---|---|---|
| equation: | (4.96) | (4.96), (5.125), (5.126) | (4.94) | (4.95) |

From left to right this sequence corresponds to the decreasing number of adjustable parameters in the theoretical equation.

We thus have constitutive equations of diffusion that agree with experimental data even better than the first-approximation DGM constitutive equations, which are formally equal to equation (4.94). [This better agreement is, however, compensated by the fact that two additional parameters are needed when using equation (4.96), and the pore-volume distribution function must be known when using equations (4.96) and (5.125), (5.126); these added parameters make the computation time longer.] Further, the application of equation (4.96) to bidisperse porous media gives a precise physical meaning to $R_1$ and $R_2$ when there are micro- and macropores, while the application of equations [like equation (4.94)] which consider the porous medium as monodispersed, although permitting treatment of the problem with a reasonable error, do not provide a correct physical picture.

A similar study was made by Patel and Butt[84] for the ternary diffusion of helium, argon, and nitrogen in twelve porous pellets prepared from seven commercial catalysts. The models tested were: equations (4.94), (4.95), b.1.4 in Table 4.3, and modified versions of the two latter models. The results obtained show that the dusty gas model [equation (4.94)] is superior to all the others both in simplicity and economy in computer time. With the exception of b.1.4, the one-parameter models were satisfactory for most materials although not as precise as their two-parameter counterparts, especially when dealing with porous media with broad macropore distributions.

## 5.4.3. Prediction of Total Diffusive Fluxes without Previous Experimental Information

The possibility of predicting total diffusive fluxes without previous experiments to determine parameters is possible starting from equation

(4.95). In fact, the integral may be evaluated *a priori* if we know the value of the tortuosity $\kappa$. The value of this parameter may be obtained theoretically in the particular case of isotropic media without dead-end branches. The resulting value is[62, 132]

$$\kappa^{-1} = 3 \qquad\qquad (5.127)$$

The validity of equation (5.127) in actual porous media, as well as the existence of some exceptions, has been discussed in related literature.[12, 49, 58, 62, 83, 88, 99] The results show that in the majority of cases the value of $\kappa$ given by equation (5.127) enables us to predict the order of magnitude of the total diffusive fluxes to within a factor of two. Thus, with only one exception, the values of $\kappa$ given in Table 5.2 agree in order of magnitude with equation (5.127).

In particular, Satterfield and Cadle[99] studied 17 commercially manufactured pellet catalysts and found that (with the exception of two catalysts which should have been calcined at high temperatures), tortuosities ranging between $\frac{1}{3}$ and $\frac{1}{7}$ (in 12 of the solids it ranged between $\frac{1}{3}$ and $\frac{1}{5}$) gives good agreement with experiment. The experimental results of Johnson and Stewart[62] on various tablets of alumina suggest a value of the tortuosity equal to $\frac{1}{4}$.

The drawback of this manner of calculating fluxes is that we cannot know *a priori* whether our solid is among the few exceptions for which the tortuosity differs significantly from $\frac{1}{3}$. In general, we may expect that anisotropic solids with a large proportion of dead-end pores (or with short lengths of large pores joined by lengths of very small pores) will have abnormally low tortuosity values while those porous solids consisting of wide and short pores will have high tortuosity values. Brown *et al.*[12] drew attention to the fact that a solid with severe pore constrictions, and consequently with a low tortuosity value, shows a high deviation ($> 100\%$) of pore distribution cumulative surface area relative to BET surface area. This may represent a criterion for deciding on the use of "normal" values of tortuosity.

Alternatively, in order to predict isobaric diffusion in a porous medium, the model proposed by Wakao and Smith[124] or some subsequent refinement[24] may be used. In spite of the initial success of this model in predicting fluxes in pellets of alumina or in Vycor glass made from pressed microporous particles,[24, 52, 124] subsequent results make it possible to verify that the model can be used, in the general case, to predict the order of magnitude of the total diffusive flux. Satterfield and Cadle[99] compared the predictions of this model with their experimental results. Table 5.3

Table 5.3.  Comparison of the Parallel Pore Model (PPM)[a] and the Random Pore Model (RPM)[b]

| Catalyst designation | % Deviation[c] | |
|---|---|---|
| | PPM | RPM |
| T 126 | 7.5 | 23 |
| T 1258 | 5 | 15 |
| T 826 | 2.5 | 19 |
| T 314 | − 77.5 | − 100 |
| T 310 | 5 | 23 |
| G 39 | − 20 | 37 |
| G 35 | − 22.5 | 37 |
| T 606 | 27.5 | − 30 |
| G 58 | 30 | − 80 |
| (T 126)′ | 10 | 8 |
| G 41 | − 10 | 10 |
| G 52 | 2.5 | − 160 |
| BASF | − 82.5 | − 100 |
| Harshaw | − 80 | − 130 |
| Haldor Topsøe | 30 | 40 |

[a] Equation (4.95) with $\kappa = 1/4$.
[b] Reprinted with permission from C. N. Satterfield and P. J. Cadle, *Ind. Eng. Chem. Proc. Des. and Dev.*, 7, 256 (1968).[99] Copyright by the American Chemical Society.
[c] See text for definition.

shows the percent deviation,

$$\% \, \text{Deviation} = 100 \, \frac{\text{observed flux} - \text{predicted flux}}{\text{observed flux}}$$

for the Wakao and Smith model (called the random pore model) as well as for the parallel pore model [equation (4.95)], with $\kappa^{-1} = 4$; the nomenclature of the catalysts corresponds to that given in the original references. It follows from the table that the parallel-pore model gives better results than the random-pore model. However, a subsequent study by Brown *et al.*[12] showed that, out of 12 porous solids studied in the pressure range 1–20 atm, 11 may be described equally well by both models (the parallel-pore model with $\kappa^{-1} = 3$); the deviation lies between 0.5 and 2.

Recently, Yang and Liu[134] have shown that the Wakao and Smith random-pore model is better than the parallel-pore model for predicting fluxes in a nuclear-reactor graphite with severe pore restrictions.

Summarizing, we may say that, in any case, the parallel-pore model [in particular, equation (4.95)], permits the prediction of total diffusive fluxes at all pressures, provided the tortuosity is experimentally measured [for example, in Table 5.2 there is good agreement with experiment when using equation (4.95)].

## 5.5.   Additional Comments

### 5.5.1.   On the Experimental Measurement of the Transport Parameters of a Porous Medium and Their Further Application

If we ask if previously measured transport parameters of a porous medium can be used in a particular case, the answer is in principle yes whenever the theoretical solution of equation (5.1) for that case fits experimental observations; this is the test of confidence of our transport parameters.

However, we may ask ourselves the following: can we apply transport parameters which have been determined under certain working conditions to other systems under other, different, working conditions? For example, assume that we have obtained our parameters from a steady state experiment without chemical reaction and that we want to apply them in a calculation of a transient system with chemical reaction; can we do this? Let us look at this question in more detail.

Assume we have a catalyst with micro- and macropores connected in parallel. In simple diffusion, the diffusion will proceed predominantly through the macropores, which offer less resistance to flow; when diffusion occurs with chemical reaction, the reaction takes place primarily in the micropores, which provide the major part of the surface area and the diffusion thus proceeds primarily along the path of higher resistance. For this reason, it is expected that the values of the effective diffusivities measured, for example, in a diffusion cell, will be greater than the values obtained (by a more cumbersome calculation) from the measurement of the effectiveness factor of the catalyst with chemical reaction.

Let us assume now that micro- and macropores are interconnected in a kind of series. For this case it is expected that both methods of measurement will give similar results. This situation has been theoretically verified by Wakao and Naruse[123] and the experimental results confirmed.[7, 59, 117]

In the particular case of a microporous catalyst it is also necessary to take account of the fact that the usual activation treatments may modify the pore radii distribution curve so that effective diffusivities are modified. For example, Stoll and Brown[112] indicate variations of the order of 11–13% in the flux of helium in a binary mixture with nitrogen in a $\gamma$-alumina catalyst with a unimodal pore distribution (average pore radius $= 47$ Å) and which had been reactivated. These authors also compared fluxes for the systems helium–nitrogen and helium–1-butene (the 1-butene isomerizes to *cis*- and *trans*-butene). In some cases the flux of helium in the reactive system is lower than that in the nonreactive system; this effect is ascribed to the coverage of the pore walls with 1-butene, resulting in a corresponding decrease in the free-pore radius.

Thus we see that, in general, the system for which we plan to apply the transport parameters should be analyzed and compared with the system used to measure these parameters. For example, data arising from a diffusion cell will be useful for systems in the steady state and without chemical reaction, while effective diffusivities measured by a chromatographic technique will be of use for transient and chemical systems. In the case of the chromatographic technique, the major difficulty will be the solution of the equation of change (5.1) to interpret the results.[26, 27, 45, 46, 50, 66, 101]

We also note that Grachev *et al.*[40] have compared the results of different methods of measurement (steady state, transient, delay time,[8] chromatographic, and with chemical reaction) in samples of an iron–molybdenum catalyst $(Fe/Mo = 1.7)$ for oxidation of methanol and a vanadium catalyst for oxidation of sulfur dioxide. Even though this work is very interesting, the analysis of the results was not very rigorous since a Fick's-type law was used.

### 5.5.2. On the Uncoupling of the Equations of Change in Multicomponent Mixtures in the Molecular Diffusion Regime

It is possible to uncouple the equations of change of the different species of a multicomponent mixture when the diffusion regime is molecular. This has been demonstrated by Cullinan[21] by applying matrix theory and a linearization method due to Toor[119] for an isobaric isothermal mixture; Cullinan obtains equations having the same form as for binary mixtures. The method of obtaining the matrix of the diffusivities was also given.[111, 120]

Sandor[96] arrived at similar results by applying Gyarmati's variation principle.[44] Turevskii *et al.*[121] solved the equations of change numerically for the same case and compared the results with those obtained by simplified procedures (such as those of Wilke,[133] Shain,[103] Toor,[119] and Konstantinov and Nikolaiev[68]) that are based on the use of expressions like the one introduced by Bird *et al.*[10] (Section 5.3.3.2), i.e., expressions having the same form as that for a binary mixture.

## Appendix. On the Use of Constitutive Diffusion Equations to Describe Total Fluxes in Nonisobaric Porous Media

In the isothermal constitutive diffusion equations,[†]

$$\frac{dc_i}{dz} = -\frac{N_i^D}{D_{ii}^K} + \sum_{\substack{j=1 \\ j \neq i}}^{\nu} \frac{x_i N_j^D - x_j N_i^D}{D_{ij}} \tag{5A.1}$$

it is a common practice to use total fluxes instead of total diffusive fluxes (i.e., as done by Hesse[53] for a nonisobaric porous catalytic pellet and Krishna[72] for a closed system). As we have seen, the total molar flux of a given species results from the addition of diffusive and viscous transport,

$$N_i = N_i^D + N_i^V \tag{5A.2}$$

where [cf. equation (4.29)]

$$N_i^V = -\frac{B_k}{\mu} x_i p \frac{dc}{dz} \tag{5A.3}$$

*For the particular case of the molecular diffusion regime* $(D_{ii}^K \gg D_{ij})$, by introducing equation (5A.3) into (5A.2) and the result into (5A.1) we find

$$\frac{dc_i}{dz} = \sum_{\substack{j=1 \\ j \neq i}}^{\nu} \frac{x_i N_j - x_j N_i}{D_{ij}} \tag{5A.4}$$

which is the Stefan–Maxwell diffusion equation written with total fluxes. Thus, with the assumption of a molecular diffusion regime as the only restriction, equation (5A.4) is valid.

However, equations such as (5A.4) are useful only if isobaric conditions are arbitrarily assumed,

$$\frac{dx_i}{dz} = \sum_{\substack{j=1 \\ j \neq i}}^{\nu} \frac{x_i N_j - x_j N_i}{c D_{ij}} \tag{5A.5}$$

---

[†]Equation (5A.1) is obtained from equation (4.12) written in terms of molar fluxes.

and it is in this way that they are actually used.[53, 72] As we have previously seen (Fig. 5.2), unless the permeability is extremely low, the pressure variation developed in the molecular diffusion regime is not very important, so we can make the simplification leading to equation (5A.5). Thus, rather fortunately, the use of equations like (5A.5) for expressing total fluxes in porous media may be regarded as correct provided: (a) the molecular diffusion regime prevails, and (b) the permeability of the porous solid is not very low. In any case one must be aware that the viscous flux, although not appearing explicitly in the equations, can be a significant fraction of the total flux.

# 6

# History of Diffusion

## PART A.  EVOLUTION OF THE KNOWLEDGE OF DIFFUSION

### 6.1.  Historical Developments

It is unusual for a book dealing with the present state of knowledge in a field of physical science to give a detailed historical analysis of the subject. But anyone undertaking a detailed study of diffusion in gases will find that some discoveries which are assumed to have been made recently were, in fact, made at the beginning of the nineteenth century, and further that some laws, experimentally verified more than a century ago, have remained in obscurity.

An historical analysis of this cultural curiosity is imperative for scientific enlightenment, and without such an analysis we feel we would not completely accomplish our objective of gaining a better understanding of the nature of diffusion.

We are aware, of course, that the history of this subject is an integral part of the history of mankind, and that every man is a prisoner of the culture of his time. Any subdivision of history is therefore artificial, but at the same time necessary, since it is the only way to analyze a problem.

We do not intend to provide a complete history, with an abundance of references, but will concentrate on the main discoveries and developments from the beginning of the nineteenth century up to our own time.

Our history has two main lines of development: the *phenomenological–experimental*[†] development began in 1826 with Thomas Graham (1805–1869), and the development of the *molecular theory* began in 1860 with

---

[†] We mean to include here *ad hoc* theoretical developments such as Graham's and Fick's laws (but not, of course, the kinetic theory).

Table 6.1.    Chronology of the Development of Our Knowledge of Diffusion

| Diffusion | Nonseparative flow |
|---|---|
| 1831. Graham's diffusion law (empirical)[43, 44] | |
| | 1839. Laminar flow in liquids (Hagen, experimental)[51] |
| | 1841. Laminar flow in liquids (Poiseuille, experimental)[92] |
| 1845. Equipartition of energy in a gas (Waterston, theory)[70] | |
| 1846. Graham's effusion law (empirical)[45] | |
| 1851. Graham's experiments on diffusion in liquids[46] | |
| 1855. Fick's law (phenomenological, for liquids)[35] | |
| | 1856. Darcy equation (empirical)[24] |
| | 1858. Laminar flow theory (Neumann)[99] |
| 1859. Maxwell's distribution of velocities in a gas[82] | |
| 1860. Maxwell's constitutive equation of diffusion (phenomenological) and formula for the diffusivity[79, 80, 82] | 1860. Laminar flow theory (Hagenbach)[99] |
| 1863. Diffusion experiments in a closed system (Graham)[47, 48] | |
| 1866. Mathematical description of a nonuniform gas (Maxwellian gas)[82] | |
| 1867. Calculation of diffusivities from experimental results (Maxwell)[81] | |
| 1870. Measurement of diffusivities (Loschmidt)[72, 73] | |
| 1871. Stefan's equation[108, 109] | |
| 1872. Boltzmann's $H$-theorem[12] | |
| 1873. Measurement of diffusivities (Stefan)[110–112] | |
| | 1875. Slip flow (Kunt and Warburg, experimental)[66] |
| 1904. Diffusion cell (Buckingham)[14] | |
| 1905. Theory of diffusion in a Lorentzian gas (Lorentz)[71] | |
| | 1909. Knudsen flow (theory and experiment)[63] |

Table 6.1—*continued*

| Diffusion | Nonseparative flow |
|---|---|
| 1912–17. Chapman's developments[17–19] | |
| 1911–17. Enskog's developments[27–29] | |
| 1922. Extension of theory to dense gases (Enskog)[30] | |
| 1924. Equation for repelling molecules (Lennard-Jones)[67] | |
| 1935. Demonstration of the appropriateness of Sonine polynomials for expansions in Chapman–Enskog's theory (Burnett)[15] | |
| | 1943. Coupling between diffusive and viscous fluxes (theory and experiment, Kramers and Kistemaker)[65] |
| 1944. Self-diffusivity in the transition regime (Bosanquet, intuitive)[13] | |
| 1948. Self-diffusivity in the transition regime (Pollard and Present, theory)[93] | |
| 1949. Constitutive equations of diffusion for multicomponent systems (Curtiss and Hirschfelder)[21] | |
| 1955. Rediscovery of Graham's law of diffusion (Hoogschagen)[57] | |
| 1961. Constitutive equations of diffusion including wall effects at uniform pressure: dusty gas model (Evans, Watson, Mason, theory)[32] | |
| 1962. Experimental verification of the dusty gas model at uniform pressure (Evans, Watson, and Truitt)[34] | |
| 1967. Constitutive equations of diffusion under pressure gradients including wall effects (Mason, Malinauskas, and Evans, theory)[77] | |
| | 1969. Constitutive equations of viscous flux in porous media (Slattery, theory)[105] |
| 1969. Experimental verification of the constitutive equations of diffusion with wall effects and viscous flow (Gunn and King)[50] | |

James Clerk Maxwell (1831–1879). We are thus deliberately skipping over Democritus, Epicurus, and Lucretius, who founded the atomic theory in ancient Greece; we are skipping over Gassendi, Hooke, and Bernoulli (1650–1750), who revived the atomic theory in the early days of modern science; and we are skipping over Harapath (1821), Waterston (1845), Krönig (1856), and Clausius (1857), who began the development of the kinetic theory. The development of the molecular theory of diffusion has been more difficult than that for the other transport phenomena, but despite some errors unbelievably repeated for more than a century [such as the erroneous coefficient $\frac{1}{3}$ Maxwell gave in his first expression for the viscosity (see Section 6.4)], we arrived, some 30 years ago, at a deeper knowledge of the subject. The phenomenological–experimental development of diffusion, on the other hand, has required more time for clarification and even in the 1960s there were unexpected surprises concerning the coupling of diffusion and nonseparative flow, a coupling that occurs in practically all laboratory experiments on diffusion.

As a guide to the reader and to help keep the historical perspective as our story develops, we give, in Table 6.1, a chronology of events in the history of diffusion. The grouping of events under the two headings "diffusion" and "nonseparative flow" reflects the fact that, as we shall see, the phenomenological–experimental studies of these phenomena were separate and independent, it being long assumed that they were completely unrelated phenomena.

### 6.1.1.  The Phenomenological–Experimental Study of Diffusion

#### 6.1.1.1.  Thomas Graham's Experiments

Until 1826 little was known about diffusion other than the experimental observation that, when two gases are kept in contact, the heavier one does not separate out at the bottom–the two gases mix uniformly.[49]

In 1826 the first[40] of Graham's many papers on gases was published; this was followed in 1829[41] by the first study of diffusion. Graham's merit lies in the fact that he was the first scientist to make laboratory experiments on diffusion and to have discovered empirical laws of diffusion that are still valid today. In his first experiments, a pure gas (hydrogen, methane, ammonia, ethylene, carbon dioxide, hydrogen sulfide, chlorine) was contained in a tube closed off by a porous plug through which it diffused into the air.[98] His first observation was that more hydrogen escaped in 2 hours than carbon dioxide in 10 hours, and that the escape rate of the gases appeared to be some function of their specific gravity, probably their square root.

Graham also developed the two-bulb closed system,[41, 42] reinvented 118 years later by Ney and Armistead,[87] for measuring diffusivities. In this experiment, Graham filled a flask with the same volumes of hydrogen and ethylene and another flask with carbon dioxide; when both flasks were connected with a tube he observed that, after 10 hours, hydrogen and ethylene had entered the second flask and had a volume ratio 12:3.1.[†]

On December 19, 1831, Graham read his paper "On the Law of Diffusion of Gases" (published two years later) before the Royal Society of Edinburgh.[43, 44] In this paper Graham stated that[49]

> The diffusion or spontaneous intermixture of two gases in contact is effected by an interchange in position of indefinitely minute volumes of gases, which volumes are not necessarily of equal magnitude, being in the case of each gas, inversely proportional to the square root of the density of that gas.

In this experiment, a gas contained in a tube closed with a porous gypsum plug at one end was allowed to escape into the atmosphere while the open end of the tube was immersed in water or mercury. Graham varied the porosity of the plug by modifying either the initial water content of the gypsum or the drying temperature. As the gas escaped from the tube, and as the air entered it, the liquid level inside the tube fell or rose, depending on whether the gas was heavier or lighter than air, respectively. Graham realized that this meant the pressure inside the tube was modified and he understood that the pressure change would complicate the interpretation of the results. To avoid this pressure gradient, he continually raised or lowered the tube so that the water level remained constant. Later scientists forgot Graham's conclusion that *diffusion can spontaneously generate a pressure gradient* and that the condition of *uniform pressure was only attainable by external imposition*, and this lapse is one of the main reasons for the confusion which arose many years later.

Table 6.2 shows that Graham's experimental results agree well with his "theory," i.e., the postulated inverse proportion of the flux to the square root of the gas density.

We must bear in mind that there were only limited tools available to Graham for interpreting his results (Fick's law was not formulated until almost 20 years later, and Maxwell's development of the molecular theory of gases was some 30 years away). Two transport laws were known: Newton's law of viscosity ("the resistance arising from 'want of lubricity' in the parts of a fluid, is, other things being equal, proportional to the velocity with which the parts of the fluid are separated from one

---

[†]Current theory gives for this ratio $(M_{C_2H_4}/M_{H_2})^{1/2} = 3.702$, which compares very well with Graham's observed value, 3.87; this gives an idea of the accuracy of Graham's measurements.

Table 6.2.    Diffusion into Air: Graham's Experimental Results vs. His Law[43, 98]

| Gas | $V_{gas}/V_{air}$ (experimental)[a] | $(M_{air}/M_{gas})^{1/2}$ (theoretical)[b] |
|---|---|---|
| $H_2$ | 3.83 | 3.7947 |
| $CH_4$ | 1.344 | 1.3414 |
| $C_2H_4$ | 1.0191 | 1.0140 |
| $CO$ | 1.0141 | 1.0140 |
| $N_2$ | 1.0143 | 1.0140 |
| $O_2$ | 0.9487 | 0.9487 |
| $H_2S$ | 0.95 | 0.9204 |
| $NO$ | 0.82 | 0.8091 |
| $CO_2$ | 0.812 | 0.891 |
| $SO_2$ | 0.68 | 0.6708 |

[a] $V_{air}$ is the volume of air that diffused into the tube (see text for a description of the experimental setup) and $V_{gas}$ is the volume of gas (listed in the left-hand column) that diffused out of the tube.
[b] $M_{air}$ and $M_{gas}$ are the molecular weights (proportional to the densities) of, respectively, air and the diffusing gas listed in the left column.

another"[97]) and what is usually known as Fourier's law of heat transfer,[36] which in fact had been suggested previously by Biot.[6, 7] In Graham's time these laws were assumed to deal with unrelated phenomena, between which there was no analogy whatsoever. The gap was bridged by intellectual abstractions such as that of Fick who drew an analogy between Ohm's law and the phenomenon of diffusion.

Graham's next study, in 1846, involved what he called "effusion": the flow of a gas into a vacuum through a small opening in a thin plate.[45] Graham studied eight gases and concluded that different gases pass through minute openings into a vacuum in times which are proportional to the square roots of their respective specific gravities, or with rates which are inversely proportional to the square roots of their specific gravities. This result may be easily derived from a thermodynamic analysis of the discharge velocity from nozzles for very low discharge pressures.[91]

We see that the statement of *Graham's laws of diffusion and effusion are identical, though these two phenomena involve quite different working conditions.* It is also interesting to note that while effusion is much more difficult to study experimentally than diffusion, it is much more easily explained theoretically.

Graham also studied the laminar flow of gases in long capillaries—a phenomenon which he called "transpiration"—and recognized that there was no simple relationship between the flux of different gases and their densities.[45]

These studies were performed some years after those carried out (with liquids) on empirical bases and independently by the German hydraulic engineer Gotthilf Heinrick Ludwig Hagen[51] in 1839 and by the French physicist Jean Louis Poiseuille[92] in 1841.

### 6.1.1.2. Fick's Law

After his study on effusion, Graham experimented with diffusion using aqueous solutions. He placed an aqueous solution in contact with pure water and measured the total amount of solute transferred to the water; Graham used hydrochloric acid and its salts of sodium, calcium, strontium, and barium; nitric acid and its salts of sodium, silver, calcium, strontium, and barium; sulfuric acid and its salts of magnesium, zinc, and thorium; acetic acid and its salts of lead and barium; sulfurous acid, ammonia, and ethanol. In his paper,[46] published in 1851, Graham reported experimental results without any further analysis for, as explained in a footnote, *every solute followed a different diffusion law.*

Four years later (1855), Adolph Fick, Demonstrator of Anatomy in Zürich, was prompted by Graham's paper on diffusion in liquids to carry out new experiments looking for a single diffusion law.[35] Fick repeated Graham's experiments, but measured density profiles with an immersion balance. There were crystals at the bottom of the solution, so that the solution at the bottom was saturated. By using Ohm's law as an analogy, Fick proposed a diffusion law, usually known as "Fick's first law" [equation (1.32)]; applying the law, Fick predicted and verified that the concentration profile must be linear for cylindrical tubes and in the steady state; he observed that the diffusivity of sodium chloride was constant throughout the solution.

Fick also predicted the concentration profiles of sodium chloride for a conical geometry and verified the profiles by working with an inverted funnel. He also studied the diffusion of sodium chloride through collodion and pig's bladder.

The successful experimental verification of Fick's law led later scientists, as we shall see below, to assume that their experiments should be described by Fick's law exclusively. This was true, but at the same time fortuitous, as we shall see in Part B.

The experimental results on laminar flow obtained by both Hagen and Poiseuille were treated analytically by Neumann and Hagenbach (independently of each other) between 1858 and 1860,[99] and they formulated what is now called the Hagen–Poiseuille law[†]; these developments, however, were not in any way related to Fick's. *Nobody suspected during those years that diffusive and viscous fluxes could be coupled.*

---

[†]So this law should really have been called the Neumann–Hagenbach law!

### 6.1.1.3.  Loschmidt's and Stefan's Studies

Graham continued to develop new laboratory techniques on diffusion and in 1863 reported the design of a two-tube closed system which consists essentially of two tubes connected by a valve that originally separates two different gases.[47, 48] Four years later, Maxwell analyzed Graham's experimental results with this system by applying Fick's law; Maxwell calculated a value for the diffusivity of carbon dioxide–air that approximates the present value very well.[81]

Loschmidt,[72, 73] in work reported in 1870, also worked with a two-tube closed system, and despite the fact that Graham preceded him in this, Loschmidt is usually credited with introducing this type of system. Von Obermayer[88] and Boardman[11] improved on this system. The tubes used by Loschmidt had a 26-mm internal diameter and those of von Obermayer, 13mm; Loschmidt's results were 6% higher than von Obermayer's.

Up to this point the types of experimental devices used for measuring diffusivities (all closed systems) and the applicability of Fick's law to them, without having to use any other equation, had given rise to a conceptual picture of diffusion as "the phenomenon by which a gaseous mixture finally reaches the state of uniform composition." While we now know this description of diffusion is valid only for a closed system, it was considered valid for any system, open or closed. This error led to several incorrect ideas (some of which are still current), for example, that diffusive fluxes are always equal and opposite.

In 1873, a few years after Loschmidt's work, Stefan developed a new technique for measuring diffusivities of mixtures of a gas and a vapor; this technique corresponds to what we have called a "semiopen system without reaction" in Section 5.3.2.[110–112] Stefan used tubes with internal diameters ranging between 0.64 and 6.16 mm. Stefan also solved the mass balance equations with Fick's law, valid for Graham's two-tube closed system experiments in the transient state, and tabulated the results.[110]

### 6.1.1.4.  Flow of Rarefied Gases (Kunt–Warburg; Knudsen)

In 1875, Kunt and Warburg[66] continued the Hagen and Poiseuille experiments, but they used gases at low pressures instead of working with liquids; they observed the so-called slip flow. Knudsen[63] (1909) continued these experiments, working at very low pressures, and observed the so-called Knudsen flow, which he explained on the basis of the kinetic theory.

### 6.1.1.5.  Other Studies

In 1904, Buckingham[14] described a diffusion cell in which it was possible to control the pressure on both faces of a porous septum or at both ends of a capillary by adjusting the flow rates of both gases. This cell, which is shown in Figure 5.15, is usually ascribed to Wicke,[122, 125] but it does not essentially differ from a device (described in Section 6.1.1.1) Graham used; in Graham's device the pressure inside a tube was controlled by varying a liquid level.

In 1923, Hertz[53] reported another technique he had developed to measure diffusivities. In Hertz's system, two gases flow through different tubes until they reach a mixing chamber, and both gases then flow out of the chamber through one tube (see Figure 6.1). Simultaneously, each component diffuses backward into both inlet streams. The trace of one gas in the inlet stream of the other can be determined by the sampling device I and from this the diffusivity of the gas can be calculated.

We thus see that, as pointed out in connection with Table 6.1, the phenomenological–experimental approach to the study of diffusion has followed two independent courses:

(a) The Graham–Fick–Loschmidt–Stefan, or "diffusion," course in which the phenomenon of diffusion is made synonymous with Fick's law exclusively and is interpreted, theoretically and experimentally, as a segregation of species totally devoid of nonseparative flow (unless that flow was specifically produced as in Hertz's technique); after the promulgation and acceptance of Fick's law, some of Graham's results were used (e.g., by Maxwell and Stefan) in order to calculate diffusivities from his results, but his other contributions were not mentioned because his law of diffusion was found not to be obeyed in the Loschmidt and Stefan experiments.

(b) The Hagen–Poiseuille–Neumann–Hagenbach–Kunt–Warburg–Knudsen course, which dealt with the nonseparative flow due to pressure gradients.

## 6.1.2.  The Development of Diffusion Theories

### 6.1.2.1.  The Developments by Maxwell, Stefan, and Boltzmann

At the end of the nineteenth century there were three theoretical approaches to diffusion:

(a) A complete development of the equations of the rigorous kinetic theory which included Boltzmann's integrodifferential equation and the transport equations developed by Maxwell[82]; the latter equations can be obtained from the Boltzmann equation.

Figure 6.1 Schematic representation of Hertz's method.[53]

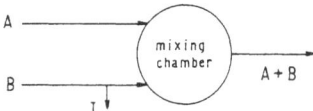

(b) A theory of momentum transport, developed independently by Maxwell[79] and Stefan,[108, 109] that led to: (b.1) an expression which correctly predicts the binary molecular diffusivity [equation (2.92)]; (b.2) an expression for diffusive fluxes in a binary system in the form of a partial momentum balance [equations (1.37) and (1.38)].

(c) A theory of diffusion based on mean free paths developed by Meyer in 1877[85]; this theory yields an expression for the binary molecular diffusivity that depends on the gas composition [equation (2.88)]; this prediction did not agree with experimental results.

As we shall immediately see, these three approaches evolved independently during the first part of the twentieth century, although the relationships between them had been demonstrated[70]: the theory of momentum transport leads to a first approximation of Chapman–Enskog's solution of Maxwell's equation of transport, and the mean free path theory leads to a solution equivalent to the first iteration of the linearized Boltzmann equation.

### 6.1.2.2.  Approximate Solutions of the Boltzmann Equation (Chapman–Enskog, Grad–Zhdanov, BGK)

In work published between 1911 and 1917, the rigorous kinetic theory equations were solved independently by Chapman,[17–19] who started from the transport equations developed by Maxwell, and by Enskog,[27–29] who started with Boltzmann's integrodifferential equation. The original Chapman–Enskog theory was valid for binary mixtures of monatomic gases and was later extended to polyatomic gaseous mixtures with quantum effects taken into account.

More rigorous solutions of the Boltzmann equation were obtained in 1949 by Grad[39] (for one-component systems) and in 1962 by Zhdanov, Kagan, and Sozykin[130] (for multicomponent systems). In 1954, Bhatnagar, Gross, and Krook[5] developed the BGK model, which uses a linearized Boltzmann equation.

### 6.1.2.3.  Partial Momentum Balances

The method of partial momentum balance was ignored for many years until Frankel used it again in 1940[37]; Furry[38] (1948), Present and de Bethune[94] (1949), and Williams[126] (1958) also used it.

### 6.1.2.4.  Mean Free Path Theory

This theory was improved by Jeans in 1921[59] by ignoring the effect of collisions between like molecules, thus obtaining an expression analogous to equation (2.92).

### 6.1.2.5.  Other Developments, Thermodynamics of Irreversible Processes

This independent method of studying transport phenomena began in 1931 with the work of Onsager.[89, 90] We have described it in Chapter 3.

## 6.2.  Present Synthesis of the Theoretical and Phenomenological–Experimental Studies

In the history of a field of science, there is very frequently a period in which many apparently unrelated experimental results and theoretical developments are obtained. This period is followed by another period of slow intellectual ripening which eventually leads to a third period characterized by a crystallization of ideas with more generalizing abstractions. These generalizations permit different aspects, hitherto isolated, to be linked together.

In our case, the molecular theories represented very important progress because these theories explained the mechanism of diffusion very well (for example, thermal diffusion was predicted theoretically prior to its experimental observation, a reversal of the usual sequence in science). However, the theory was valid for systems without walls, i.e., systems that cannot really be represented by any laboratory device, and, as we now know, walls are responsible for the coupling between the diffusive and viscous fluxes (furthermore, it is nonsense to speak of a viscous flux in a system without walls). There was thus a very-well-built theoretical frame but there were experimental results that could not be explained within that frame.

### 6.2.1.  The Work of Kramers and Kistemaker and of Hoogschagen

The first experimental evidence (provided we forget Graham!) of the coupling between diffusive and viscous fluxes was obtained in 1943 when Kramers and Kistemaker[65] connected two bulbs containing hydrogen and air to both ends of a capillary 0.920 mm in diameter and detected a spontaneously developed pressure gradient (a maximum pressure difference of 0.02 mm Hg was observed). The authors explained their results by momentum transfer arguments and the use of Fick's law.

In the period 1945–46, Skinner and Beeck,[104] as cited by Wheeler,[121] detected the pressure drop for hydrogen–ethane diffusing in the Knudsen regime in a silica gel pellet with pores 30 Å in diameter, and Bernstein[4] in 1949 detected it for hydrogen–nitrogen and hydrogen–argon in the Knudsen regime in a capillary 0.0472 mm in diameter.

By that time, the analysis of the influence of diffusion on a chemical reaction occurring in a catalyst pellet[23, 113, 114, 129]† lent interest to the study of diffusion within porous media. Thus Wicke (1941) studied the surface diffusion of carbon dioxide diluted in nitrogen within activated carbon, porous clay, and fritted glass,[125] and he studied (1949) the variation of catalytic activity with pellet size, which indirectly gives an effective diffusivity[123]; Weisz[120] (1957) measured the interdiffusion of nitrogen and hydrogen within pellets of silica–alumina and chromia–alumina, calculating the effective diffusivity, and repeated the study with carbon pellets, calculating axial and radial effective diffusivities from material provided by Walker et al.[116, 117]

In all these studies the diffusion cell was used and uniform pressure was maintained. As in Loschmidt's experiments, it was assumed that Fick's law described the phenomenon of diffusion completely. Consequently, only one diffusive flux was measured, with the other flux assumed to be equal and opposite.

However, Hoogschagen[57] (1955) studied the diffusion of the pairs hydrogen–oxygen, nitrogen–oxygen, and carbon dioxide–oxygen in tablets of carbonyl iron with pores 0.8 $\mu$m in diameter; he measured both countercurrent fluxes. Hoogschagen was surprised to observe that the values of the two molar fluxes were not equal but were inversely proportional to the square root of their molecular weight. This meant that, in spite of the uniform pressure, there was a net molar flux.

Hoogschagen, who overlooked Graham's law of diffusion, thought that he had discovered a new law (which began to be called the "square root law"), which he explained by applying momentum transfer arguments. He used the constitutive equation of diffusion [equation (1.37)] developed originally by Maxwell[82] (and which was given in textbooks[58, 102] at the time of Hoogschagen's work ) and which had usually been applied letting $N_{zA} = -N_{zB}$, therefore reducing it to Fick's law. Hoogschagen solved equation (1.37) using $N_{zA}/N_{zB} = -(M_B/M_A)^{1/2}$.

Kramers and Kistemaker, on the one hand, and Hoogschagen on the other, thus showed that diffusion was not as simple as was heretofore believed.

### 6.2.2. Systems with Diffusion and Chemical Reaction

Despite the above developments, diffusion was still poorly understood. This is evident from the fact that the conclusions obtained by Kramers and Kistemaker and by Hoogschagen were considered valid only

---

†These studies were in fact preceded by those of Juttner in 1909 as quoted by Mercer and Aris[84] and of Sir James Jeans in 1921.[59]

for systems without reaction. For example, there is no essential difference between Thiele's pioneer study in 1939[113] and those theoretical papers appearing after Hoogschagen's work that dealt with the effectiveness factor for reactions in which equation (5.73) is not verified: all of them assume uniform pressure.

Thiele applied Fick's law exclusively and, among his working hypotheses, there is one which states that[113] "There is no draft [net molar flow] through the catalyst grains, all transfer being by diffusion, or resulting from a change in volume during reaction." Thiele expressed the volume change by a coefficient on which the effectiveness factor depended. Using Fick's law, he analytically solved the equation of change for a first-order irreversible reaction for slab geometry. Twenty-six years later, Weekman and Gorring[119] applied Hoogschagen's equations for the reaction $A = bB$, letting $bN_{rA} = -N_{rB}$ and assuming uniform pressure; they obtained the effectiveness factor numerically for spherical geometry and reaction orders 0, 1, and 2. The analytical solution for the zeroth-order reaction under identical conditions was obtained by Lin and Lih.[69] In 1961, from the experimental results for the effectiveness factor, Hawthorn[52] obtained an empirical formula for volume changes as a function of a Thiele modulus. The use of a Thiele modulus was theoretically justified by Bischoff[10] in 1965 by introducing an effective diffusivity as a function of a volume change coefficient (this coefficient has also been used recently[68]); however, this solution is not rigorous since diffusive and viscous effects are not given separately by their respective constitutive equations.

Solutions for the multicomponent mixture were also obtained for the transition regime using a constitutive equation such as equation (4.12), but the fluxes involved were assumed to be the total fluxes (diffusive plus viscous)[16] and uniform pressure was assumed. In 1965, Wakao, Otani, and Smith[115] calculated the pressure gradient, but again used the constitutive equations of diffusion with total fluxes (this error will be considered in Section 6.3).

### 6.2.3.  Attempts at Explaining Wall Effects

In Loschmidt's and Stefan's experiments the frequency of the molecule–wall collisions was negligible compared to that of molecule–molecule collisions; in Knudsen's experiments at $p \to 0$, the reverse situation prevailed. The diffusion experiments with porous media may represent a state intermediate between the two extremes considered.

It seems that Bosanquet[13]† was the first (in 1944) to consider the combined effects of molecule–molecule and molecule–wall collisions. He

---

†Bosanquet's report has been classified and is not available.[106]

considered the diffusion of a pure gas in a capillary as a random walk phenomenon in which the flight of a molecule is modified by colliding either with another molecule or with the wall. In such a process the frequencies of both types of collisions are additive. The relationship between the mean free path $\lambda^R$ of this random walk phenomenon, the mean free path $\lambda_p$ of the molecule–wall collisions, and the mean free path $\lambda$ of the molecule–molecule collisions, assuming that the total collision frequency is given by $\bar{v}/\lambda^R$, where $\bar{v}$ is the average velocity of a molecule, is

$$\frac{\bar{v}}{\lambda^R} = \frac{\bar{v}}{\lambda_p} + \frac{\bar{v}}{\lambda} \tag{6.1}$$

The diffusivities associated with each of these mechanisms can be considered to be proportional to $\lambda^R \bar{v}$, so that

$$\frac{1}{D_{AA}^R} = \frac{1}{D_{AA}^K} + \frac{1}{D_{AA}} \tag{6.2}$$

In 1948, Pollard and Present[93] developed a complex expression for $D_{AA}^R$ founded on kinetic theory and which involved an integral that could be numerically solved for a few values of the Knudsen number. The quantity $F_{zA}/\Delta p$ was plotted as a function of the Knudsen number and the result was compared with that arising from the much simpler (and intuitive) equation (6.2) assuming that $D_{AA}^K = \frac{2}{3} R \bar{v}_A$ and $D_{AA} = \frac{1}{3} \lambda \bar{v}_A$ (the expression for $D_{AA}$ being erroneous, as we shall see in Section 6.4); the agreement between Pollard and Present's results and Bosanquet's was very good. Later, in 1951, Wheeler[121] proposed that

$$D_{AB}^R = D_{AB}\left[ 1 - \exp\left(-D_{AA}^K/D_{AB}\right) \right] \tag{6.3}$$

This was the state of knowledge of the influence of molecule–wall collisions on diffusion when effective diffusivities in porous media began to be measured. This knowledge was not in contradiction with the generally held belief that $N_A^D = J_{AM}$.

However, the interpretation of the effective diffusivity became more complex as a consequence of Hoogschagen's findings. The constitutive equation of diffusion used by Hoogschagen [equation (1.37)] involved the molecular diffusivity alone (this is justified for Hoogschagen's results since he used a macroporous solid). But this equation is incompatible with equations (6.2) or (6.3) since, if we introduce $D_{AB}^R$ into equation (1.37), the resulting expression cannot be reduced to the correct form for the Knudsen regime. This incompatibility showed the absence of a rigorous development capable of extending the Chapman–Enskog (or Grad–Zhdanov) equations to systems with walls.

### 6.2.4.  The Development of the Dusty Gas Model by E. A. Mason et al.

The study of nuclear reactors, a subject alien to heterogeneous catalysis, seems to be the stimulus that, at the beginning of the 1960s, led to the extension of the Chapman–Enskog equations to systems with walls.

Nuclear reactors cooled by helium have a central body of graphite and uranium (indicated by II in Figure 5.16); by fission of the uranium, the radioactive gases B are generated. To keep the reactor cooled and, at the same time, to remove the radioactive contaminants as rapidly as possible, the central body is surrounded by a jacket of a low-permeability graphite leaving an annular space to let the gases B flow at 20 atm pressure; outside the jacket the cooling gas A flows at 21 atm pressure and 600–800°C.[34] The interdiffusion of A and B, under a pressure drop of 1 atm, takes place within the jacket. Curiously, this device is in essence very similar to the diffusion cell described in Section 5.3.4.1 with the uranium core as the source of B.

The problem of diffusion in the above setup led to studies on the interdiffusion of helium and argon in macroporous graphites by Evans, Truitt, and Watson.[31] The first study, completed at the end of 1960, used equations similar to those of Hoogschagen's to interpret experimental results in graphite with pore diameters of approximately 3 $\mu$m. The "square root law" was verified and the effective diffusivity was measured at different pressures (1–6 atm) and temperatures (room temperature to 100°C).[31]

About the same time (1961), Wicke and Hugo[124] also verified the "square root law" for the pairs carbon monoxide–nitrogen (whose molecular weights are the same) and carbon dioxide–hydrogen in fritted glass at room temperature and atmospheric pressure, and new studies on the Kramers–Kistemaker effect appeared.[25, 83]

Simultaneously, E. A. Mason, together with R. B. Evans III and G. M. Watson, developed the dusty gas model for uniform pressures[32] in 1961 and subsequently extended it by computing the effect of pressure gradients.[33]

The dusty gas model for isobaric conditions was verified for the interdiffusion of helium and argon in a low-permeability graphite.[34]

The theoretical developments of the dusty gas model were completed in 1967 in a paper in which second-approximation diffusivities are introduced, slip flow is interpreted as a Knudsen mechanism (the variation of $[D_{AA}^K]_2$ as the pressure varies; cf. Section 4.3.1.3), and pressure diffusion appears as a mechanism of transport distinct from viscous transport.[75] These equations were experimentally verified in 1969.[50]

The extension of the Chapman–Enskog equations was thus finally obtained by a fundamental theoretical study of the effects of walls on diffusion.

## PART B.    THE ERRORS, CONFUSION, AND MISUNDERSTANDINGS

### 6.3.    The Fluxes Involved in the Phenomenon

#### 6.3.1.    The Measurements of Graham, Loschmidt, and Stefan

If we return to the nineteenth century work, we observe that Loschmidt and Stefan determined diffusivities exclusively by applying Fick's law (which, we must remember, was obtained for diffusion in liquids) to gaseous systems under different working conditions. The diffusivities measured in this way agree very well with modern measurements by other techniques. Graham, before Fick's law was known, obtained two phenomenological laws, one for diffusion and the other for effusion.

If we now consider the early work from the point of view of our present knowledge, many questions arise: Is the accuracy of Loschmidt's and Stefan's results merely fortuitous? How is it possible that they obtained correct results with only the aid of a law valid for liquids and that ignored the existence of nonequimolar and viscous fluxes? Why, in Loschmidt's and Stefan's time, was Graham's law taken as only a rough approximation if his measurements agree with the present-day ones? What relationship is there, if any, between Hoogschagen's, Loschmidt's, and Stefan's results and the Kramers–Kistemaker observations? Why do many textbooks confuse Graham's laws of diffusion and effusion? These questions show that the study of diffusion has traveled a tortuous road with many bypasses and dead ends. We intend now to explain these (real and apparent) contradictions on the basis of our present knowledge (or ignorance). A very important part of the task of explaining the contradictions was carried out by Professor E. A. Mason and others in articles written with just this objective.[74–76]

Graham's results led to a law for the ratio of the fluxes of two species interdiffusing in an isobaric porous medium. What Graham measured was the total diffusive flux:

$$N_A^D = J_{AM} + N_A^N \qquad (6.4)$$

Obviously, Graham did not know whether he was measuring $N_A^D$ or $J_{AM}$ and so his law remained as the expression of a ratio of the fluxes of species which "are diffusing," without specifying (in fact, without suspecting) whether such a ratio belonged to $N_A^D/N_B^D$ or $J_{AM}/J_{BM}$, since at that time (and for many years afterward) nothing was known about the possible existence of two kinds of diffusive fluxes.

From his experiments on diffusion in liquids (which led eventually to Fick's law), Graham could not obtain any simple law. This is understandable from the way the experiment was performed. Graham measured the macroscopic transfer of solute from the solution during a given interval (he did not measure the mass transfer rate but the total amount of mass transferred). The calculation of a diffusivity from such a result would involve a calculation which was almost impossible in his day, requiring: (1) the formulation of a constitutive equation of diffusion, an equation unknown at that time; (2) writing the differential equation for the mass balance of solute (equation of change of the solute in the transient state); (3) integration of the equation for several values of the diffusivity, checking the predicted with the experimental results to find the appropriate value for the diffusivity.

We can therefore understand why it was generally accepted that, as Graham stated in his paper,[46] every solute has a different diffusion law.

Similarly, when Fick's law was postulated, it was impossible to establish a difference between $N_A^D$ and $J_{AM}$. But if we let our fancy wander, we can imagine that, had Graham been able to apply Fick's law to his results with gases, he would very probably have concluded, erroneously, that

$$D_{AB}/D_{BA} = (M_B/M_A)^{1/2} \qquad (6.5)$$

The phenomenon of diffusion is often misunderstood even today as shown by the fact that equation (6.5) has been cited in contemporary studies.[62, 64]

In contrast to Graham's experiments, Loschmidt's were made in a closed system, as were Fick's; consequently, in their systems the net molar flux was zero and, from equation (1.9), it follows that

$$N_A = J_{AM} \qquad (6.6)$$

Thus, the total flux, which here is the observed flux, is the same as the diffusive flux. Furthermore, since Loschmidt did not work with narrow capillaries, it was impossible to detect a pressure gradient and, consequently, to even consider the possibility of any flux other than the diffusive one.

The equation of change for component A on a molar basis in Loschmidt's experiment is given by [cf. equation (5.1)]

$$- \partial N_{zA}/\partial z = \partial c_A/\partial t \tag{6.7}$$

If Fick's law is introduced into equation (6.7), we obtain what is commonly called "Fick's second law"; the solution of the resulting equation enabled Loschmidt to calculate diffusivities from his results. Since wall effects were negligible, the diffusivities involved were the molecular ones and so Loschmidt's results were correct,[†] and this explains the agreement with modern results, although they also revealed, as Mason has stated,[78] how difficult it is to measure diffusivities accurately.

Evidently, the ratio of the observed fluxes, $N_A/N_B$, in Loschmidt's experiments was the ratio of the total (diffusive plus viscous) fluxes rather than that of the total diffusive fluxes, $N_A^D/N_B^D$, as was the case in Graham's experiment. Loschmidt, assuming that fluxes were only diffusive, had originally intended to verify Graham's law of diffusion; he obviously could not and, obviously too, he ascribed the error to Graham. Graham's law of diffusion was thus discredited and was for the most part forgotten; only his law of effusion was considered to be valid. Subsequently, many people thought that there was only *one* Graham's law: For example, a theoretical explanation of Hoogschagen's results included the statement[26]: "This result [the 'square root law,' i.e., Graham's law of diffusion!] should not be confused with what has been known as Graham's law of diffusion for over a hundred years."

On the other hand, Graham's law of diffusion predicts that if we have the gas pairs A–C and B–C, the ratio of the total diffusive fluxes of A and B is given by

$$N_{A-C}^D/N_{B-C}^D = (M_B/M_A)^{1/2} \tag{6.8}$$

where $N_{A-C}^D$ is the flux of A when this component interdiffuses with C, and similarly for $N_{B-C}^D$. However, if we assume that the fluxes observed by Graham are described by Fick's law and we remember that $D_{ij} = D_{ji}$, we

---

[†]The solution of equation (6.7) together with Fick's law is based on solving equation (6.7) for a semiinfinite medium (or assuming that $c_A = 0$ as a boundary condition). The solution yields $D_{AB} = \pi N_{MA}^2/S^2 c_{A0}^2 t$, where $N_{MA}$ is the number of moles that have diffused into a given volume of the chamber of B at time $t$ and $S$ is the cross section of the tube.[60] It is important to observe that, in order to satisfy the boundary condition $c_A = 0$, there must be no natural convection. Natural convection can be avoided by placing the heaviest gas in the lower chamber and working with tubes that are not very wide (to avoid high Grashof numbers, which depend on $R^3$) and that are long enough. The fact that Loschmidt's results are 6% higher than those of von Obermayer's can be explained by the natural convection present in Loschmidt's experiments.[60]

conclude that

$$D_{CA}/D_{CB} = (M_B/M_A)^{1/2} \qquad (6.9)$$

which is false. But, if $M_C$ is much greater than both $M_A$ and $M_B$, equation (6.9) approximates a correct result, as is seen from the application of equation (2.92) to calculate $D_{CA}$ and $D_{CB}$. Graham's law of diffusion was therefore at least treated as a rough approximation.

For Loschmidt to perform "Graham type" experiments in a closed system, it would have been necessary to connect the bulbs containing A and B by means of a tube provided with a frictionless piston (see Figure 5.6). The piston would move as a consequence of the nonequimolar diffusion and restore uniform pressure.

Some years later, Stefan developed his method for determining diffusivities (see Section 5.3.2.1 and Figure 5.7). If the evaporating liquid is A, then $N_B = 0$ and the total flux of A is given by

$$N_A = \frac{J_{AM}}{1 - x_A} \qquad (6.10)$$

As in Loschmidt's experiments, the Stefan tube is not narrow, so pressure gradients are undetectable. Furthermore, the liquid density is higher than the vapor–gas density by approximately three orders of magnitude, so the descent rate of the liquid interface is very slow compared with that of the gaseous component A as it leaves the interface. Hence, a quasi-steady-state can be assumed in which $\partial c_A/\partial t = 0$.[9] Consequently, the equation of change for A on a molar basis is given by

$$D_{AB} \frac{d}{dz} \left( \frac{1}{1 - x_A} \frac{dc_A}{dz} \right) = 0 \qquad (6.11)$$

where usually $x_A \ll 1$ for a vapor–gas system, which means that in Stefan's experiments $N_A \approx J_{AM}$. The solution of equation (6.11) provides a means of calculating $D_{AB}$ from experimental results.[†]

We observe so far that in two different experiments, Loschmidt and Stefan measured molecular diffusivities correctly by applying Fick's law exclusively; in both cases the observed flux corresponded to the diffusive

---

[†]Stefan solved equation (6.11) with $1 - x_A = 1$ (in fact, he ignored the existence of this factor) assuming that $c_A = 0$ in the upper part of the tube and that $c_A = c_{Aeq}$ in the gas–liquid interface. The solution shows that $D_{AB} = N_{MA}L/Sc_{Aeq}t$, where $L$ is the average length for diffusion during the time $t$ (namely, the average distance between the liquid interface and the mouth of the tube). As in Loschmidt's experiment, it is necessary to avoid natural convection.

flux in spite of the existence of viscous and nonequimolar fluxes. It is therefore easy to understand why the latter two fluxes remained ignored. As a consequence, the phenomenon of diffusion was interpreted as a separative flux with respect to fixed coordinates with no other fluxes unless they were intentionally produced, as in Hertz's method, where the equation of change of A on a molar basis is[†]

$$D_{AB}\frac{d^2c_A}{dz^2} + \vartheta_M\frac{dc_A}{dz} = 0 \qquad (6.12)$$

This wrong picture of diffusion as a purely separative flux can still be found nowadays[62] and generates doubts and confusion (see, for example, the footnote on p. 1969 of Reference 127 and the section titled "The Hartly–Crank Theory" in Reference 3).

From this analysis of the Loschmidt and Stefan experiments, we see that, by extension, we can find a great variety of diffusion problems in which $N_A = J_{AM}$, with $J_{AM}$ described by Fick's law[20, 60]; but in these problems either wall effects are negligible or the binary interdiffusion is equimolar, and these two conditions severely restrict the usefulness of these solutions for porous media.

The coupling between the viscous and diffusive fluxes caused confusion for many years. The original equations of the dusty gas model[33] contained total (viscous plus diffusive) fluxes although the form of the equations was the same for the nonisobaric as for the isobaric case. The error was corrected by the authors themselves[77] but has since been repeated by others.[128] Because this error has been particularly important in the evolution of the knowledge of diffusion, we will devote the next section to it.

### 6.3.2.  Meaning of the Fluxes Involved in the Constitutive Equations of Diffusion

The development of the dusty gas model for pressure gradients and uniform temperature began with the Chapman–Enskog equations with coefficients calculated in the first approximation. Some terms in these equations involve the relative velocities of two species and it does not matter whether these relative velocities are calculated from the diffusive or total velocities [see equation (1.22)]. But in the term that describes the molecule–wall collisions, there is no difference of velocities since the particles are at rest in the reference frame; only the velocity of the gaseous

---

[†]The solution of equation (6.12) with the condition $c_A = c_{A0}$ at $z = 0$ yields $D_{AB} = \vartheta_M z \ln^{-1}(c_{A0}/c_A)$, where $c_A$ is the concentration of A at $z$.

species appears in this term, and this is the total diffusion velocity. Nevertheless, in the original article,[33] the authors assumed that this was the total velocity, and, as a consequence, they stated that:

> The beauty of the model is that it has buried the forced flow problem in the total flux N, such as its dependence on pressure drop or gas viscosity; .... It is somewhat surprising that [the equation obtained] has exactly the same form when pressure gradients exist as when the pressure is uniform.

The authors had obtained a constitutive equation for the total flux in terms of diffusivities exclusively. This is as if we assumed that equation (4.11) provides the total flux. Obviously, this error was not ignored by the authors, who, when referring to their equation (9) [analogous to our equation (4.11) written on a molar basis and with $j = p$, $\Delta' = 0$, $\nabla T = 0$] stated:

> We also see that the factor [which multiplies $N_i$] corresponds to a flow resistance, but contains only diffusion coefficients and thus the model possesses only a diffusion mechanism for flow. As remarked previously, this is a weakness of the model which we must try to circumvent.

To circumvent the problem, the authors replaced the diffusivity by an adjustable coefficient. But in fact, the model does not possess only a diffusion mechanism, as the authors themselves showed five years later[77]; the error did not reside in a weakness of the model but in a misinterpretation of the fluxes (we will discuss this further below). Nevertheless, the authors experimentally verified their original equations by applying them to results obtained for the interdiffusion of helium and argon under pressure gradients in macroporous graphite.[31] The equations were consequently accepted as correct. For example, Wakao, Otani, and Smith[115] used the original DGM equation and said:

> Evans and colleagues have shown that the (total) flux $N_A$ with a pressure gradient can still be represented by [equation (4.46) on a molar basis with $\nabla n = 0$ and with the total flux N replacing the $N_i^D$] provided the effect of pressure on $D_{AB}$ is taken into account.

Wakao *et al.* rewrote this equation as

$$N_{zA} = \frac{-D_{AB}(dc_A/dz) + (N_{zA} + N_{zB})x_A}{1 + (D_{AB}/D_{AA}^K)} \tag{6.13}$$

where, according to the authors,

$$N_z = N_{zA} + N_{zB} = N_{zA}^D + N_{zB}^D + N_{zA}^{SV} + N_{zB}^{SV} \tag{6.14}$$

and $N_{zi}^{SV}$ are the slip plus viscous contributions.

If equation (6.14) is introduced into equation (6.13) and also into the analogue of (6.13) for $N_{zB}$, and the resulting expressions added, the result obtained is not consistent with equation (6.14) since it contains a viscous contribution that depends on diffusivities [see the discussion following equation (6.19)]. In spite of this inconsistency, equations (6.13) and (6.14) were supported by two arguments: (1) for pure systems the equations reduce to a result very similar to that obtained theoretically by Scott and Dullien[101]; (2) the authors[115] verified their equations experimentally. In more detail these arguments are:

(1) For pure systems, equation (6.13), with the aid of equation (6.14) reduces to

$$N_{zA} = -D_{AA}^R \frac{dc}{dz} + \frac{N_z}{1 + D_{AA}/D_{AA}^K} \tag{6.15}$$

where [cf. equation (6.2)]

$$D_{AA}^R \equiv D_{AA}^K \Phi \tag{6.16}$$

and the Wakao, Otani, and Smith expression $\Phi_{WOS}$ for $\Phi$ is

$$\Phi_{WOS} \equiv \frac{D_{AA}}{D_{AA}^K + D_{AA}} = \frac{Kn}{1 + Kn} \tag{6.17}$$

Writing equation (6.15) in terms of $\Phi_{WOS}$ and $N_z^{SV}$, we obtain

$$N_{zA} = -D_{AA}^K \Phi_{WOS} \frac{dc}{dz} + N_z^{SV}(1 - \Phi_{WOS}) \tag{6.18}$$

This expression is similar to those obtained by Scott and Dullien[101] and by Weber[118]; in fact, Weber [see equation (1.74)] proposed that $\Phi$ in equation (6.16) is given by $\Phi_W = 0$ for $N^V$ (pure viscous flow) and his result therefore became correct. Wakao and colleagues plotted $\Phi_{WOS}$ as a function of Kn (letting $D_{AA} = \frac{1}{3}\lambda\bar{v}_A$) and compared it with the plot of Scott and Dullien's function $\Phi_{SD}$ vs. Kn, where $\Phi_{SD}$ is a complex function of Kn. The agreement between both functions was very good.

(2) Wakao, Otani, and Smith[115] verified equation (6.18) for the results obtained by Knudsen for the flow of carbon dioxide in his capillary No. 4 (24 tubes, $3.3 \times 10^{-5}$ cm radius, and 2 cm long at 25°C); to do this, they integrated equation (6.18) assuming that $\Delta p \ll p$, thus replacing $dc/dz$ by $\Delta c/\Delta z$ and introducing $\bar{p}$ in the viscous contribution. They also verified their own results for the flow of nitrogen in a capillary with a 12.44-cm radius at 23.5°C.

Summarizing our discussion up to this point we observe that:

(a)     Mason and colleagues obtained a constitutive equation of diffusion with wall effects starting from an equation that was valid for systems without walls and in which it did not matter whether $N_i$ or $N_i^D$ was used.

(b)     As a consequence of a misinterpretation of the fluxes involved, incompatible equations were obtained.

(c)     For pure systems the constitutive equation obtained reduces to a result obtained by another method.

(d)     The constitutive equation was experimentally verified for pure as well as for binary systems.

Let us see in more detail how (d) was possible.

If equation (6.14) is introduced into equation (6.13), we obtain

$$N_{zA} = -D_{AB}^R \frac{dc_A}{dz} + \frac{N_z x_A}{1 + D_{AB}/D_{AA}^K} \tag{6.19}$$

We know that equation (6.19) is correct only if $N_{zA}$ and $N_z$ are replaced by $N_{zA}^D$ and $N_z^D$, respectively. The correct expression, instead of equation (6.19), is then

$$N_{zA} = -D_{AB}^R \frac{dc_A}{dz} + \frac{N_z x_A}{1 + D_{AB}/D_{AA}^K} + N_z^V x_A \left(1 - \frac{1}{1 + D_{AB}/D_{AA}^K}\right) \tag{6.20}$$

A comparison of equations (6.19) and (6.20) shows that the error in equation (6.19) is the lack of an addend—the third term on the right-hand side of equation (6.20). Nevertheless, we observe that:

(i) In the Knudsen regime, $D_{AB}/D_{AA}^K \to \infty$, $N_z^V \to 0$ and equations (6.19) and (6.20) reduce to

$$N_{zA} = -D_{AA}^K(dc_A/dz) \tag{6.21}$$

(ii) In the molecular diffusion regime, equation (6.19) and equation (6.20) both reduce to

$$N_{zA} = -D_{AB} \frac{dc_A}{dz} + (N_{zA} + N_{zB})x_A \tag{6.22}$$

(iii) If $dp/dz$ is such that $N_z^V \approx 0$, the missing addend in equation (6.19), $N_z^V x_A\{1 - [1 + D_{AB}/D_{AA}^K]^{-1}\}$, is negligible in comparison to the first two terms of equation (6.19).

We thus see that in three particular cases the false equation (6.19) reduces to the same expression as the correct equation (6.20). The experimental results verified by Mason and colleagues correspond to case (ii) and those verified by Wakao *et al.* correspond to case (iii).

Some five years after the original work on the DGM, Mason and colleagues realized that the fluxes had been misinterpreted[77]:

> In the earlier work it was assumed that [equation (6.13)] applied to the total flux ... . We now realize that [equation (6.11)] applies only to the diffusive component.

Gunn and King[50] took these statements into account and verified the correct expression with the aid of other experimental results.[55]

### 6.3.3. The Nonisobaric Case in Porous Media

At the time Thiele submitted his work,[113] it was assumed that the diffusive contribution was completely determined by Fick's law; we now know that this is true for equimolar counterdiffusion in the molecular diffusion regime and involving reactions without change in the total number of moles. Thiele did not refer to a change in moles but to a volume change, which produces an isobaric net flux.

After Thiele's work, the experiments of Kramers and Kistemaker[65] and of Hoogschagen[57] proved that a coupling between the diffusive and viscous fluxes existed; however, this result was not applied immediately to a reacting system in a porous medium, probably because the application of the new concepts was not straightforward, and all three systems (Thiele's Kramers–Kistemaker's, and Hoogschagen's) were treated as corresponding to three different phenomena instead of as three particular cases of the same phenomenon.[†]

The misinterpretation of the fluxes discussed earlier only added to the confusion regarding porous media. The correct solution for porous media was obtained recently[1, 61] and has explained the results obtained by Hawthorn[52] as well as the solution proposed by Bischoff,[10] as we have seen in Section 5.3.3.1.

---

[†]We can see why from the following: For all three systems we have diffusive fluxes (described by their respective constitutive equations, which in turn provide the ratio of fluxes) and viscous fluxes (described by the Darcy equation). For systems with a chemical reaction of the type $A = bB$, the ratio of the total fluxes (diffusive plus viscous) is given by $-N_B/N_A = b$; $b = 1$ means uniform pressure and absence of viscous flux, while for $b \neq 1$ the lighter component diffuses in the same direction as the viscous flux. In a closed system, such as that used by Kramers and Kistemaker, we have the same diffusive and viscous phenomena but with the restriction that $-N_B/N_A = 1$; the heavier component therefore diffuses in the same direction as the viscous flux, and, as in the case of chemical reaction, if $M_A = M_B$, there will be uniform pressure. In the diffusion cell used by Hoogschagen, when the pressure is uniform the ratio $N_B^D/N_A^D$ is given by Graham's law of diffusion.

### 6.3.4.  Nature of the Total Diffusive Flux

The results obtained by Hoogschagen showed an unexpected phenomenon: that in an isothermal, isobaric binary gaseous mixture, without gravity effects, the mere existence of composition gradients generates a net molar flux. We have seen in Chapters 1 and 4 that the constitutive equations of diffusion predict this net molar flux in a very simple manner. Today, this appears to be logical when we realize that the constitutive equations of diffusion are nothing else but a momentum balance of a diffusing species and that, if we sum up the constitutive equations over all the species, the momentum transferred to the walls will be obtained. It seems that this concept was not clear some years ago for, on the one hand, constitutive equations of restricted validity were obtained by writing down a momentum balance[96, 100, 103] while, on the other hand, attempts were made to explain Graham's law of diffusion using a different momentum balance.[2, 54, 57, 96, 100, 124] As none of these derivations started from the Boltzmann equation, many doubts remained about the equations obtained.

It is therefore not surprising that other explanations were proposed to elucidate the nature of the total diffusive flux, such as the statement[26] that the isobaric net flux is due to the coupling between inertial forces (mass density gradients) and composition gradients (partial molar density gradients). The results of the study in which this statement appeared showed that Graham's law of diffusion was a limiting case of the ratio $-N_A/N_B = (M_B/M_A)^n$, with $\cdot 0.5 \leqslant n \leqslant 1$.[26] Unfortunately, the derivation is faulty since the starting equations predict a net mass flux opposite in direction to the experimentally observed direction.

## 6.4.  Maxwell's Two Viscosity Coefficients

In Chapter 2 we made an elementary prediction of the transport coefficients by means of what is called the "simplified kinetic theory." This theory gives the viscosity of a gas as

$$\mu = \tfrac{1}{2} m n \bar{v} \lambda \tag{6.23}$$

However, in 1860, Maxwell[80] calculated the viscosity of an ideal gas by applying a simplified kinetic theory and obtained

$$\mu = \tfrac{1}{3} m n \bar{v} \lambda \tag{6.24}$$

The same ratio $\tfrac{2}{3}$ of the coefficients in these two equations also occurs for the expressions obtained in Chapter 2 and by Maxwell for the self-diffusivity, the binary diffusivity, and the thermal conductivity.

Equations (6.23) and (6.24) have endured independently up to the present. However, very few papers refer to both of these expressions,[93, 101] but in those that do, equation (6.24) is either dismissed because it is considered as the more simplified form, or the difference between the two equations is correctly ascribed to an error introduced by Maxwell in his original development.[86]

The origin of the different numerical coefficients in equations (6.23) and (6.24) is to be found in a careful study of the derivations leading to these equations. The derivation of equation (6.23) takes into account the molecules coming from within a distance $\pm\lambda$ of the reference plane and which, therefore undergo one collision at that plane, thus modifying their velocities. On the other hand, the derivations (including Maxwell's) leading to equation (6.24) include all the molecules, colliding or not, which cross the reference plane (so that a fraction of these molecules do not exchange momentum) and this is an error since it can be demonstrated that, for the latter conditions, the molecules transport, on the average, the gas properties corresponding to a distance $\pm\frac{2}{3}\lambda$ from the reference plane.[95] This is the origin of the ratio $\frac{2}{3}$ by which the two expressions (6.23) and (6.24) differ.

If we consider the actual momentum transfer in the reference plane, equation (6.24) seems to be describing a more simplified case although its derivation is more complicated. Furthermore, in the Chapman–Enskog theory, the calculation of the viscosity for colliding rigid spheres leads to an expression which is practically the same as equation (6.23). It is difficult therefore to understand why equation (6.24) has persisted along with equation (6.23). Undoubtedly, the reason lies in the high regard for Maxwell. The intellectual necessity of obtaining the coefficient $\frac{1}{3}$ (to arrive at the same result as Maxwell's!) not only induced the use of Maxwell's working hypotheses, but also led to the statement of incorrect hypotheses, such as the assertion that for a gas in equilibrium the molecular flux in one direction is given by[56] $F_z = \frac{1}{6} n\bar{v}$ (instead of the true value $F_z = \frac{1}{4} n\bar{v}$), or another assertion that the molecules undergo their last collision, on an average, at a distance located $\pm\frac{2}{3}\lambda$ from the reference plane.[8]

It is of interest to note that it seems Maxwell himself had noticed his error but he did not point it out explicitly. For example, when Maxwell calculated the slip friction coefficient[82] (see Section 2.2.2), he used the coefficient $\frac{1}{2}$. This obviously increased the confusion and so we can find derivations which start from Maxwell's original work in 1860 and obtain a slip friction coefficient $\frac{2}{3}$ of that obtained by Maxwell himself some years later through what is called "a more refined analysis"[95] (in fact, it is not more refined but correct and simpler!).

These are not, however, the only misinterpretations in this theoretical field. For example, there are common fallacies in the derivation of the Boltzmann equation,[22] especially regarding the collision terms.

# References

## References for Chapter 1

1. Adzumi, H., *Bull. Chem. Soc. Japan*, **12**, 292 (1937).
2. Adzumi, H., *Bull. Chem. Soc. Japan*, **14**, 343 (1939).
3. Ash, R., Baker, R. W., and Barrer, R. M., *Proc. Roy. Soc. (London)*, **A229**, 434 (1967).
4. Ash, R., and Barrer, R. M., *Surface Sci.*, **8**, 461 (1967).
5. Ash, R., Barrer, R. M., and Lowson, R. T., *Surface Sci.*, **21**, 265 (1970).
6. Aylmore, L. A. G. and Barrer, R. M., *Proc. Roy. Soc. (London)*, **A290**, 477 (1966).
7. Babbit, J. D., *Can. J. Res.*, **28A**, 449 (1950).
8. Bangham, D. H., and Fakhoury W., *Proc. Roy. Soc. (London)*, **A130**, 81 (1930).
9. Barrer, R. M., *Disc. Far. Soc.*, **3**, 61 (1948).
10. Barrer, R. M., *Appl. Mat. Res.*, **2**, 129 (1963).
11. Barrer, R. M., *Can. J. Chem.*, **41**, 1768 (1963).
12. Barrer, R. M., *Proc. Brit. Cer. Soc.*, **5**, 21 (1965).
13. Barrer, R. M., in: *The Solid–Gas Interface* (E. A. Flood, ed.), Vol. II, Marcel Dekker Inc., New York (1967).
14. Barrer, R. M. and Barrie, J. A., *Proc. Roy. Soc. (London)*, **A213**, 250 (1952).
15. Barrer, R. M., and Gabor, T., *Proc. Roy. Soc.*, **A251**, 356 (1959).
16. Barrer, R. M., and Gabor, T., *Proc. Roy. Soc.*, **A256**, 267 (1960).
17. Barrer, R. M., and Strachan, E., *Proc. Roy. Soc.*, **A231**, 52 (1955).
18. Bartholomew, R. F., and Flood, E. A., *Can. J. Chem.*, **43**, 1968 (1965).
19. Bennett, C. O., and Myers, J. E., *Momentum, Heat, and Mass Transfer*, McGraw-Hill Book Co., New York (1962).
20. Bird, R. B., *Advances in Chemical Engineering*, Vol. I, Academic Press Inc., New York (1956).
21. Bird, R. B., Stewart, W. E., and Lightfoot, E. N., *Transport Phenomena*, John Wiley & Sons Inc., New York (1960).
22. Brinkman, H. C., *Appl. Sci. Res.*, **A1**, 81 (1947).
23. Callihan, A. D., Manhattan Project Report, M-1157 (September 4, 1944).
24. Carman, P. C., *Proc. Roy. Soc. (London)*, **A211**, 526 (1952).
25. Carman, P. C., *Flow of Gases through Porous Media*, Butterworths Scientific Publications, London (1956).
26. Carman, P. C., and Malherbe, P. le R., *Proc. Roy. Soc. (London)*, **A203**, 165 (1950).
27. Carman, P. C., and Raal, F. A., *Proc. Roy. Soc. (London)*, **A201**, 38 (1951).
28. Carman, P. C., and Raal, F. A., *Trans. Far. Soc.*, **50**, 842 (1954).

29. Carman, P. C., and Stein, L. H. S., *Trans. Far. Soc.*, **52**, 619 (1956).
30. Christiansen, C., *Wied. Ann.*, **41**, 565 (1890).
31. Darcy, H. P. G., *Les Fontaines Publiques de la Ville de Dijon*, Victor Dalmont, Paris (1856).
32. de Boer, J. H., *The Dynamical Character of Adsorption*, 2nd ed., Oxford University Press, London (1968).
33. Duncan, J. B., and Toor, H. L., *A.I.Ch. E. Journal*, **8**, 38 (1962).
34. Engel, H. H., Sc.D. Thesis, Dept. Chem. Eng., M.I.T., Cambridge, Massachusetts (1967).
35. Flood, E. A., and Heyding, R. D., *Can. J. Chem.*, **32**, 660 (1964).
36. Flood, E. A., and Huber, M., *Can. J. Chem.*, **33**, 203 (1955).
37. Flood, E. A., and Tomlinson, R. H., *Can. J. Res.*, **B26**, 38 (1948).
38. Flood, E. A., Tomlinson, R. H. and Leger, A. E., *Can. J. Chem.*, **30**, 389 (1952).
39. Gilliland, E. R., and Engel, H. H., *A.I.Ch.E. Journal*, **8**, 530 (1962).
40. Gilliland, E. R., Baddour, R. F., and Russell, J. L., *A.I.Ch.E. Journal*, **4**, 90 (1958).
41. Gunn, R. D., and King, J. C., *A.I.Ch.E. Journal*, **15**, 507 (1969).
42. Haul, R., *Angew. Chem.*, **621**, 10 (1950).
43. Haul, R., *Nature*, **171**, 519 (1953).
44. Hellund, E. J., *Phys. Rev.*, **57**, 737 (1940).
45. Higashi, K., Ito, H., and Pishi, J., *J. Japan Atom En. Soc.*, **5**, 846 (1963).
46. Higashi, K., Ito, H., and Pishi, J., *J. Nucl. Sci. Tech.*, **1**, 293 (1964).
47. Hirschfelder, J. O., Curtiss, C. F., and Bird, R. B., *Molecular Theory of Gases and Liquids*, John Wiley and Sons Inc., New York (1954).
48. Horiguchi, Y., Hudgins, R. R., and Silveston, P. L., *Can. J. Chem. Eng.*, **49**, 76 (1971).
49. Kammermeyer, K., and Rutz, L. O., *Chem. Eng. Progr. Symp. Ser.*, No. 24, **55**, 163 (1959).
50. Knudsen, M., *Ann. Phys.*, **28**, 75 (1909).
51. Krückels, W. W., *Chem. Eng. Sci.*, **28**, 1565 (1973).
52. Kuhn, H., Manhattan Project Report, M-1157 (March 8, 1944).
53. Kunt, A., and Warburg, E., *Pogg. Ann.*, **155**, 337 (1875).
54. Mason, E. A., and Evans III, R. B., *J. Chem. Educ.*, **46**, 358 (1969).
55. Mason, E. A., and Marrero, T. R., *Advances in Atomic and Molecular Physics*, Vol. VI, Robert Maxwell & Co., London (1970).
56. Maxwell, J. C., *Scientific Papers*, Cambridge University Press, London (1890).
57. McCarty, K. P., and Mason, E. A., *Phys. Fluids*, **3**, 908 (1960).
58. Nicholson, D., and Petropoulos, J. H., *J. Coll. Interface Sci.*, **45**, 459 (1973).
59. Opfell, B., and Sage, B. H., *Ind. Eng. Chem.*, **47**, 918 (1955).
60. Poiseuille, J. L., *Comptes Rendus*, **11**, 961 (1840).
61. Poiseuille, J. L., *Comptes Rendus*, **12**, 112 (1841).
62. Rohsenow, W. M. and Choi, H. Y., *Heat, Mass, and Momentum Transfer*, Prentice-Hall Inc., Englewood Cliffs, N.J. (1961).
63. Rothfeld, L. B., and Watson, Ch. C., *A.I.Ch.E. Journal*, **9**, 19 (1963).
64. Russell, B., *Our Knowledge of the External World*, George Allen and Unwin Ltd., London (1964).
65. Russell, J. L., Sc.D. Thesis, Dept. Chem. Eng., M.I.T., Cambridge, Massachusetts 1955.
66. Schleicher, K., Manhattan Project M-1472 (January 5, 1945).
67. Schrödinger, E., *What is Life?*, Cambridge University Press, London (1958).
68. Scott, D. S., and Dullien, F. A. L., *A.I.Ch.E. Journal*, **8**, 113 (1962).
69. Sladek, K. J., Sc.D. Thesis, Dept. Chem. Eng., M.I.T., Cambridge, Massachusetts (1967).
70. Sladek, K. J., Gilliland, E. R., and Baddour, R. F., *Ind. Eng. Chem. Fund.*, **13**, 100 (1974).
71. Slattery, J. C., *Momentum, Energy, and Mass Transfer in Continua*, McGraw-Hill Book Co. Inc., New York (1972).
72. Spiegler, K. S. S., *Ind. Eng. Chem. Fund.*, **5**, 529 (1966).

73. Stefan, J., *Sitzber. Akad. Wiss. Wien*, **63**, 63 (1871).
74. Stefan, J., *Sitzber. Akad. Wiss. Wien*, **65**(2), 323, (1872).
75. Taylor, J. B., and Langmuir, I., *Phys. Rev.*, **44**, 423 (1933).
76. Toor, H. L., *A.I.Ch.E. Journal*, **3**, 198 (1957).
77. Volmer, M., and Adhikari, G., *Z. Physik.*, **35**, 170 (1925).
78. Volmer, M., and Adhikari, G., *Z. physikal. Chem.*, **119**, 46 (1926).
79. Warburg, E., *Pogg. Ann.*, **159**, 399 (1876).
80. Weaver, J. A., Ph.D. Thesis, Dept. Chem. Eng., University of Delaware, Newark, Delaware 1965.
81. Weber, S., *Kgl. Danske Videnskab. Slebskab. Mat. Fyz. Medd.*, **28** (1954).
82. Welty, J. R., Wicks, C. E., and Wilson, R. E., *Fundamentals of Momentum, Heat, and Mass Transfer*, John Wiley & Sons Inc., New York (1969).
83. Wentworth, C. K., *Amer. J. Sci.*, **242**, 478 (1944).
84. Wicke, E., and Kallenbach, R., *Kolloidzohr*, **97**, 135 (1941).
85. Wicke, E., and Voigt, W., *Angew. Chem.*, **B19**, 94 (1947).

# References for Chapter 2

1. Carman, P. C., *Flow of Gases through Porous Media*, Butterworths Scientific Publications, London (1956).
2. Clausing, P., Dissertation, University of Amsterdam, Amsterdam, The Netherlands (1918).
3. Hirsch, E. H., *J. Appl. Phys.*, **32**, 977 (1961).
4. Knudsen, M., *Ann. Phys.*, **28**, 75 (1909).
5. Loeb, L. B., *Kinetic Theory of Gases*, 2nd ed., McGraw-Hill Book Co., Inc., New York (1934).
6. Millikan, R., *Phys. Rev.*, **21**, 217 (1923).
7. Moelwyn-Hughes, E. A., *Physical Chemistry*, 2nd ed., Pergamon Press, London (1965).
8. Pollard, W. G., and Present, P. D., *Phys. Rev.*, **73**, 762 (1948).
9. Tsien, H. S., *J. Aero. Sc.*, **13**, 653 (1946).
10. von Smoluchowsky, M., *Ann. Phys.* (*Leipzig*), **33**, 1559 (1910).
11. von Smoluchowsky, M., *Ann. Phys.* (*Leipzig*), **35**, 983 (1911).

# References for Chapter 3

1. Amdur, I., and Schatsky, T. F., *J. Chem. Phys.*, **29**, 1425 (1958).
2. Andrussow, L., *Z. Elektrochem.*, **54**, 566 (1950).
3. Arnold, J. H., *Ind. Eng. Chem.*, **22**, 1091 (1930).
4. Bailey, R. G., and H. T. Chen, *Chem. Eng.*, **82**, 86 (March 17, 1975).
5. Barker, T., and Everett, D., *Trans. Far. Soc.*, **58**, 1608 (1962).
6. Bhatnagar, P. L., Gross, E. P., and Krook, M., *Phys. Rev.* **94**, 511 (1954).
7. Cercignani, C., and Daneri, A., *J. Appl. Phys.*, **34**, 3509 (1963).
8. Cercignani, C., and Pagani, C. D., in: *Rarefied Gas Dynamics* (C. L. Brundin, ed.), Vol. 1, p. 555, Academic Press Inc., New York (1967).
9. Cercignani, C., and Sernagiotto, F., *Phys. Fluids*, **9**, 40, (1966).
10. Chapman, S., *Phil. Trans. Roy. Soc.*, **A211**, 433 (1912).
11. Chapman, S., *Phil. Trans. Roy. Soc.*, **A216**, 279 (1916).
12. Chapman, S., *Phil. Trans. Roy. Soc.*, **A217**, 115 (1917).

13. Chapman, S., and Cowling, T. G., *Proc. Roy. Soc. (London)*, **A179**, 159 (1941).
14. Chapman, S., and Cowling, T. G., *The Mathematical Theory of Non-Uniform Gases*, 3rd ed., Cambridge University Press, London (1970).
15. Chapman, S., and Dootson, F. W., *Phil. Mag.*, **33**, 248 (1917).
16. Chen, N. H., and Othmer, D. F., *J. Chem. Eng. Data*, **7**, 37 (1962).
17. Cohen, E. G. D., in: *Transport Phenomena in Fluids* (H. J. M. Hanley, ed.), pp. 157, Marcel Dekker, New York (1969).
18. Curtiss, C. V., and Hirschfelder, J. O., *J. Chem. Phys.*, **17**, 550 (1949).
19. De Groot, S. R., *Thermodynamics of Irreversible Processes*, North-Holland Publishing Company, Amsterdam (1951).
20. de Troyer, A., van Itterbeek, A., and Rietveld, A. O., *Physica*, **17**, 938 (1951).
21. Duncan, J. B., and Toor, H. L., *A.I.Ch.E. Journal*, **8**, 38 (1962).
22. Enskog, D., *Physik. Z.*, **12**(56), 533 (1911).
23. Enskog, D., *Ann. Phys.*, **38**, 731 (1912).
24. Enskog, D., *Archiv för Matematik, Astronomi och Fysik*, **16**, Sect. 16 (1922) in German; Kinetische Theorie der Vorgänge in mässig verdünnten Gasen, Inaugural Dissertation, Uppsala (Sweden), Almqvist and Wiksell (1917).
25. Ernst, M. H., Haines, L. K., and Dorfman, J. R., *Rev. Mod. Phys.*, **41**, 296 (1969).
26. Ferziger, J. H., *Phys. Fluids*, **10**, 1448 (1967).
27. Fuller, E. N., Schettler, P. D., and Giddings, J. C., *Ind. Eng. Chem.*, **58**(5), 19 (1966).
28. Gilliland, E. R., *Ind. Eng. Chem.*, **26**, 681 (1950).
29. Goldstein, H., *Classical Mechanics*, Addison-Wesley, Reading, Massachusetts (1959).
30. Grad, H., *Comm. Pure Appl. Math.*, **2**, 325 (1949).
31. Grad, H., *Comm. Pure Appl. Math.*, **2**, 331 (1949).
32. Gross, E. P., and Krook, M., *Phys. Rev.*, **102**, 593 (1956).
33. Hamel, B. B., Ph.D. Thesis, Princeton University, Princeton, New Jersey (1963).
34. Hamel, B. B., *Phys. Fluids*, **8**, 418 (1965).
35. Hamel, B. B., *Phys. Fluids*, **9**, 12 (1966).
36. Hellund, E. J., *Phys. Rev.*, **57**, 737 (1940).
37. Hirschfelder, J. O., Curtiss, C. F., and Bird, R. B., *Molecular Theory of Gases and Liquids*, John Wiley & Sons Inc., New York (1954).
38. Holway Jr., L. H., *Phys. Fluids*, **9**, 1658 (1966).
39. Holway Jr., L. H., in: *Rarefied Gas Dynamics*, (C. L. Brundin, ed.), Vol. 1, p. 759, Academic Press Inc., New York (1967).
40. Huang, K., *Statistical Mechanics*, p. 72, John Wiley, New York, (1963).
41. Kihara, T., *Imperfect Gases*, Asakura Bookstore, Tokyo (1949). (Translated into English by the United States Office of Air Research, Wright-Patterson Air Force Base.)
42. Lang, H., *Z. Angew. Math. Mech.*, **48**, 208 (1968).
43. Lang, H., *Mitt. Max-Planck-Inst. Strömungsforsch. Aerodyn. Versuchanst.*, No. 43 (1968).
44. Lang, H., *Chem. Eng. Sci.*, **14**, 21 (1971).
45. Loyalka, S. K., *Phys. Fluids*, **14**, 21 (1971).
46. Loyalka, S. K., *Phys. Fluids*, **14**, 2291 (1971).
47. Loyalka, S. K., *Phys. Fluids*, **14**, 2599 (1971).
48. Marrero, T. R., and Mason, E. A., *J. Phys. Chem. Ref. Data*, **1**, 1 (1972).
49. Marrero, T. R., and Mason, E. A., *A.I.Ch.E. Journal*, **19**, 498 (1973).
50. Mason, E. A., *J. Chem. Phys.*, **27**, 75 (1957).
51. Mason, E. A., and Marrero, T. R., *Advances in Atomic and Molecular Physics*, Vol. VI, Robert Maxwell & Co., London (1970).
52. Mason, E. A., Weissman, S., and Wendt, R. P., *Phys. Fluids*, **7**, 174 (1964).
53. Mathur, G. P., and Thodos, G., *A.I.Ch.E. Journal*, **11**, 613 (1965).
54. Maxwell, J. C., *Phil. Trans. Roy. Soc.*, **157**, 49 (1867); reprinted in *Scientific Papers*, Vol. II, Dover, New York (1962).

55. Morse, T. F., *Phys. Fluids*, **7**, 2012 (1964).
56. Muckenfuss, C., *J. Chem. Phys.*, **59**, 1747 (1973).
57. Oguchi, H., in: *Rarefied Gas Dynamics* (C. L. Brundin, ed.), Vol. 1, p. 745, Academic Press Inc., New York (1967).
58. Onsager, L., *Phys. Rev.*, **37**(I), 405 (1931).
59. Onsager, L., *Phys. Rev.*, **38**(II), 2265 (1931).
60. Onsager, L., *Ann. N.Y. Acad. Sci.*, **46**, 241 (1945).
61. Othmer, D. F., and Chen, N. H., *Ind. Eng. Chem. Proc. Des. and Dev.*, **1**, 249 (1962).
62. Popielawski, J., *Molec. Phys.*, **10**, 583 (1966).
63. Popielawski, J., *Molec. Phys.*, **12**, 97 (1967).
64. Popielawski, J., and Baranowski, B., *Molec. Phys.*, **9**, 59 (1965).
65. Shendalman, L. H., *J. Chem. Phys.*, **51**, 2483 (1969).
66. Simons, S., *Proc. Roy. Soc.* (*London*), **A301**, 387 (1967).
67. Simons, S., *Proc. Roy. Soc.* (*London*), **A301**, 401 (1967).
68. Sinagoglu, O., and Pitzer, K., *J. Chem. Phys.*, **32**, 1279 (1960).
69. Sirovich, L., *Phys. Fluids*, **5**, 908 (1962).
70. Sitarski, M., and Popielawski, J., *Molec. Phys.*, **19**, 741 (1970).
71. Sitarski, M., and Popielawski, J., *Molec. Phys.*, **23**, 365 (1972).
72. Slattery, J. C. and Bird, R. B., *A.I.Ch.E. Journal*, **4**, 137 (1958).
73. Sone, Y., and Yamamoto, K., *Phys. Fluids*, **11**, 1672 (1968).
74. Stefan, J., *Sitzber. Akad. Wiss. Wien*, **63**, 63 (1871).
75. Stefan, J., *Sitzber. Akad. Wiss. Wien*, **65**, 323 (1872).
76. Toor, H. L., *A.I.Ch.E. Journal*, **8**, 38 (1962).
77. van Eekelen, H. A. M., and Smit, W., *Phys. Fluids*, **14**, 2295 (1971).
78. Waldmann, L., *Handbuch der Physik* (S. Flugge, ed.), Vol. XII, pp. 295–514, Springer Verlag, Berlin (1958).
79. Walker, E. L., and Tanenbaum, B. S., *Phys. Fluids*, **11**, 1951 (1968).
80. Wang Chang, C. S., Uhlenbeck, G. E., and de Boer, J. H., *Studies Stat. Mech.*, **2**, 241 (1964).
81. Welander, P., *Arkiv Physik*, **7**, 507 (1954).
82. Wilke, C. R., and Lee, C. Y., *Ind. Eng. Chem.*, **47**, 1253 (1955).
83. Zhdanov, V., Kagan Yu., and Sazykin, A., *Zh. Eksperim. Teor. Fiz.*, **42**, 857 (1962) [English translation: *Soviet Phys.–JETP*, **15**, 596 (1962)].

# References for Chapter 4

1. Adzumi, H. B., *Bull. Chem. Soc. Japan*, **12**, 304 (1937).
2. Annis, B. K., and Mason, E. A., *Phys. Fluids*, **13**, 1452 (1970).
3. Arnell, J. C., and Hennenberry, G. O., *Can. J. Res.*, **A26**, 29 (1948).
4. Azzam, M. I. S., and Dullien, F. A. L., *Ind. Eng. Chem. Fund.*, **15**, 281 (1976).
5. Barrer, R. M., and Barrie, R. A., *Proc. Roy. Soc.* (*London*), **A213**, 250 (1952).
6. Barrer, R. M., and Grove, D. M., *Trans. Far. Soc.*, **47**, 826 (1951).
7. Barrer, R. M., and Nicholson, D., *Brit. J. Appl. Phys.*, **17**, 1091 (1966).
8. Barrer, R. M., and Strachan, E., *Proc. Roy. Soc.* (*London*), **A231**, 52 (1955).
9. Beavers, G. S., and Joseph, D. D., *J. Fluid Mech.*, **30**, 197 (1967).
10. Bird, R. B., Stewart, W. E., and Lightfoot, E. N., *Transport Phenomena*, John Wiley & Sons Inc., New York (1960).
11. Breton, J. P., *Phys. Fluids*, **12**, 2019 (1969).
12. Breton, J. P., *J. Physique*, **31**, 613 (1970).
13. Breton, J. P., *Physica*, **50**, 365 (1970).
14. Brinkman, H. C., *Appl. Sci. Res.*, **A1**, 27 (1947).

15. Brock, J. R., *J. Colloid Sci.*, **18**, 489 (1963).
16. Brown, L. F., and Haynes, H. W., *Ind. Eng. Chem. Fund.*, **8**, 601 (1969).
17. Brown, L. F., and Haynes, H. W., and Manogue, H. W., *J. Catal.*, **14**, 220 (1969).
18. Brugeman, D. A. G., *Ann. Phys.*, **24**, 636 (1935).
19. Carman, P. C., *Flow of Gases through Porous Media*, Butterworths Scientific Publications, London (1956).
20. Chapman, S., and Cowling, T. G., *The Mathematical Theory of Non-Uniform Gases*, Cambridge University Press, London (1952).
21. Childress, S., *J. Chem. Phys.*, **56**, 2527 (1972).
22. Childs, E. C., and Collis-George, N., *Proc. Roy. Soc. (London)*, **A201**, 392 (1950).
23. Derjaguin, B. F., and Bakanov, S. P., *Soviet Physics–Doklady*, **2**, 326 (1957); *Disc. Far. Soc.*, **30**, 130 (1960).
24. Eger, K., *Mitt. Max-Planck-Inst. Strömungsforsch. Aerodyn. Versuchanst.*, No. 51 (1971).
25. Evans, III, R. B., Watson, G. M., and Mason E. A., *J. Chem. Phys.*, **35**, 2076 (1961).
26. Evans, III, R. B., Watson, G. M., and Truitt, J., *J. Appl. Phys.*, **33**, 2682 (1962).
27. Evans, III, R. B., Watson, G. M., and Truitt, J., *J. Appl. Phys.*, **34**, 2020 (1963).
28. Fatt, I., *A.I.M.E. Pet. Trans.*, **207**, 144 (1956).
29. Feng, Ch., and Stewart, W. E., *Ind. Eng. Chem. Fund.*, **12**, 143 (1973).
30. Foster, R. N., Butt, J. B., and Bliss, H., *J. Catal.*, **7**, 179 (1967).
31. Foster, R. N., Butt, J. B., and Bliss, H., *J. Catal.*, **7**, 191 (1967).
32. Gaede, W., *Ann. Phys.*, **41**, 289 (1913).
33. Graham, T., *Phil. Mag.*, **2**, 175, 269, 351 (1833).
34. Grew, K. E., and Ibls, T. L., *Thermal Diffusion in Gases*, Cambridge University Press, New York (1952).
35. Grove, D. M., and Ford, M. G., *Nature*, **182**, 999 (1958).
36. Gunn, R. D., and King, J., *A.I.Ch.E. Journal*, **15**, 507 (1969).
37. Happel, J., and Brenner, H., *Low Reynolds Number Hydrodynamics*, Prentice-Hall, Englewood Cliffs, New Jersey (1965).
38. Haring, R. E., and Greenkorn, R. A., *A.I.Ch.E. Journal*, **16**, 477 (1970).
39. Hasse, H. R., and Cooke, W. R., *Proc. Roy. Soc. Ser.*, **A125**, 196 (1929).
40. Haynes, Jr., H. W., M.S. Thesis, University of Colorado, Boulder, Colorado (1966).
41. Haynes, Jr., H. W., Ph.D. Thesis, University of Colorado, Boulder, Colorado (1969).
42. Haynes, Jr., H. W., and Brown, L. F., *A.I.Ch.E. Journal*, **17**, 491 (1971).
43. Henry, Jr., J. P., Cunningham, R. S., and Geankoplis, C. J., *Chem. Eng. Sci.*, **22**, 11 (1967).
44. Hewitt, G. F., and Sharratt, E. W., *J. Nucl. Mat.*, **13**, 197 (1964).
45. Higby, J. W., and Pahl, M., *Z. Naturforsch.*, **7a**, 533, 542 (1952).
46. Hill, T. L., *An Introduction to Statistical Thermodynamics*, Addison-Wesley, Reading, Massachusetts (1960).
47. Hirschfelder, J. O., Curtiss, Ch. F., and Bird, R. B., *Molecular Theory of Gases and Liquids*, John Wiley & Sons Inc., New York (1954).
48. Hoogschagen, J., *J. Chem. Phys.*, **21**, 2096 (1953).
49. Hoogschagen, J., *Ind. Eng. Chem.*, **47**, 906 (1955).
50. Horák, Z., and Schneider, P., *Chem. Eng. J.*, **2**, 26 (1971).
51. Hwang, S. T., and Kammermeyer, K., *Can. J. Chem. Eng.*, **44**, 82 (1966).
52. Hwang, S. T., and Kammermeyer, K., *Sep. Sci.*, **1**, 629 (1966).
53. Hwang, S. T., and Kammermeyer, K., *Ind. Eng. Chem. Fund.*, **7**, 671 (1968); **8**, 601 (1969).
54. Johnson, M. F. L., and Stewart, W. E., *J. Catal.*, **4**, 252 (1965).
55. Kawazoe, K., Sujiyama, I., and Fukuda, Y., *Kagaku Kogaku*, **30**, 1007 (1966).
56. Kim, K. K., and Smith, J. M., private communication (1973).

57. Klinkenberg, L. J., *A.P.I. Drilling and Production Practice*, American Petroleum Institute, New York (1941).
58. Knudsen, M., *Ann. Phys.*, **28**, 75 (1909).
59. Kramers, H. A., and Kistemaker, J., *Physica*, **10**, 699 (1943).
60. Kucherov, R. A., and Rikenglaz, L. E., *Zhur. Eksp. Teor. Fiz.*, **36**, 1758 (1959) [English translation: *Soviet Phys.–JETP* **9**, 1253, (1959)].
61. Kuwabara, S., *J. Phys. Soc. Japan*, **14**, 257 (1959).
62. Lang. H., *Mitt. Max-Planck-Inst. Strömungsforsch. Aerodyn. Versuchanst.*, No. 43 (1968).
63. Lang, H., and Eger, K., *Z. physikal. Chem.*, *Neue Folge*, **68**, 130 (1969).
64. Lang, H., and Loyalka, S. K., *Z. Naturforsch*, **27a**, 1307 (1972).
65. Laranjeira, M. F., *Physica*, **26**, 417 (1960).
66. Loyalka, S. K., *Phys. Fluids*, **14**, 2291 (1971).
67. Loyalka, S. K., *Phys. Fluids*, **14**, 2599 (1971).
68. Lundgren, T. S., *J. Fluid Mech.*, **51**, 273 (1972).
69. Marshall, T. J., *J. Soil Sci.*, **9**, 1 (1958).
70. Mason, E. A., *J. Chem. Phys.*, **39**, 522 (1963).
71. Mason, E. A., and Chapman, S., *J. Chem. Phys.*, **36**, 627 (1962).
72. Mason, E. A., Evans, III, R. B., and Watson, G. M., *J. Chem. Phys.*, **38**, 1808 (1963).
73. Mason, E. A., Malinauskas, A. P., and Evans, III, R. B., *J. Chem. Phys.*, **46**, 3199 (1967).
74. Mason, E. A., and Marrero, R. T., *Advances in Atomic and Molecular Physics*, Vol. VI, Robert Maxwell and Co., London (1970).
75. Mason, E. A., and Monchick, L., *J. Chem. Phys.*, **36**, 1622 (1962).
76. Maxwell, J. C., *A Treatise on Electricity and Magnetism*, 2nd ed., Vol. I, Oxford University Press, London (1881).
77. Michaels, A. S., *A.I.Ch.E. Journal*, **5**, 270 (1959).
78. Mood, A. F., and Graybill, F. A., *Introduction to the Theory of Statistics*, 2nd ed., McGraw-Hill Book Co. Inc., New York, (1963).
79. Neale, G. H., and Nader, W. K., *A.I.Ch.E. Journal*, **19**, 112 (1973).
80. Neale, G. H., and Nader, W. K., *A.I.Ch.E. Journal*, **20**, 530 (1974).
81. Omata, H., and Brown, L. F., *A.I.Ch.E. Journal*, **18**, 967 (1972).
82. Pakula, R. J., and Greenkorn, R. A., *A.I.Ch.E. Journal*, **17**, 1265 (1971).
83. Petersen, E. E., *A.I.Ch.E. Journal*, **17**, 1265 (1971).
84. Pollard, W. G., and Present, R. D., *Phys. Rev.*, **73**, 762 (1948).
85. Prager, S., *Physica*, **29**, 129 (1963).
86. Rothfeld, L. B., and Watson, Ch., *A.I.Ch.E. Journal*, **9**, 19 (1963).
87. Saffman, P. G., *Stud. Appl. Math.*, **50**, 93 (1971).
88. Sandler, S. I., *Ind. Eng. Chem. Fund.*, **11**, 424 (1972).
89. Satterfied, Ch. N., *Ind. Eng. Chem. Fund.*, **8**, 175 (1969).
90. Satterfield, Ch. N., and Cadle, P. J., *Ind. Eng. Chem. Fund.*, **7**, 202 (1968).
91. Satterfield, Ch. N., and Cadle, P. J., *Ind. Eng. Chem. Proc. Des. and Dev.*, **7**, 257 (1968).
92. Satterfield, Ch. N., and Saraf, S. K., *Ind. Eng. Chem. Fund.*, **4**, 451 (1965).
93. Scheidegger, A. E., *The Physics of Flow through Porous Media*, University of Toronto Press, Toronto (1960).
94. Schmitt, K., and Waldmann, L. W., *Z. Naturforsch.*, **15a**, 843 (1960).
95. Scott, D. S., and Cox, K. E., *Can. J. Chem. Eng.*, **38**, 201 (1960).
96. Scott, D. S., and Dullien, F. A. L., *A.I.Ch.E. Journal*, **8**, 113 (1962).
97. Shendalman, L. H., *J. Chem. Phys.*, **51**, 2483 (1969).
98. Shimizu, M., Watanabe, F., and Sujiyama, S., *J. Chem. Eng. Japan*, **4**, 331 (1971).
99. Silveston, P. L., *A.I.Ch.E. Journal*, **10**, 132 (1964).
100. Slattery, J. C., *A.I.Ch.E. Journal*, **13**, 1066 (1967).
101. Slattery, J. C., *A.I.Ch.E. Journal*, **15**, 866 (1969).

102. Slattery, J. C., *Momentum, Energy and Mass Transfer in Continua*, McGraw-Hill Book Co. Inc., New York (1972).
103. Spiegler, K. S., *Ind. Eng. Chem. Fund.*, 5, 529 (1966).
104. Steeley, W. A., and Halsey, Jr., G. D., *J. Phys. Chem.*, 59, 57 (1955).
105. Wakao, N., and Smith, J. M., *Chem. Eng. Sci.*, 17, 825 (1962).
106. Waldmann, L., *Z. Naturforsch.*, 14a, 589 (1959).
107. Waldmann, L., and Schmitt, K. H., *Z. Naturforsch.*, 16a, 1343 (1961).
108. Weisz, P. B., *Chem. Tech.*, 498 (August, 1973).
109. Weisz, P. B., and Schwartz, A. B., *J. Catal.*, 1, 399 (1962).
110. Wheeler, A., *Advances in Catalysis*, Vol. III, Academic Press Inc., New York (1951).
111. Wyllie, M. R. J., and Gardner, G. H. F., *World Oil*, 149, 210 (1958).
112. Zhdanov, V. M., *Zh. Tekh. Fiz.*, 37, 192 (1967) [English translation: *Soviet Phys.–Tech. Phys.* 12, 134 (1967)].

# References for Chapter 5

1. Abed, R., Ha, K. D., and Rinker, R. G., *A.I.Ch.E. Journal*, 20, 391 (1974).
2. Abed, R., and Rinker, R. G., *A.I.Ch.E. Journal*, 19, 618 (1973).
3. Abed, R., and Rinker, R. G., *J. Catal.*, 34, 246 (1974).
4. Apecetche, M. A., González, M. G., Williams, R. J. J., and Cunningham, R. E., *J. Catal.*, 29, 451 (1973).
5. Ash, R., Barrer, R. M., and Lowson, R. T., *J. Chem. Soc., Far. Trans.*, 169, 2166 (1973).
6. Ash, R., Barrer, R. M., and Pope, C. G., *Proc. Roy. Soc. (London)*, A271, 1, 19 (1963).
7. Balder, J. R., and Petersen, E. E., *J. Catal.*, 11, 195, 202 (1968).
8. Barrer, R. M., *J. Phys. Chem.*, 57, 35 (1953).
9. Barrer, R. M., *Proc. A.I.Ch.E.* (I. Chem. E. Joint Meeting, London), 1, 112 (1965).
10. Bird, R. B., Stewart, W. E., and Lightfoot, E. N., *Transport Phenomena*, John Wiley & Sons Inc., New York (1960).
11. Bosanquet, C. H., British T.A. Report Br-507, London (1944).
12. Brown, L. F., Haynes, Jr., H. W., and Manogue, W. H., *J. Catal.*, 14, 220 (1969).
13. Buckingham, E., *U.S. Dept. Agr., Bur. Soils, Bull.* No. 25 (1904).
14. Cadle, P. J., Ph.D. Thesis, M.I.T., Cambridge, Massachusetts, (1966).
15. Campbell, F. R., Hills, A. W. D., and Paulin, A., *Chem. Eng. Sci.*, 25, 929 (1970).
16. Carberry, J. J., *Ind. Eng. Chem. Fund.*, 14, 131 (1975).
17. Carty, R., and Schrodt, T., *Ind. Eng. Chem. Fund.*, 14, 276 (1975).
18. Chen, N. Y., Lucki, S. J., and Mower, E. B., *J. Catal.*, 13, 329 (1969).
19. Chen, N. Y., and Weisz, P. B., *Chem. Eng. Progr. Symp. Ser.*, 63, 86 (1967).
20. Chen, O. T., and Rinker, R. G., *Chem. Eng. Sci.*, 34, 51 (1979).
21. Cullinan Jr., H. T., *Ind. Eng. Chem. Fund.*, 4, 133 (1965).
22. Cunningham, R. S., Ph.D. Thesis, The Ohio State University, Columbus, Ohio (1966).
23. Cunningham, R. S., and Geankoplis, C. J., *Ind. Eng. Chem. Fund.*, 7, 429 (1968).
24. Cunningham, R. S., and Geankoplis, C. J., *Ind. Eng. Chem. Fund.*, 7, 535 (1968).
25. Curtiss, C. F., University of Wisconsin, Naval Research Laboratory, Report CM-476 (1948).
26. Davis, R. B., and Scott, D. S., Symposium on Fundamentals of Heat and Mass Transfer, 58th Annual Meeting, *A.I.Ch.E.*, Philadelphia, Pa., p. 21, (1965).
27. Davis, R. B., and Scott, D. S., *Symposium on Porous Structure and Catalytic Transport Processes in Heterogeneous Catalysis*, (G. K. Boreskov, ed.), Akad. Kiado, Budapest (1972).

28. Di Napoli, N. M., Williams, R. J. J., and Cunningham, R. E., *Lat. Am. J. Chem. Eng. and Appl. Chem.*, **5**, 101 (1975).
29. Evans, III, R. B., Watson, G. M., and Mason, E. A., *J. Chem. Phys.*, **35**, 2076 (1961).
30. Evans, III, R. B., Watson, G. M., and Truitt, J., *J. Appl. Phys.*, **33**, 2682 (1962).
31. Evans, III, R. B., Watson, G. M., and Truitt, J., *J. Appl. Phys.*, **34**, 2020, (1963).
32. Feng, C. F., Kostrov, V. V., and Stewart, W. E., *Ind. Eng. Chem. Fund.*, **13**, 5 (1974).
33. Feng, C. F., and Stewart, W. E., *Ind. Eng. Chem. Fund.*, **12**, 143 (1973).
34. Foster, R. N., Ph.D. Thesis, Yale University, New Haven, Connecticut (1966).
35. Foster, R. N., Bliss, H., and Butt, J. B., *Ind. Eng. Chem. Fund.*, **5**, 580 (1966).
36. Foster, R. N., Butt, J. B., and Bliss, H., *J. Catal.*, **7**, 179, 191 (1967).
37. Frisch, H. L., *Polym. Lett.*, **3**, 13 (1965).
38. Gorring, R. L., *J. Catal.*, **31**, 13 (1973).
39. Gotoh, S., Manner, M., Sorensen, J. P., and Stewart, W. E., *Ind. Eng. Chem. Fund.*, **12**, 119 (1973).
40. Grachev, G. A., Veskov, V. S., Ione, K. G., Malinovskaya, O. A., and Slin'ko, M. G., *Kinetics and Catalysis*, **12**, part 2, 1152 (English translation) (1971).
41. Gros, J. B., and Bugarel, R., *Chem. Eng. Sci.*, **29**, 1465 (1974).
42. Gunn, R. D., Ph.D. Thesis, University of California, Berkeley, California (1967).
43. Gunn, R. D., and King, C. J., *A.I.Ch.E. Journal*, **15**, 507 (1969).
44. Gyarmati, I., *On the Principles of Thermodynamics*, Budapest (1957); *Zhur. Fiz. Khim.*, **39**, 1489 (1965) [English translation: *Russ. J. Phys. Chem.*, **39**(6), 788 (1965)]; *Acta Chim. Acad. Scien. Hung.*, **43**, 353 (1965); *Nonequilibrium Thermodynamics*, Budapest (1957).
45. Hashimoto, N., and Smith, J. M., *Ind. Eng. Chem. Fund.*, **12**, 353 (1973).
46. Hashimoto, N., and Smith, J. M., *Ind. Eng. Chem. Fund.*, **13**, 115 (1974).
47. Hawtin, P., *Chem. Eng. Sci.*, **26**, 1783 (1971).
48. Haynes, Jr., H. W., Ph.D. Thesis, University of Colorado, Boulder, Colorado (1969).
49. Haynes Jr., H. W., and Brown, L. F., *A.I.Ch.E. Journal*, **17**, 491 (1971).
50. Haynes, Jr., H. W., and Sarma, P. N., *A.I.Ch.E. Journal*, **19**, 1043 (1973).
51. Heinzelmann, F. J., Wasan, D. T., and Wilke, C. R., *Ind. Eng. Chem. Fund.*, **4**, 55 (1965).
52. Henry, Jr., J. P., Cunningham, R. S., and Geankoplis, C. J., *Chem. Eng. Sci.*, **22**, 11 (1967).
53. Hesse, D., *Chem. Eng. Sci.*, **32**, 427 (1977).
54. Hesse, D., and Köder, J., *Chem. Eng. Sci.*, **28**, 807 (1973).
55. Hills, A. W. D., *Chem. Eng. Sci.*, **23**, 297 (1968).
56. Hite, R. H., and Jackson, R., *Chem. Eng. Sci.*, **32**, 703 (1977).
57. Hoogschagen, J., *Ind. Eng. Chem.*, **47**, 906 (1955).
58. Horák, Z., and Schneider, P., *Chem. Eng. J.*, **2**, 26 (1971).
59. Jackson, R., *Chem. Eng. Sci.*, **29**, 1413 (1974).
60. Jiratová, K., and Horák, Z., *J. Chem. Prum. (Czechoslovakia)*, **27**, 5 (1977); *Int. Chem. Eng.*, **18**, 297 (1978).
61. Johns, L. E., and De Gance, A. E., *Ind. Eng. Chem. Fund.*, **14**, 237 (1975).
62. Johnson, M. F. L., and Stewart, W. E., *J. Catal.*, **4**, 248 (1965).
63. Jost, W., *Diffusion in Solids, Liquids and Gases*, Academic Press Inc., New York (1960).
64. Kapral, R., Hudson, S., and Ross, J., *J. Chem. Phys.*, **53**, 4387 (1970).
65. Kehoe, J. P., and Aris, R., *Chem. Eng. Sci.*, **28**, 2094 (1973).
66. Kirillov, V. A., Matros, Yu. Sch., Kuzin, V. Z., and Slin'ko, M. G., *Kinetika i Kataliz.*, **12**, part 2, 185 (English translation) (1971).
67. Kirkendall, E. O., *Trans. A.I.M.E.*, **147**, 104 (1942).
68. Konstantinov, E. N., and Nikolaiev, A. M., *Izv. Vyssh. Uchebn. Zaved., Neft. Gaz.*, No. 1 (1964).

69. Kosov, L. D., and Kurlapov, L. I., *Soviet Physics–Tech. Phys.*, **10**, 1623 (1966).
70. Kotousov, L. S., *Soviet Physics–Tech. Phys.*, **9**, 1679 (1965).
71. Kramers, H., and Kistemaker, J., *Physica (Utrecht)*, **10**, 699 (1943).
72. Krishna, R., *Ind. Eng. Chem. Fund.*, **16**, 228 (1977).
73. Lee, C. Y., and Wilke, C. R., *Ind. Eng. Chem.*, **46**, 2381 (1954).
74. Lin, K., and Lih, M. M., *A.I.Ch.E. Journal*, **17**, 1234 (1971).
75. Maa, J. R., *Ind. Eng. Chem. Fund.*, **9**, 283 (1970).
76. Marcellin, P., *Rev. Pract. Froid Condit. l'air*, **10**, 23 (1972).
77. Marrero, T. R., and Mason, E. A., *J. Phys. Chem. Ref. Data*, **1**, 3 (1972).
78. Mason, E. A., Malinauskas, A. P., and Evans, III, R. B., *J. Chem. Phys.*, **46**, 3199 (1967).
79. Mason, E. A., and Marrero, T. R., *Advances in Atomic and Molecular Physics*, Vol. VI, Robert Maxwell and Co., London (1970).
80. McCarty, K. P., and Mason, E. A., *Phys. Fluids*, **3**, 908 (1960).
81. Miller, L., and Carman, P. C., *Nature*, **186**, 549 (1960); **191**, 375 (1961).
82. Omata, H., and Brown, L. F., *A.I.Ch.E. Journal*, **18**, 967 (1972).
83. Omata, H., and Brown, L. F., *A.I.Ch.E. Journal*, **18**, 1063 (1972).
84. Patel, P. V., and Butt, J. B., *Ind. Eng. Chem. Proc. Des. and Dev.*, **14**, 298 (1975).
85. Pommersheim, J. M., and Ranck, B. A., *Ind. Eng. Chem. Fund.*, **12**, 246 (1973).
86. Prager, S., and Long, F. A., *J. Amer. Chem. Soc.*, **73**, 4072 (1951).
87. Pyun, C. W., and Ross, J., *J. Chem. Phys.*, **40**, 2572 (1964).
88. Raghavan, N. S., and Doraiswamy, L. K., *Ind. Eng. Chem. Proc. Res. and Dev.*, **16**, 519 (1977).
89. Rao, S. S., and Bennett, C. O., *Ind. Eng. Chem. Fund.*, **5**, 573 (1966).
90. Remick, R. R., and Geankoplis, C. J., *Ind. Eng. Chem. Fund.*, **9**, 206 (1970).
91. Remick, R. R., and Geankoplis, C. J., *Ind. Eng. Chem. Fund.*, **12**, 214 (1973).
92. Remick, R. R., and Geankoplis, C. J., *Chem. Eng. Sci.*, **29**, 1447 (1974).
93. Ross, J., and Mazur, P., *J. Chem. Phys.*, **35**, 19 (1961).
94. Rothfeld, L. B., Ph.D. Thesis, University of Wisconsin, Madison, Wisconsin (1961).
95. Rothfeld, L. B., *A.I.Ch.E. Journal*, **9**, 19 (1963).
96. Sandor, J., *Zh. Fiz. Khim.*, **44**, 1552 (1970).
97. Sarmasaev, M. T., *Vop. Obshch. Prikl. Fiz. Tr.*, **2**, 164 (1969).
98. Satterfield, C. N., and Cadle, P. J., *Ind. Eng. Chem. Fund.*, **7**, 202 (1968).
99. Satterfield, C. N., and Cadle, P. J., *Ind. Eng. Chem. Proc. Des. and Dev.*, **7**, 256 (1968).
100. Schmitt, K. H., and Waldmann, L., *Z. Naturforsch.*, **15a**, 843 (1960).
101. Schneider, P., and Smith, J. M., *A.I.Ch.E. Journal*, **14**, 762 (1968).
102. Scott, D. S., and Dullien, F. A. L., *A.I.Ch.E. Journal*, **8**, 113 (1962).
103. Shain, S. A., *A.I.Ch.E. Journal*, **7**, 17 (1961).
104. Shizgal, B., *J. Chem. Phys.*, **55**, 76 (1971).
105. Shizgal, B., and Karplus, M., *J. Chem. Phys.*, **52**, 4262 (1970); **54**, 4345, 4357 (1971).
106. Simons, J., *Chem. Phys. Letters*, **12**, 454 (1972).
107. Smikelsgras, A. D., and Kirkendall, E. O., *Metals Techn. Tech. Publ.*, **13**, 2071 (1946).
108. Smikelsgras, A. D., and Kirkendall, E. O., *Trans. A.I.M.E.*, **171**, 130 (1947).
109. Snider, N. S., and Ross, J., *J. Chem. Phys.*, **44**, 1087 (1966).
110. Stewart, W. E., Gotoh, S., and Sorensen, J. P., *Ind. Eng. Chem. Fund.*, **12**, 114 (1973).
111. Stewart, W. E., and Probe, R., *Ind. Eng. Chem. Fund.*, **3**, 224 (1964).
112. Stoll, D. R., and Brown, L. F., *J. Catal.*, **32**, 37 (1974).
113. Suetin, P. E., and Volobuev, P. V., *Soviet Phys.–Tech. Phys.*, **9**, 859 (1964).
114. Suetin, P. E., and Volobuev, P. V., *Zh.T.E.*, **34**, 66 (1964).
115. Suetin, P. E., and Volobuev, P. V., *Tr. Ural. Polytekh. Inst.*, **189**, 69 (1971).
116. Suetin, P. E., and Volobuev, P. V., *Izv. Vyssh. Uchebn. Zaved. Fiz.*, **9**, 11 (1972).
117. Toei, R., Okazaki, M., Nakanishi, K., Kondo, Y., Hayashi, M., and Shiozaki, Y., *J. Chem. Eng. Japan*, **6**, 50 (1973).

118. Toor, H. L., *A.I.Ch.E. Journal*, **3**, 198 (1957).
119. Toor, H. L., *A.I.Ch.E. Journal*, **10**, 448 (1964).
120. Toor, H. L., and Arnold, K. R., *A.I.Ch.E. Journal*, **11**, 746 (1965).
121. Turevskii, E. N., Aleksandrov, I. A., and Dvoiris A. D., *Khim. i Tekh. Top. i Masel*, No. 4, 36 (1971) [English translation: *Chem. Technol. Fuels Oil*, **7**(4), 289 (1971)].
122. Volobuev, P. V., and Suetin, P. E., *Soviet Phys.-Tech. Phys.*, **10**, 269 (1965).
123. Wakao, N., and Naruse, Y., *Chem. Eng. Sci.*, **29**, 1304 (1974).
124. Wakao, N., and Smith, J. M., *Chem. Eng. Sci.*, **17**, 825 (1962).
125. Waldmann, L., and Schmitt, K. H., *Z. Naturforsch.*, **16a**, 1343 (1961).
126. Weekman, Jr., V. W., and Gorring, R. L., *J. Catal.*, **4**, 260 (1965).
127. Weisz, P. B., *Chem. Tech.*, 498 (August,1973).
128. Wicke, E., *Kolloid Z.*, **93**, 129 (1940).
129. Wicke, E., and Hugo, P., *Z. physikal. Chem. (Neue Folge)*, **28**, 401 (1961).
130. Wicke, E., and Kallenbach, R., *Kolloid Z.*, **97**, 135 (1941).
131. Widom, B., *J. Chem. Phys.*, **34**, 2050 (1961).
132. Wiggs, P. K. C., in: *Structure and Properties of Porous Materials* (D. H. Everett and F. S. Stone, eds.), Academic Press, New York (1958).
133. Wilke, C. R., *Chem. Eng. Progr.*, **46**, 95 (1950).
134. Yang, R. T., and Liu, R. J., *Ind. Eng. Chem. Proc. Res. and Dev.*, **18**, 245 (1979).

## References for Chapter 6

1. Apecetche, M. A., González, M. G., Williams, R. J. J., and Cunningham, R. E., *J. Catal.*, **29**, 451 (1973).
2. Barrer, R. M., *Appl. Mat. Res.*, **2**, 129 (1963).
3. Bearman, R. J., *J. Phys. Chem.*, **65**, 1961 (1961).
4. Bernstein, R. B., *J. Chem. Phys.*, **17**, 209 (1949).
5. Bhatnagar, P. L., Gross, E. P., and Krook, M., *Phys. Rev.*, **94**, 511 (1954).
6. Biot, J. B., *Bibliothèque Britannique*, **27**, 310 (1804).
7. Biot, J. B., *Traité de Physique*, Vol. VI, Paris (1816).
8. Bird, R. B., Stewart, W. E., and Lightfoot, E. N., *Transport Phenomena*, John Wiley & Sons., New York (1960).
9. Bischoff, K. G., *Chem. Eng. Sci.*, **18**, 711 (1963).
10. Bischoff, K. G., *A.I.Ch.E. Journal*, **11**, 351 (1965).
11. Boardman, L. E., and Wild, N. E., *Proc. Roy. Soc. (London)*, **A162**, 511 (1937).
12. Boltzmann, L., *Wien. Ber.*, **66**, 275 (1872).
13. Bosanquet, C. H., *British TA. Report B.R.* 507 (Sept. 27, 1944).
14. Buckingham, E., *U.S. Dept. Agr., Bur. Soils, Bull.*, No. 25 (1904).
15. Burnett, D., *Proc. Lond. Math. Soc.*, **39**, 385 (1935); **40**, 382 (1935).
16. Butt, J. B., *A.I.Ch.E. Journal*, **9**, 707 (1963).
17. Chapman, S., *Phil. Trans. Roy. Soc.*, **A211**, 433 (1912).
18. Chapman, S., *Phil. Trans. Roy. Soc.*, **A216**, 279 (1916).
19. Chapman, S., *Phil. Trans. Roy. Soc.*, **A217**, 115 (1917).
20. Crank, J., *The Mathematics of Diffusion*, Oxford University Press, Cambridge (1956).
21. Curtiss, C. F., and Hirschfelder, J. O., *J. Chem. Phys.*, **17**, 550 (1949).
22. Dahlbert, I., *J. Phys. A.*, **6**, 1800 (1973).
23. Damköhler, G., *Der Chemie Ingenieur*, Vol. III, Pt. 1, Akademische Verlagsgesellschaft M. B. H., Leipzig (1937).
24. Darcy, H. P. G., *Les Fontaines Publiques de la Ville de Dijon*, Victor Dalmont, Paris (1856).

25. Derjaguin, B. V., and Batova, D. A., *Doklady Akad. Nauk S.S.S.R.*, **128**, 323 (1959).
26. Dullien, F. A. L., and Scott, D., *Chem. Eng. Sci.*, **17**, 771 (1962).
27. Enskog, D., *Physik Z.*, **12**, 56, 533 (1911).
28. Enskog, D., *Ann. Physik*, **38**, 731 (1912).
29. Enskog, D., *Archiv für Matematik, Astronomi och Fysik*, **16**, Section 16, Paragraph 16, in German (1912); Kinetische Theorie der Vorgänge in mässig verdünnten Gasen, Inaugural Dissertation, Upsala, Sweden, Almqvist and Wilksell (1917).
30. Enskog, D., *Svensk. Akad. Handl.*, **4**, 63 (1922).
31. Evans, III, R. B., Truitt, J., and Watson, G. M., *J. Chem. Eng. Data*, **6**, 522 (1961).
32. Evans, III, R. B., Watson, G. M., and Mason, E. A., *J. Chem. Phys.*, **35**, 2076 (1961).
33. Evans, III, R. B., Watson, G. M., and Mason, E. A., *J. Chem. Phys.*, **36**, 1894 (1962).
34. Evans, III, R. B., Watson, G. M., and Truitt, J., *J. Appl. Phys.*, **33**, 2682 (1962).
35. Fick, A., *Pogg. Ann.*, **94**, 59 (1855); abridged in *Phil. Mag.*, **10**, 30 (1855).
36. Fourier, J. B. J., *Theorie Analytique de la Chaleur*, Paris, (1822).
37. Frenkel, S. P., *Phys. Rev.*, **57**, 661 (1940).
38. Furry, W. H., *Am. J. Phys.*, **16**, 63 (1948).
39. Grad, H., *Comm. Pure and Appl. Math.*, **2**, 331 (1949).
40. Graham, T., *Ann. Phil.*, **12**, 69 (1826).
41. Graham, T., *Quart. J. Sci.*, **2**, 74 (1829).
42. Graham, T., *Ann. Phys.*, **17**, 341 (1829).
43. Graham, T., *Phil. Mag.*, **2**, 175, 269, 351 (1833).
44. Graham, T., *Ann. Phys.*, **28**, 331 (1833).
45. Graham, T., *Phil. Trans.*, 573 (1846).
46. Graham, T., *Liebig's Ann.*, **77**, 56 (1851); abstracted on page 197.
47. Graham, T., *Phil. Trans. Roy. Soc.*, **153**, 385 (1863).
48. Graham, T., *Ann. Phys.*, **120**, 415 (1863).
49. Graham, T., *Chemical and Physical Researches*, University Press, Edinburgh (1876).
50. Gunn, R. D., and King, C. J., *A.I.Ch.E. Journal*, **15**, 507 (1969).
51. Hagen, G. J. L., *Ann. Phys. Chem.*, **46**, 423, 442 (1839).
52. Hawthorn, R. D., *National Meeting, A.I.Ch.E.*, Dec. 2–7, New York (1961).
53. Hertz, G., *Z. Physik*, **19**, 35 (1923).
54. Hewitt, G. F., and Sharratt, E. W., *Nature*, **198**, 952 (1962).
55. Hewitt, G. F., and Sharratt, E. W., *J. Nucl. Mat.*, **13**, 206 (1964).
56. Hirschfelder, J. O., Curtiss, C. F., and Bird, R. B., *Molecular Theory of Gases and Liquids*, John Wiley & Sons Inc., New York (1954).
57. Hoogschagen, J., *Ind. Eng. Chem.*, **47**, 906 (1955).
58. Hougen, O. A., and Watson, K. M., *Chemical Process Principles*, John Wiley & Sons Inc., New York (1948).
59. Jeans, J., *Dynamical Theory of Gases*, 3rd ed., Cambridge University Press, Cambridge (1921).
60. Jost, W., *Diffusion in Solids, Liquids and Gases*, Academic Press Inc., New York (1960).
61. Kehoe, J. P. G., and Aris, R., *Chem. Eng. Sci.*, **28**, 2094 (1973).
62. Kirk, A. D., *J. Chem. Educ.*, **44**, 745 (1967).
63. Knudsen, M., *Ann. Phys.*, (*Leipzig*), **28**, 75 (1909).
64. Kosov, N. D., and Kurlapov, L. I., *Zh. Tekh. Fiz.*, **35**, 2120 (1965) [English translation: *Soviet Physics–Tech. Phys.*, **10**, 1623 (1966)].
65. Kramers, H., and Kistemaker, J., *Physica*, **10**, 699 (1943).
66. Kunt, A., and Warburg, E., *Pogg. Ann.*, **195**, 337, 525 (1875).
67. Lennard-Jones, J. E., *Proc. Roy. Soc.*, **A106**, 441 (1924).
68. Lih, M. M., and Lin, K., *A.I.Ch.E. Journal*, **19**, 832 (1973).
69. Lin, K., and Lih, M. M., *A.I.Ch.E. Journal*, **17**, 1234 (1971).

70. Loeb, L. B., *Kinetic Theory of Gases*, 2nd ed., McGraw-Hill Book Company, New York (1934).
71. Lorentz, H. A., *Proc. Amsterdam Acad.*, **7**, 438, 585, 684 (1905).
72. Loschmidt, J., *Wien. Ber.*, **61**, 367 (1870).
73. Loschmidt, J., *Wien. Ber.*, **62**, 468, (1870).
74. Mason, E. A., and Evans, III, R. B., *J. Chem. Educ.*, **46**, 358 (1969).
75. Mason, E. A., Evans, III, R. B., and Love, L. D., *J. Chem. Educ.*, **46**, 423 (1969).
76. Mason, E. A., and Kronstadt, B., *J. Chem. Educ.*, **44**, 740 (1967).
77. Mason, E. A., Malinauskas, A. P., and Evans, III, R. B., *J. Chem. Phys.*, **46**, 3199 (1967).
78. Mason, E. A., and Marrero, T. R., *Advances in Atomic and Molecular Physics*, Vol. VI, Robert Maxwell & Co., London (1970).
79. Maxwell, J. C., *Phil. Mag.*, **20**, 21 (1860); reprinted in Reference 82.
80. Maxwell, J. C., *Phil. Mag.*, **19**, 9 (1860).
81. Maxwell, J. C., *Phil. Trans. Roy. Soc.*,.**157**, 49 (1867).
82. Maxwell, J. C., *Scientific Papers*, Cambridge University Press, Cambridge, (1890); reprinted by Dover Publications Inc., New York, (1962).
83. McCarty, K. P., and Mason, E. A., *Phys. Fluids*, **3**, 908 (1960).
84. Mercer, M. C., and Aris, R., *Lat. Am. J. Chem. Eng. and Appl. Chem.*, **1**, 149 (1971).
85. Meyer, O. E., *Kinetic Theory of Gases*, 2nd ed. (translated by R. E. Baynes), Longmans Green, New York (1899).
86. Moelwyn-Hughes, E. A., *Physical Chemistry*, 2nd ed., Pergamon Press Ltd., London (1965).
87. Ney, E. P., Armistead, F. C., *Phys. Rev.*, **71**, 14 (1947).
88. Obermayer, A. von, *Wien. Ber.*, **81**(II), 1102 (1880).
89. Onsager, L., *Phys. Rev.*, **37**(I), 405, 426 (1931).
90. Onsager, L., *Phys. Rev.*, **38**(II), 2265, 2279 (1931).
91. Perry, R. H., and Chilton, C. H., *Chemical Engineers Handbook*, 5th ed., McGraw-Hill Kokakusha Ltd., Tokyo (1973), p. 5–29, equations (5.71) and (5.73).
92. Poiseuille, J. L., *Comptes Rendus* **12**, 112 (1841).
93. Pollard, W. G., and Present, R. D., *Phys. Rev.*, **73**(II), 762 (1948).
94. Present, R. D., and de Bethune, A. J., *Phys. Rev.*, **75**, 1050 (1949).
95. Rohsenow, W. M., and Choi, H., *Heat, Mass and Momentum Transfer*, Prentice-Hall Inc., Englewood Cliffs, New Jersey (1963).
96. Rothfeld, L. B., and Watson, C. C., *A.I.Ch.E. Journal*, **2**, 129 (1963).
97. Rouse, H., and Ince, S., *History of Hydraulics*, Dover Publications Inc., New York (1957).
98. Ruckstuhl, A., *J. Chem. Educ.*, **28**, 594 (1951).
99. Schiller, L., *Drei Klassiker der Strömungslehre: Hagen, Poiseuille, Hagenbach*, Leipzig, (1933).
100. Scott, D., and Dullien, F. A. L., *A.I.Ch.E. Journal*, **8**, 113 (1962).
101. Scott, D., and Dullien, F. A. L., *A.I.Ch.E. Journal*, **8**, 293 (1962).
102. Sherwood, T. K., *Absorption and Extraction*, McGraw-Hill Book Company Inc., New York (1937).
103. Silveston, P. L., *A.I.Ch.E. Journal*, **10**, 132 (1964).
104. Skinner, L., and Beeck, O., *Shell Development Laboratories*, Emeryville, California (1945–1946).
105. Slattery, J. C., *A.I.Ch.E. Journal*, **15**, 866 (1969).
106. Spiegler, K. S., *Ind. Eng. Chem. Fund.*, **5**, 529 (1966).
107. Stefan, J., *Phil. Trans. Roy. Soc.*, **138**, 224 (1871).
108. Stefan, J., *Sitzber. Akad. Wiss. Wien*, **63**, 63 (1871).
109. Stefan, J., *Sitzber. Akad. Wiss. Wien*, **65**, 323 (1872).

110. Stefan, J., *Sitzber. Akad. Wiss. Wien*, **II68**, 385 (1873).
111. Stefan, J., *Sitzber. Akad. Wiss. Wien*, **II79**, 161 (1879).
112. Stefan, J., *Sitzber. Akad. Wiss. Wien*, **II98**, 1418 (1889).
113. Thiele, E. W., *Ind. Eng. Chem.*, **31**, 916 (1939).
114. Wagner, C., *Z. physikal. Chem. Abt.* **A193**, 1 (1943).
115. Wakao, N., Otani, S., and Smith, J. M., *A.I.Ch.E. Journal*, **11**, 435 (1965).
116. Walker Jr., P. L., and Rusinko Jr., F. J., *J. Phys. Chem.*, **59**, 241 (1955).
117. Walker Jr., P. L., Rusinko, Jr., F. J., and Raats, E., *Nature*, **176**, 1167 (1955).
118. Weber, S., *Kgl. Danske Videnskab. Selskab. Mat. Fys. Medd.*, **28**, (1954).
119. Weekman Jr., V. W., and Gorring, R. L., *J. Catal.*, **4**, 260 (1965).
120. Weisz, P. B., *Z. physikal. Chem.*, *N.F.*, **11**, 1 (1957).
121. Wheeler, A., *Advances in Catalysis*, Vol. III, Academic Press Inc., New York (1951).
122. Wicke, E., *Kolloid Z.*, **93**, 129 (1940).
123. Wicke, E., and Brötz, W., *Chem. Ing. Tech.*, **21**, 219 (1949).
124. Wicke, E., and Hugo, P., *Z. physikal. Chem.*, *N.F.*, **28**, 401 (1961).
125. Wicke, E., and Kallenbach, R., *Kolloid Z.*, **97**, 135 (1941).
126. Williams, F. A., *Am. J. Phys.*, **26**, 467 (1958).
127. Wright, P. G., *Far. Trans. II*, **68**, 1959 (1972).
128. Youngquist, G. R., *Ind. Eng. Chem.*, **62**, 52 (1970).
129. Zel'dovich, Ya. B., *Acta Physicochimica U.S.S.R.*, **10**, 583 (1939).
130. Zhdanov, V., Kagan, Yu., and Sozykin, A., *Zh. Eksperim. Teor. Fiz.*, **42**, 857 (1962) [English translation: *Soviet Phys. JETP* **15**, 596 (1962)].

# Index